Springer
Tokyo
Berlin
Heidelberg
New York
Hong Kong
London
Milan
Paris

T. Sekimura · S. Noji
N. Ueno · P.K. Maini (Eds.)

Morphogenesis and Pattern Formation in Biological Systems

Experiments and Models

With 175 Figures, Including 16 in Color

Springer

Toshio Sekimura, D.Sc.
Professor
Department of Biological Chemistry
College of Bioscience and Biotechnology, Chubu University
1200 Matsumoto-cho, Kasugai, Aichi 487-8501, Japan

Sumihare Noji, D.Sc.
Professor
Department of Biological Science and Technology
Faculty of Engineering, University of Tokushima
2-1 Minami-Josanjima, Tokushima 780-8506, Japan

Naoto Ueno, Ph.D.
Professor
National Institute for Basic Biology
38 Nishigonaka, Myodaiji-cho, Okazaki, Aichi 444-8585, Japan

Philip K. Maini, Ph.D.
Professor
Centre for Mathematical Biology
Mathematical Institute, University of Oxford
24-29 St. Giles', Oxford OX1 3LB, UK

ISBN 978-4-431-00644-2 Springer-Verlag Tokyo Berlin Heidelberg New York

Library of Congress Cataloging-in-Publication data applied for.

Printed on acid-free paper

Typesetting: Camera-ready by the editors and authors

SPIN: 10905630

Foreword

Although it is really only during the past 15 years or so that mathematical, or theoretical, biology has become an accepted discipline in its own right, the diversity of modelling applications and the resulting increase in our knowledge of basic biological processes has far exceeded my expectations and, I know, those of many others, including die-hard experimentalists initially hostile to theory. The days of the mathematical biologist polymath are over, and it is an indication of the scientific sophistication and exciting development of the field that specialised subfields have arisen and are continually increasing. There are now few biological fields in which theoreticians are not involved. You never now hear from mathematical or physical colleagues such comments as "He started to work in biology because he couldn't make it in mathematics" or other equally ridiculous comments. The clear implication was that biology is an easy field. Biomedical science is so obviously *the* science of the foreseeable future it would have been tragic for mathematicians to have remained outside. For many of us who work in biology, Henri Poincaré (1854–1912), however, has summed up one of the major reasons: "The scientist does not study nature because it is useful: he studies it because he delights in it because it is beautiful. If nature were not beautiful it would not be worth knowing, and if nature would not be worth knowing, life would not be worth living."

Other unrealistic views still persist and even flourish in some places, namely, that only through rigorous mathematical analysis can we understand biological processes. I do not mean to imply that what these purely analytical people do is necessarily trivial, only that it is almost always biologically irrelevant. Few genuine (from my point of view) theoretical biologists, and certainly no biologist, take this view seriously. Their attitude was succinctly put forward by Auguste Comte (1798–1857) in his *Positive Philosophy*: "In mathematics we find the primitive source of rationality; and to mathematics must biologists resort for means to carry out their researches." Much more recently, in the early days of catastrophe theory hype, it was again expounded and in fact did much harm to the cause of interdisciplinary mathematics biology collaboration. Those days are now far behind us, but these attitudes still arise from time to time. Before ending this brief polemic I firmly believe that our work in biology can only really be judged by biologists whose standards are often very much higher particularly for those in other disciplines who move into this field! I should add here that I feel the greatest growth of new areas of involvement will be in medical applications where many fewer mathematicians are currently involved. Among those medical scientists seeking collaboration there is, in my experience, a more open approach to genuine collaboration and a willingness to invest their time.

Although I believe it is an objective view (although many would say it could not be more subjective), the involvement of theoreticians in the biological sciences is among the most exciting and fascinating research a scientist can do. This truly interdisciplinary international meeting, commemorating the 50-year anniversary of Turing's monumental paper on the chemical basis of morphogenesis, on "Morphogenesis and Pattern Formation in Biological Systems—Experiments and Models" organised by Professor Toshio Sekimura (Chubu University), Professor Philip Maini (University of Oxford) and their colleagues and the articles in these proceedings exemplify the now-accepted view that good mathematical biology is closely related to real biology. They are an impressive offering of the current state of many of these new areas of application and give an excellent introduction to some of the subtleties and challenges that the biomedical sciences pose. What also makes these proceedings so interesting is the diversity of problems in which theory has made a significant contribution to our biological understanding: they are a valuable contribution to this interdisciplinary field.

Understanding spatial pattern formation processes whether in development, ecology, epidemiology, and so on is certainly not the be-all and end-all. However, if there was a competition for that slot, pattern generation would unquestionably be on the shortlist, and it is testament to this view that the articles in this volume deal with patterning. Elucidating the development of spatio-temporal pattern and form is clearly one of the fundamental questions and challenges of the twenty-first century whether it is morphogenesis, the dynamics of interacting populations in ecology, the complex species interactions in epidemiology, neuronal connectivity and information processing, the growth of tumours, and so on. Enormous progress has been made in the past two decades in a wide spectrum of areas, but the detailed mechanisms in practically all pattern-formation situations are still unknown, which of course is another reason the field is so fascinating. Notwithstanding this fact, it is not necessary to have such a complete knowledge for simpler model systems, which capture the key elements, to be enormously useful in applications; this is particularly apposite in medicine.

Mathematical (theoretical) models for pattern formation in a general sense can suggest possible mechanisms, and these have played a seminal role in the study of pattern formation in a wide range of problems and significantly influenced the thinking of experimentalists. Analysis of the models can provide biological insight into which elements are essential for pattern development and can be useful in predicting the effect of altering given parameters (and experimental conditions) on the resultant pattern. Mathematical modelling can also suggest experiments to further elucidate the underlying pattern-formation process. I would stress, however, that mathematical descriptions of patterning phenomena are not explanations.

In spite of the seemingly endless series of exciting new discoveries in other areas of biology from mapping the genome to cloning animals, the major problems in pattern generation are still unknown in detail. In an interesting survey conducted by the journal *Science* in 1994 (Barinaga, Science 266:561–564, 1994) more than 100 leading developmental biologists were asked what they thought were the most important unanswered problems in development and where they thought the biggest breakthroughs would come in the following 5 years. Among the 66 responses, the most important unanswered question was that of how the body's specialised organs and tissues are formed, namely, morphogenesis. Coming second after morphogenesis was how the actual mechanisms evolved and how evolution acted on the mechanisms to effect change and generate new species. By a significant margin the largest number of votes on "Development's greatest unsolved mysteries" was "What are the molecular mechanisms of morphogenesis?" Also high up on their lists was the question "How are patterns established in the early embryo?"

Whatever the specific discipline, the principal use of any theory is in its predictions, and even though different models might be able to create similar spatio-temporal patterns, they are mainly distinguished by the different experiments they suggest and, of course, how closely they relate to the real biology. The best mathematical models help to unravel the underlying biological mechanisms and show how a patterning process works and, importantly, then make biological predictions. From a theoretical point of view, the art of good modelling in biology relies not only on sufficient mathematical expertise (often not at all sophisticated), but also on: (1) a sound understanding and appreciation of the biological problem; (2) a realistic mathematical representation of the important biological phenomena; and (3) a biological interpretation of the mathematical analysis and results in terms of insights and predictions. It is interesting to compare the proceedings of this conference in the light of these comments and those of many such conferences 10 years ago. The practicality and genuine biological relevance in the articles in this volume clearly bear out this view as to biological relevance.

An important point arising from theoretical models is that any pattern contains its own history. Consider a simple engineering analogy which we have given before (Murray et al., J. Franklin Inst., 335B:303–328, 1998) of our role in trying to understand a biological process. It is one thing to suggest that a bridge requires a thousand tons of steel, that any less will result in too weak a structure, and any more will result in excessive rigidity. It is quite another matter to instruct the workers on how best to put the pieces together. In morphogenesis, for example, it is conceivable that the cells involved in tissue formation and deformation have enough expertise that given the right set of ingredients and initial instructions they could be persuaded to construct whatever element one wants. This is the goal of many who are searching for a full and predictive understanding. However, it seems very likely that the global effect of all this sophisticated cellular activity would be critically sensitive to the sequence of events occurring during development. As scientists we should concern ourselves with how to take advantage of the limited opportunities we have for communicating with the workforce so as to direct experiments towards an acceptable end-product. Although this is a touch philosophical, even a cursory look at many theories in the literature reveals a fixation on simplistic explanations.

None of the individual models that have been suggested for any biological patterning process, and not even all of them put together, could be considered a complete model. For example, in the case of several specific spatial pattern-formation situations, each model has shed light on different aspects of the process and we can now say what the most important conceptual elements have to be in a complete model. The articles here help to highlight where our knowledge is deficient and suggest directions in which fruitful experimentation might lead us. As we have said, a critical test of these theoretical constructs is in their impact on the experimental community.

The last 25 years have seen a blossoming of the field of theoretical or mathematical biology from the relatively few working in the field in the 1970s to the several thousand now, albeit with varying degrees of biological relevance.

Early interdisciplinary centres, such as the Centre for Mathematical Biology in Oxford, promoted the genuine integration of experimentalists and theoreticians by bringing them together, along with many young researchers for extended periods. The express aim of the organisers of this meeting was to bring together experimentalists and theoreticians in a meaningful way; in this they have been eminently successful.

The ability of a model to mimic patterns observed in a developmental process is only the first, but necessary, step to its candidacy as a possible mechanism. A model must be able to make testable predictions. The processes that occur in development are, of course, highly complex, and mathematical models are, necessarily, based on gross simplification. However, they do focus on particular elements of the process and help determine the role played by such elements in the overall process. The role of theory is to provide a conceptual framework for the experimenter and to make predictions which encourage, or provoke (also very informative), experimentation. The articles in this volume provide a good view of the current state of pattern-formation modelling in several different areas.

Pattern-formation studies are sometimes criticized for their lack of inclusion of genes in the models. But then criticism can be levelled at any modelling abstraction of a complex system to a relatively simple one. It should be remembered that the generation of pattern and form, particularly in development, is usually a long way from the level of the genome. Of course genes play crucial roles in development but they do not actually create patterns. Many of the evolving patterns could hardly have been anticipated solely by genetic information.

A frequently asked question is why we use mathematics to study something as intrinsically complicated and ill-understood as development, angiogenesis, wound healing, interacting population dynamics, regulatory networks, epidemics, and so on. We suggest, as we have said often before, that mathematical modelling must be used if we ever hope to genuinely and realistically convert an understanding of the underlying mechanisms into a predictive science. Mathematics is required to bridge the gap between the level on which most of our knowledge is accumulating (cellular and below) and the macroscopic level of the patterns we see. A mathematical approach lets us explore the logic of pattern formation. Even if the mechanisms were well understood—and they certainly are far from it at this stage—mathematics would be required to explore the consequences of manipulating the various parameters associated with any particular scenario. In the case of such things as wound healing, and it will be increasingly so in angiogenesis with the cancer connection, the number of options that are fast becoming available to wound and cancer managers will become overwhelming unless we can find a way to simulate particular treatment protocols before applying them in practice. The latter has already been of use in understanding the efficacy of various treatment scenarios with brain tumours.

There is no doubt that we are a long way from being able to reliably simulate actual patterning mechanisms notwithstanding the multitude of theories that abound. In morphogenesis the active cellular control of key processes is poorly understood. Despite such limitations, we argue that exploring the logic of biological processes is worthwhile, in some current situations even essential in our present state of knowledge. It allows us to take a hypothetical mechanism and examine its consequences in the form of a mathematical model, make predictions, and suggest experiments that would verify or invalidate the model; the latter is frequently biologically informative. In fact, the very process of constructing a mathematical model can be useful in its own right. Not only must we commit to a particular mechanism, we are also forced to consider what is truly essential to the process and what the key players (variables) are. We are thus involved in constructing frameworks on which we can hang our understanding. The equations, the mathematical analysis, and the numerical simulations that follow serve to reveal quantitatively, as well as qualitatively, the consequences of that logical structure.

In conclusion, theoretical modelling has been proven to be useful in the study of a remarkably wide spectrum of biological problems of which these proceedings show. The best integrative biology studies have served to highlight where our knowledge is deficient and to suggest directions in which fruitful experimentation might lead us. A crucial aspect of this research is the interdisciplinary content, and a critical test of all theoretical constructs should be in their impact on the experimental community. The field of mathematical, or theoretical, biology has now achieved some level of maturity, and we believe that future dialogue between experimentalists and theoreticians will lead us most rapidly towards a fuller understanding, if not a complete one, of several important biomedical processes.

October 2002
James D. Murray
Department of Applied Mathematics
Box 352420
University of Washington
Seattle, Washington 98195-2420, USA

Preface

Understanding the mechanisms that underlie the processes of morphogenesis and pattern formation is one of the central goals in current biology, and is intensively studied in a wide range of research fields from molecular genetics to ecology to medicine. Since the discovery of *Hox* genes in the 20th century, the problem has become a realistic research theme in both experimental and theoretical biology. Morphogenesis refers to the processes involved in causing changes in form during embryonic development. Pattern formation, on the other hand, refers to the processes that generate spatially inhomogeneous structures in biological systems. However, both processes are often highly interconnected. Pattern formation is widely observed not only at microscopic levels but also at the macroscopic level in, for example, forest structures in population biology.

The methods for studying morphogenesis and pattern formation in different fields tend to be specific and are often almost independent of each other. No one has yet developed a method for dealing with the complexity of multiscale interactions that occur in almost all biological patterning processes. Given the latest advances in technology, we decided that the time was ripe to bring together experimentalists and theoreticians to begin addressing these issues.

A meeting entitled "Morphogenesis and Pattern Formation in Biological Systems— Experiments and Theories" (**mpb2002**) was held at Chubu University, Japan, September 24–27, 2002, and was co-chaired by Professor Philip K. Maini of the University of Oxford and myself. As 2002 is the 50th anniversary of Alan Turing's seminal paper, "The Chemical Basis for Morphogenesis", we decided that the meeting should commemorate this anniversary. Turing's original idea of so-called diffusion-driven instability established the notion of self-organization and crystallized the idea of chemical morphogen later extended by Lewis Wolpert (1969). Half of the participants at **mpb2002** were experimental biologists while the other half were theoretical or mathematical biologists. The meeting provided a stimulating interdisciplinary forum and there was much interaction among all participants.

This volume of the proceedings of **mpb2002** includes papers presented by invited speakers of the meeting. We hope that the reader will enjoy the papers and feel the atmosphere of **mpb2002** through this book. Scientific sessions of the meeting covered a wide range of research fields from genes to ecology and medicine in both experiments and theories as follows: "Morphogenesis and Pattern Formation in Animals", "Morphogenesis and Pattern Formation in Plants", "Morphogenesis and Pattern Formation Viewed from the Behavior of Individual Cells", "Models for Pattern Formation and Experiments", "Spatial Pattern and Structure Formation in Ecological Systems", "Spatio-temporal Pattern Formation in Epidemiology", "Morphogenesis and Pattern Formation in Medicine", "Diversity of Biological Patterns in the Fossil Record and Their Meaning in Morphological Evolution", and "Relationship Between Ontogeny and Phylogeny".

We hope that **mpb2002** and this book will prove to be stepping-stones or triggers to create a new integrative research field in the near future. Finally, I have to say that without the strong support of Professors Sumihare Noji of the University of Tokushima and Naoto Ueno of the National Institute for Basic Biology, Okazaki, Japan, this volume would not have been produced so smoothly and promptly.

Acknowledgments

First of all, I would like to thank President Atsuo Iiyoshi of Chubu University for his generous and continuous support. I would like to thank Professor James D. Murray of the University of Washington for his kind support. I also acknowledge Professors Okitsugu Yamashita and Tadashi Noguchi of Chubu University for their strong support. I express my sincere thanks to all members of the **mpb2002** organizing committee for their kind cooperation and continuous assistance: Philip K. Maini (Oxford, UK), Sumihare Noji (Tokushima, Japan), Masayuki Kakehashi (Hiroshima, Japan), Mark A.J. Chaplain (Dundee, UK), Kiyotaka Okada (Kyoto, Japan), Naoto Ueno (Okazaki, Japan), Kei Inouye (Kyoto, Japan), Toshihiko Hara (Hokkaido, Japan), and Rihito Morita (Chiba, Japan).

The conference was co-hosted by the Japanese Association for Mathematical Biology (JAMB), the Society for Mathematical Biology (SMB), and a Research Project on Dynamics of Developmental Systems funded by the Ministry of Education, Culture, Sports, Science and Technology, Japan. Financial support was provided by the Japan Society for the Promotion of Science, a Research Project on Dynamics of Developmental Systems funded by the Ministry of Education, Culture, Sports, Science and Technology, Japan; the Wellcome Trust, UK; the Daiko Foundation, Nagoya, Japan; and Chubu University. The conference was also supported by the Society for Science on Form, Japan.

December 2002
Toshio Sekimura
Chubu University
Japan

Contents

Contributors

Alexander R.A. Anderson
Department of Mathematics, University of Dundee
23 Perth Road, Dundee DD1 4HN, UK
E-mail: anderson@maths.dundee.ac.uk

Vincenzo Capasso
Milan Research Centre for Industrial and Applied Mathematics, University of Milan
Via Saldini 50, 20131 Milan, Italy
E-mail: capasso@mat.unimi.it

Mark A.J. Chaplain
The SIMBIOS Centre, Department of Mathematics, University of Dundee
23 Perth Road, Dundee DD1 4HN, UK
E-mail: chaplain@maths.dundee.ac.uk

Liam Dolan
Department of Cell and Developmental Biology, John Innes Centre
Norwich, NR4 7UH, UK
E-mail: liam.dolan@bbsrc.ac.uk

Shigeo Hayashi
Center for Developmental Biology, RIKEN
2-2-3 Minatojima-minami, Chuo-ku, Kobe 650-0047, Japan
E-mail: shayashi@cdb.riken.go.jp

Tomáš Herben
Institute of Botany, Academy of Sciences of the Czech Republic, CZ-252 43 Pruhonice
Department of Botany, Faculty of Science, Charles University, Prague, Czech Republic
E-mail: herben@site.cas.cz

Hisao Honda
Department of Health Science, Hyogo University
2301 Shinzaike, Hiraoka-cho, Kakogawa, Hyogo 675-0101, Japan
E-mail: hihonda@humans-kc.hyogo-dai.ac.jp

Hisashi Inaba
Department of Mathematical Sciences, University of Tokyo
3-8-1 Komaba, Meguro-ku, Tokyo 153-8914, Japan
E-mail: inaba@ms.u-tokyo.ac.jp

Kei Inouye
Department of Botany, Kyoto University
Kitashirakawa-Oiwake-cho, Sakyo-ku, Kyoto 606-8502, Japan
E-mail: inoue@cosmos.bot.kyoto-u.ac.jp

Masayuki Kakehashi
Faculty of Medicine, Hiroshima University
Kasumi, Minami-ku, Hiroshima 734-8551, Japan
E-mail: mkake@hiroshima-u.ac.jp

Shigeru Kuratani
Center for Developmental Biology, RIKEN
2-2-3 Minatojima-minami, Chuo-ku, Kobe 650-0047, Japan
E-mail: saizo@cdb.riken.go.jp

Chiyoko Machida
College of Bioscience and Biotechnology, Chubu University
1200 Matsumoto-cho, Kasugai, Aichi 487-8501, Japan
E-mail: cmachida@isc.chubu.ac.jp

Anotida Madzvamuse
Department of Numerical Analysis and Mathematical Biology, Oxford University
Computer Laboratory, Wolfson Build, Parks Rd, Oxford OX1 3QD, UK
E-mail: Anotida.Madzvamuse@comlab.ox.ac.uk

Hans Meinhardt
Max-Planck-Institut fuer Entwicklungsbiologie
Spemannstr. 35, Tübingen, D-72076, Germany
E-mail: hans.meinhardt@tuebingen.mpg.de

Sasuke Miyazima
College of Bioscience and Biotechnology, Chubu University
1200 Matsumoto-cho, Kasugai, Aichi 487-8501, Japan
E-mail: miyazima@isc.chubu.ac.jp

Rihito Morita
Department of Earth Sciences, Natural History Museum and Institute, Chiba
955-2 Aoba-cho, Chuo-ku, Chiba 260-8682, Japan
E-mail: morita@chiba-muse.or.jp

Vidyanand Nanjundiah
Developmental Biology and Genetics Laboratory, Indian Institute of Science
Bangalore 560012, India
E-mail: vidya@ces.iisc.ernet.in

Sumihare Noji
Department of Biological Science and Technology, University of Tokushima
2-1 Minami-Josanjima, Tokushima 780-8506, Japan
E-mail: noji@bio.tokushima-u.ac.jp

Kiyotaka Okada
Department of Botany, Graduate School of Science, Kyoto University
Kitashirakawa-Oiwake-cho, Sakyo-ku, Kyoto 606-8502, Japan
E-mail: kiyo@ok-lab.bot.kyoto-u.ac.jp

Hans G. Othmer
Department of Mathematics, University of Minnesota
206 Church Street S.E., Minneapolis, MN 55455, USA
E-mail: othmer@math.umn.edu

Enrico Savazzi
Hagelgrand 8, 75646 Uppsala, Sweden
E-mail: enrico@savazzi.net

Nanako Shigesada
Department of Information and Computer Sciences, Nara Women's University
Kita-Uoya Nishi-machi, Nara 630-8506, Japan
E-mail: sigesada@ics.nara-wu.ac.jp

Akihiro Sumida
Institute of Low Temperature Science, Hokkaido University
North 19 West 8, Sapporo 060-0819, Japan
E-mail: asumida@lowtem.hokudai.ac.jp

Ryuji Takaki
Department of Mechanical Systems Engineering
Tokyo University of Agriculture and Technology
2-24-16 Nakamachi, Koganei, Tokyo 184-8588, Japan
E-mail: takaki@cc.tuat.ac.jp

Masao Tasaka
Graduate School of Biological Sciences, Nara Institute of Science and Technology
8916-5 Takayama, Ikoma, Nara 630-0101, Japan
E-mail: m-tasaka@bs.aist-nara.ac.jp

Keiko U. Torii
Department of Botany, University of Washington
1521 NE Pacific, Seattle, WA 98195-5325, USA
E-mail: ktorii@u.washington.edu

Takao Ubukata
Institute of Geosciences, Shizuoka University
836 Oya, Shizuoka 422-8529, Japan
E-mail: sbtubuk@ipc.shizuoka.ac.jp

Naoto Ueno
Department of Developmental Biology, National Institute for Basic Biology
38 Nishigonaka, Myodaiji-cho, Okazaki, Aichi 444-8585, Japan
E-mail: nueno@nibb.ac.jp

Yoshiyuki Usami
Institute of Physics, Kanagawa University
3-27-1 Rokkakubashi, Yokohama 221-8686, Japan
E-mail: usami-yoshiyuki@nifty.com

Alfried P. Vogler
Department of Entomology, The Natural History Museum
Cromwell Road, SW7 5BD, London, UK
E-mail: apv@nhm.ac.uk

Masayuki Yokozawa
Department of Earth Resources Science
National Institute for Agro-Environmental Sciences
3-1-3 Kannondai, Tsukuba, Ibaraki 305-8604, Japan
E-mail: myokoz@niaes.affrc.go.jp

Part I

Models for Morphogenesis and Pattern
Formation

Pattern Forming Reactions and the Generation of Primary Embryonic Axes

Hans Meinhardt

Max-Planck-Institut für Entwicklungsbiologie, Spemannstraße 35,
D-D-72076 Tübingen, Germany
e-mail: hans.meinhardt@tuebingen.mpg.de

1.1 Necessary extensions of simple pattern forming mechanisms to generate the primary body axes

In his pioneering paper Turing [37] discovered a mechanism that allows the generation of patterns even when starting from more or less homogeneous initial situations. To account for essential steps in the early patterning of higher organisms, several extensions have to be made to overcome problems inherent in simple Turing reaction-diffusion type mechanisms:

1. **The wavelength problem:** Patterns generated by the Turing reaction-diffusion mechanism have a characteristic wavelength. Upon growth a transition from a polar pattern into a symmetric and ultimately into a periodic pattern is expected. However, in many developing systems substantial growth is possible without losing the polar character. It is proposed that the maintenance of a polar pattern is accomplished by a feedback of the pattern on the ability of the cells to perform the patterning reaction, i.e., on their competence. Cells distant to a once formed organizing region lose the competence to form additional maxima, making the first formed maximum dominate.

2. **The midline problem:** The formation of a coordinate system for a bilaterally symmetric organism requires the formation of a midline, i.e., a reference *line* and not a reference point for the mediolateral patterning. The formation of a single straight line requires the cooperation of a spot-like and a stripe-like system. In vertebrates the midline is formed by a local elongation under the influence of the organizer. In contrast, in insects an inhibitory influence from a local dorsal organizer allows midline formation only ventrally.

3. **The left-right pattern:** It is proposed that the midline signal induces the 'left' signal but higher levels of the midline signal repress the 'left' signal. Therefore, the 'left'-signal has to escape from the midline to a lateral position. This mechanism needs only a minute asymmetry for a

reproducible shift to the left. It accounts for a random left-right patterning if the systematic asymmetry is lost.

1.2 Pattern formation in the radial-symmetric Hydra: an ancient pattern forming system organizing a single axis

In discussing biological systems to which his theory could apply Turing mentioned explicitly the formation of a single maximum on a sphere for generating the first embryonic axis for driving gastrulation. As an example for a periodic pattern he discussed the formation of the tentacle ring of the small freshwater polyp Hydra. Hydra, however, is not only interesting for the periodic tentacle formation. Recently it has turned out that Hydra can be regarded as a living fossil that documents axis formation before bilaterality and trunk formation had been invented during evolution. A comparison of expression patterns of homologous genes in higher organisms and Hydra suggests that, counter-intuitively, the foot of the Hydra has the same positional address as the forebrain and heart in higher organisms. The foot of a Hydra represents, therefore, the most anterior structure. Conversely, the gastric opening, conventionally called the head of the Hydra, is in fact the most posterior pole and corresponds to the anus. The trunk is a later invention in evolution that became inserted into a narrow region close to the gastric opening (Fig. 1.1; [23]). In other words, the major portion of the ancestral organism evolved into the heart and brain. A Hydra can be reformed even after dissociation into individual cells and re-aggregation [7]. This clearly indicates that *de novo* patterning is possible along its single ancestral axis. In the following, models for this axis formation will be discussed. The mediolateral and left-right patterning require not only separate pattern forming systems, but these patterns must also obtain the correct orientation with the primary anteroposterior axis. Corresponding models will be discussed in subsequent sections.

1.3 Local self-enhancement and long-range inhibition as the driving force for primary pattern formation

Frequently, pattern-forming reactions are generally called 'reaction-diffusion systems'. However, interacting substances with different diffusion rates will not necessarily form a pattern. We have shown that pattern formation occurs only if a short-ranging self-enhancing reaction is coupled with a long-ranging substance that reacts in an antagonistic way [8, 18, 21, 22, 24]. Small local elevations above the average grow further due to the local positive feedback while the spreading antagonistic effect leads to a de-activation at larger distances. Knowing that local autocatalysis and long-range inhibition is the driving force facilitates substantially the design and an intuitive understanding of pattern forming reactions.

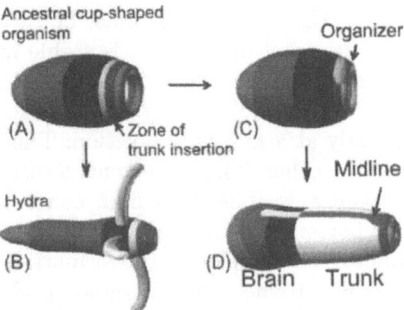

Fig. 1.1. Relation of the axis in the freshwater polyp Hydra to those of higher organisms. A common ancestor (A) of Hydra (B) and vertebrates (D) is assumed to have a cup-shaped radial-symmetric geometry. The bottom of the cup-shaped organism (left) gave rise to the foot in Hydra and to the most anterior part of vertebrates, the forebrain and heart (Nkx2.5 expression [9, 29,40]; light grey at left in each panel). In contrast, the opening of the gastric cavity gave rise to the so-called mouth opening in Hydra and to the anus of higher organisms (*Wnt*-expression [11, 39]). An essential intermediate step is assumed to be a widening of the blastoporus. The posterior organizer, originally involved in marking the whole blastoporus, became restricted to a hot spot on this ring (C). This Spemann-type organizer [10] attracts cells of the blastoporus [13]. After being once close to the organizer, the cells leave the organizer region as a unified band that forms the midline. Thus, midline formation is proposed to depend on the conversion of a ring into a rod. A patch-like signal causes the local elongation of the stripe-like midline ([22, 23]; Fig. 1.4).

Pattern formation from initially more or less homogeneous situations is not restricted to living systems. The formation of sand dunes, patterns of erosion, lightning, stars and galaxies are examples of pattern formation in non-animated systems. It is easy to see that patterning in these processes is also based on the mechanism we have proposed: local perturbations grow further due to the local self-enhancement but become spatially confined due to the long-range antagonistic effects that limit the self-enhancing process.

A simple molecular realization of this concept consists of a substance that has an autocatalytic feedback on its own synthesis. We have called this substance the activator $a(x)$. The production rate of the activator is slowed down by a long ranging molecule, the inhibitor $h(x)$, which is produced under the control of the activator:

$$\frac{\partial a}{\partial t} = \frac{\rho a^2}{h} - \mu_a a + D_a \frac{\partial^2 a}{\partial x^2} + \sigma_a \qquad (1.1)$$

$$\frac{\partial h}{\partial t} = \rho a^2 - \mu_h h + D_h \frac{\partial^2 h}{\partial x^2} + \sigma_h. \qquad (1.2)$$

A necessary condition for the formation of a stable pattern is that the inhibitor diffuses much faster and has a shorter half-life than the activator, i.e., the conditions $D_h \gg D_a$ and $\mu_h > \mu_a$ must be satisfied. The latter condition makes sure that the resulting pattern is stable in time. Whenever the field size exceeds the range of the activator, a homogeneous distribution of both substances becomes unstable (see Fig. 1.2). At this critical extension a maximum can appear only at a marginal position. Thus, one side of the field becomes different from the other. Therefore, such a mechanism is appropriate to generate an embryonic axis. The local high concentration can act as an organizing region. The pattern can be initiated by small fluctuations. Maternally supplied asymmetries can accelerate the formation of a single maximum. A stable situation is reached when the surrounding cloud of inhibition balances the activator increase. The resulting pattern is independent of the mode of initiation for a wide range of parameter values.

There are many other realizations of the same principle. In most inorganic patterning systems mentioned above the antagonistic effect results from a depletion of ingredients that are required for the self-enhancing process. Correspondingly the activator autocatalysis can be antagonized by the depletion of a substrate that is consumed during the autocatalysis. An example of such a system is

$$\frac{\partial a}{\partial t} = \rho s a^2 - \mu a + D_a \frac{\partial^2 a}{\partial x^2} + \rho_0. \tag{1.3}$$

$$\frac{\partial s}{\partial t} = \sigma - \rho s a^2 - \nu s + D_s \frac{\partial^2 s}{\partial x^2}. \tag{1.4}$$

This reaction has properties more similar to an activator-inhibitor reaction with saturation [24]. This is because the substrate concentration cannot become smaller than zero. This leads to an inherent upper boundary level for the activator production. In the literature, this interaction is sometimes referred to as the Schnakenberg reaction [33] but was part of our original proposal (Eq. 11 in [8]). Further, the autocatalysis need not be direct but can be realized by two other substances which inhibit each other [18]. In the patterning within insect segments the long-range inhibition of the same feedback loop is replaced by a long-range activation of a different feedback loop that locally represses the first. In this way, at least two cell states appear in close proximity. A stripe-like pattern is especially stable since a long common boundary of the two cell states allows an efficient mutual stabilization. The unravelling of the *engrailed-wingless* interaction has provided strong support for this view [21].

In addition to stable gradients a system of a self-enhancing and an antagonistic reaction can generate many other essential elementary patterns that are frequently required during development: periodic patterns in space if the range of the antagonist is much smaller than the field size, stripes if the autocatalysis saturates [19], oscillations and travelling waves if the antagonistic reaction has a longer time constant [18, 24].

1.4 Comparison with the Turing mechanism

Turing exemplified the mechanism he envisaged by the following set of equations ([37], p. 42).

$$\frac{dx}{dt} = 5x - 6y + 1 \qquad (1.5)$$

$$\frac{dy}{dt} = 6x - 7y + 1 \quad (+ \text{ diffusion}). \qquad (1.6)$$

Both Eqs. (1.5) and (1.6) look very similar and it is not immediately obvious why such a reaction leads to pattern formation. It is easy to see, however, that this interaction satisfies our conditions: x has a positive feedback on its own production rate while the long-ranging y molecule, produced under x control, acts antagonistically by destroying the x molecules. Therefore, Turing's mechanism can generate basically the same types of pattern as the lateral inhibition mechanism, i.e., graded concentration profiles and isolated maxima [2, 14]. In his paper, Turing did not mention anything like 'lateral inhibition'. It seems, however, that later he recognized this crucial point. In unpublished notes found after his death the following sentence occurred "The amplitude of the waves is largely controlled by the concentration V of 'poison'" (see [12], p. 494).

The particular mechanism proposed by Turing had an essential drawback: its molecular basis is unreasonable. According to Eq. (1.5), the number of x molecules disappearing per time unit is assumed to be proportional to the number of y molecules but independent of the number of x molecules present. In other words, x molecules can disappear even if no x molecules are left. This can lead to negative concentrations. Turing has seen this problem and proposed to ignore negative concentrations. One could repair this defect by assuming a degradation also proportional to x. This requires, however, a nonlinear autocatalytic activator production. A possible interaction is similar to Eq. (1.1) except that the inhibitor does not lower the production but increases the destruction of the activator (the equation for h would be the same as Eq. 1.2).

$$\frac{\partial a}{\partial t} = \rho a^2 - \mu_a ah + D_a \frac{\partial^2 a}{\partial x^2} + \sigma_a. \qquad (1.7)$$

If, as in the Turing-type reactions, the half-life of the activator declines with its peak height, at high concentrations a transition into an oscillating mode may occur and the maxima start to 'breathe'.

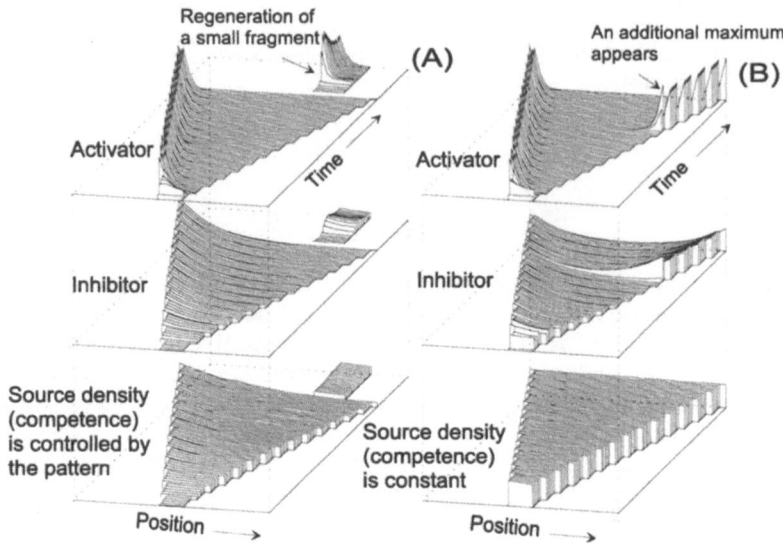

Fig. 1.2. The maintenance of a polar pattern during growth: the solution of the wavelength problem. (A) An activator - inhibitor system with an additional feedback of the inhibitor on the ability of the cells to perform the pattern forming reaction (ρ in Eqs. 1.1, 1.2 and 1.8; source density, bottom distribution). At small field size, only a marginal maximum can be formed. This leads on a long time scale to a graded ρ distribution. Due to the reduced ρ level in regions distant to the activated (organizing) region, a further activation is efficiently suppressed. After removal of the head and thus of the inhibitor-producing region, pattern regeneration is possible and occurs due to the ρ gradient according to the original polarity, in agreement with observations in many systems. Since the wavelength of the activator-inhibitor system is much smaller than the field size, small fragments are also able to regenerate, in agreement with experimental observations. (B) Without this feedback, i.e., if the source density remains unchanged, secondary maxima can arise.

1.5 The wavelength problem: how to avoid multiple maxima in growing fields

A simple pattern forming reaction has a particular wavelength. A graded concentration profile can be maintained only over a range of about a factor of two. With a further increase in the field size, the range of the antagonistic reaction becomes insufficient. Transitions from the polar into symmetric and ultimately into periodic distributions would occur, either by insertion of new, or by splitting of existing, maxima. This is inappropriate if the graded concentration profile should be used in the growing embryo as positional information for the determination of the primary body axes. Multiple maxima could lead to severe malformations such as the formation of several partially fused embryos instead of one. Observations clearly demonstrate that nature was able to solve this problem. Again Hydra is a good example. Its polar nature is maintained over a wide range of sizes.

A single maximum can be stabilized if cells distant to an established maximum lose the capability to perform the pattern forming reaction [19]. We have termed this capability of the cells "source density" (bottom distribution in Fig. 1.2A), corresponding to the observable feature of competence. Cells at larger distances become unable to compete with the primary maximum. To achieve a smoothly graded competence it is assumed that either the activator or the inhibitor has a positive feedback on the competence. Together with Eqs.(1.1) and (1.2) the feedback of the pattern forming system on the ability to perform the autocatalysis can have the following form:

$$\frac{\partial \rho}{\partial t} = c \, \frac{h}{\rho} - \mu_\rho. \qquad (1.8)$$

In this case the inhibitor, with its shallow gradient, leads to a similarly distributed source density ρ. In the example given, h increases with increasing ρ. Therefore, this feedback must be slowed down at higher ρ levels to avoid an overall explosion. This is the reason for the factor $1/\rho$ in Eq. (1.8). In a region distant to the primary activation, the initiation of secondary maxima becomes less likely due to the reduced competence. Thus, the maximum first formed at a small size becomes dominant.

In many systems, regeneration occurs in such a way that the polarity is maintained. Again, Hydra is a well-known example. This is a straightforward consequence of the model. The graded source density keeps track of the polarity. A small fragment regenerates a pattern according to the original polarity since the graded source density is always higher in those cells originally closer to the organizing region (Fig. 1.2A). These cells have a head start in the competition to form the new organizing region. Regeneration can proceed faster since no symmetry breaking is required.

According to this model, an organizing region has two opposing influences on the surrounding tissue. The long-ranging inhibitory effect prevents the formation of additional organizing regions in the surrounding tissues. A positive feedback keeps only the surrounding cells competent. Why do both effects not simply cancel each other? The two effects have different time constants. After (partial) removal of an organizing region, the inhibition has to decay rapidly in order for regulation to occur. In contrast, competence should have a much longer time constant. It has to remain almost unchanged on the timescale required for pattern regulation (Fig. 1.2). Thus, the formation of a pattern is not the only important step. To make development reproducible and to suppress malformations it is also essential that the capability to form a particular pattern fades away at the correct later stage.

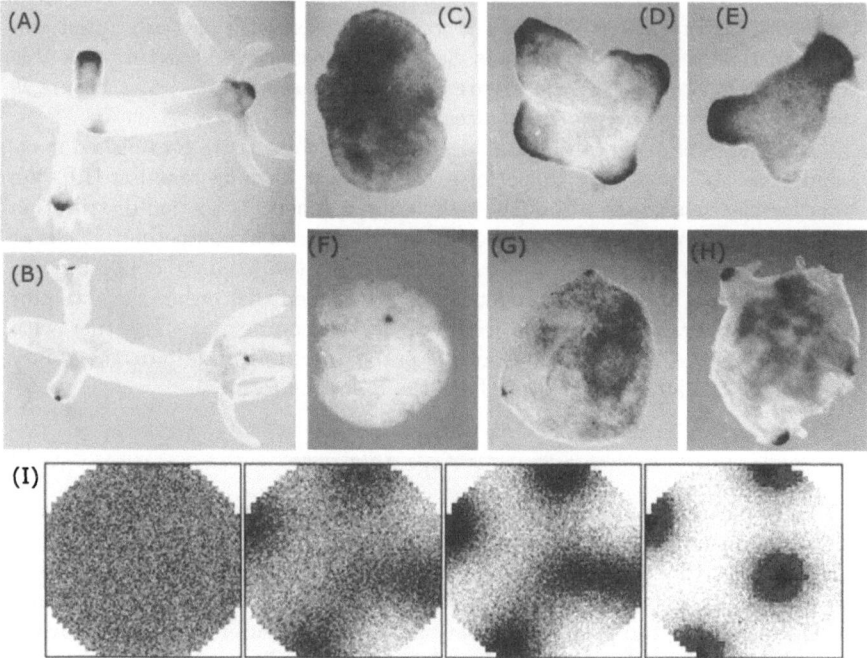

Fig. 1.3. Patterning with different wavelengths in Hydra. (A, B) While $Hy\beta$-cat (A) and Hy-Tcf expression are more smoothly distributed in the head region, Hy-Wnt expression is sharply confined to the oral opening (B). (C-E) During pattern formation in re-aggregated cells, $Hy\beta$-cat and hy-Tcf appear first uniformly distributed and subsequently become more restricted to regions that eventually form the heads. (F-H) In contrast, Hy-Wnt appears directly in sharp spots that form the future oral opening [11]. (I) The nested pattern formation can be accounted for by the assumption that $Hy\beta$-cat / Hy-Tcf concentration is the precondition to trigger Hy-Wnt (black) that becomes more localized than the initiating $Hy\beta$-cat / Hy-Tcf pattern (Photographs courtesy of B. Hobmayer, T. Holstein and colleagues; see [11]).

1.6 Evidence for patterning by multiple systems of different wavelength

If Hydra tissue is dissociated into individual cells and these cells are allowed to re-aggregate, within 4 days new heads and feet are formed [7]. This is a striking example of the formation of a pattern from an initially uniform situation. By probing for molecules that are known to play a role in organizer formation of higher organisms, gene homologues to β-catenin, wingless, Tcf [11] and Brachyury [36] have been found. All these genes are expressed in the so-called head, i.e., around the opening of the gastral column (Figs 1.1 and 1.3). Upon regeneration these molecules behave as expected from our theory: the activity reappears about 2 hours after head removal in the region where the

regenerating head is expected. Ectopic expression of *Hyβ-catenin* can induce
a secondary embryonic axis in *Xenopus*, indicating that the crucial pathways
for organizer formation are well preserved in evolution. Although all these
genes are expressed in the head region, there are differences in the expression
pattern: *Hyβ-catenin* and *Hy-Tcf* have a somewhat wider distribution. During
bud formation, these are distributed over the entire bud in a graded fashion.
Later, high levels become more restricted to the proper head region. In contrast, from the beginning *Hy-Wnt* is more confined to the surrounding of the
opening of the gastric column. Thus, both systems have different wavelength.
These different expression patterns suggest that these molecules are not just
part of the same autoregulatory loop.

Both *β-catenin/Tcf* on the one hand and *Hy-Wnt* on the other have different expression patterns in aggregates (Fig. 1.3). Taking cells from the gastric
column that do not express these genes, *Hyβ-catenin/Tcf* become first active
in a homogeneous way (Fig. 1.3 C-E; [11]). Later, isolated local high concentrations appear at spots that eventually give rise to head formation. This is
what is expected from the theory. In contrast, no such initial homogeneous
distribution has been found in *Hy-Wnt*. The activity emerges directly in very
sharp spots that later form the oral opening. This suggests that first relatively
smooth *Hyβ-catenin/Tcf* distributions are formed by a genuine patterning
process. The highest levels of these peaks are used to trigger the sharper
Hy-Wnt activity. This behaviour can be reproduced in simulations (Fig. 1.3).

This sequence of events seems to suggest a somewhat different mechanism
of using different wavelengths as suggested above: a pattern with a longer
wavelength is formed first and is used to localize a second pattern with a
shorter wavelength. However, in the adult organism the system with the longer
wavelength, *Hyβ-catenin/Tcf*, is still confined to the head region, while experiments indicate that the competence has a graded distribution over the entire
animal. The molecular nature of this gradient is not yet clear. It is known,
however, how this gradient can be modified. Diacyl-glycerol (DAC) can cause
the tissue of the whole animal to take on properties characteristic for near-
head cells [27]. Tentacles can appear anywhere along the body column, and
after foot removal, for instance, a second head regenerates instead of a foot.
This suggests that at least three systems are involved in the formation of
the head region: (i) the primary system (*Hyβ-catenin/Tcf*) that sharpens in
the course of time, (ii) an even sharper signal is triggered by high levels of
the primary system, providing the signal for the oral opening (*Hy-Wnt*). (iii)
A very long-ranging system - modifiable by DAG - that mines competence.
While arguments can be given that *Hy-Wnt* has an autoregulatory element,
so far no molecule has been found that has the characteristic properties of
the long-ranging inhibitor. The regulatory features clearly indicate that this
long-range inhibition exists but its molecular basis is unknown. The communication over the entire axis, i.e., the spread of the antagonist, must be a rapid
process since about two hours are sufficient for the re-establishment of a new
head signal after head removal. In contrast, it takes about two days to revert

the polarity of Hydra tissue, indicating that the polarity-determining gradient must have a much longer time constant.

1.7 The problem of midline formation

To proceed from a radial by symmetric to a bilateral body plan, the crucial invention was the generation of a midline, i.e., the reference point (or, better, a reference line) for mediolateral patterning. In vertebrates, this is the notochord. It controls the formation of many other structures such as the spinal chord. Theoretical considerations revealed that the generation of a single long extended structure is a difficult task that cannot be accomplished by a single pattern-forming reaction. What generates the asymmetry such that the midline can have a long extension in one dimension but only a small extension in the other direction? Early midline removal in the chick embryo leads to its regeneration [30], showing that the midline is maintained by a dynamic pattern-forming process. Saturation in the autocatalysis can lead to stripe-like patterns. However, when initiated by random fluctuations, the stripes bend, bifurcate and their width is of the same order as the distance between the stripes (Fig. 1.4A; [19]). A strengthening of the lateral inhibition is inappropriate to increase the spacing between the stripes since this would lead to a disintegration of the stripes into separate patches. A single straight stripe can be generated by the cooperation of a spot-producing and a stripe-producing system. The spot-forming system makes sure that only a single stripe is formed. Experimental observations suggest that two completely different strategies are evolved in vertebrates and insects (Figs. 1.4, 1.5; [23]).

1.7.1 Midline formation in vertebrates

One way to generate a single long extended straight stripe consists of a spot-forming system that triggers a stripe-forming system while the stripe system repels the spot system. The maximum of the spot system becomes shifted in front of the tip of the incipient stripe. In this way, the spot system elongates the stripe system [22]. This is what happens during midline formation in vertebrates (Fig. 1.4): the organizer attracts cells of the ring-shaped margin that forms the opening of the cup-shaped organism (or the ectoderm/endoderm border; i.e., the marginal zone in the frog or the germ ring in the fish). After being once close to the organizer, these cells become reprogrammed such that they are no longer attracted. They are left behind the moving organizer or Hensen's node as the midline. Thus, the midline forms sequentially from anterior to posterior by local elongation accomplished by the local organizer.

For the simulation shown in Fig. 1.4 two activator-inhibitor systems are assumed. The system a, b produces a stripe-like pattern due to saturation (via κ); the system c, d produces a spot-like pattern. The spot-activator c triggers

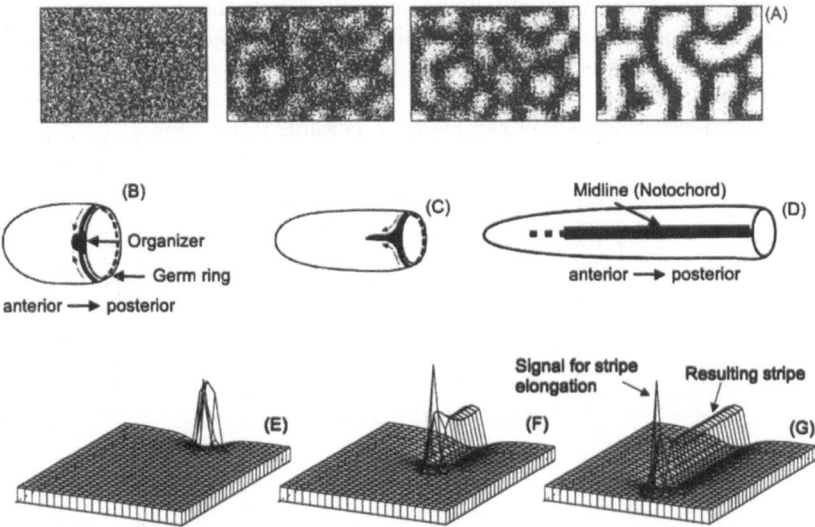

Fig. 1.4. Model for midline formation in vertebrates. (A) Saturation of autocatalysis leads to a stripe-like pattern. However, the stripes are neither straight nor is only a single stripe formed. (B-D) Proposed mechanism [22]: the spot-like Spemann-type organizer at the opening of the cup-shaped gastrula attracts cells from all around the opening (blastoporus; germ ring in the fish, marginal zone in amphibians). Cells leaving this zone form a unified band (the notochord). (E-G) Stages in the formation of a single straight line by local elongation. Two activator-inhibitor systems are involved Eqs. (1.9-1.13). The a, b-system forms stripes due to saturation ($\kappa > 0$). This system is induced by the c, d-system that has a patch-forming characteristic (no saturation). The induction results from the term $\sigma_a c$. The baseline inhibitor production σ_b makes sure that no spontaneous trigger occurs. After induction, the a system repels the c-system by removing a factor e that is required for c-autocatalysis ($-\nu e a$ in Eq. 1.13). This causes a shift of the c-signal into that neighbouring region that has the highest e-concentration. This is the region in front of the stripe since there fewer cells of the stripe system contribute to the e-removal; e is produced by all cells with the rate σ_e. Calculated with Eqs. (1.9-1.13) and $D_a = 0.001 \mu_a = 0.01; \sigma_a = 0.01; \kappa = 0.3; \rho_a = 0.01 \pm 1\%$ random fluctuation; $D_b = 0.2; \mu_b = 0.015; \sigma_b = 0.003; D_c = 0.002; \mu_c = 0.015; \sigma_c = 0.001; D_d = 0.2; \mu_d = 0.01; \sigma_d = 0.001; \sigma_e = 0.005; \mu_e = 0.005; \nu_e = 0.04; D_e = 0.03$.

the stripe activator a while the stripe activator repels the spot-activator, enforcing them to escape into an adjacent region. To orient the elongation of the stripe, it is assumed that all cells produce a factor e that is removed by the midline a; BMP would be a good candidate for the factor. The following set of equations provides an example for a possible interaction:

$$\frac{\partial a}{\partial t} = \frac{\rho_a(a^2 + \sigma_a c)}{b(1 + \kappa a^2)} - \mu_a a + D_a \Delta_a \tag{1.9}$$

$$\frac{\partial b}{\partial t} = \rho_a a^2 + \sigma_b - \mu_b b + D_b \Delta_b \tag{1.10}$$

$$\frac{\partial c}{\partial t} = \frac{\rho_c e(c^2 + \sigma_c a)}{d} - \mu_c c + D_c \Delta_c \tag{1.11}$$

$$\frac{\partial d}{\partial t} = \mu_d(ec^2 - d) + D_d \Delta_d + \sigma_d \tag{1.12}$$

$$\frac{\partial e}{\partial t} = \sigma_e - \mu_e e - \nu_e ea + D_e \Delta_e. \tag{1.13}$$

Since the autocatalysis of the signal is enhanced by e, the shift occurs towards those neighbouring cells that have the highest e concentration. Usually these are the cells at the tip of the elongating line since there fewer cells contribute to the removal of e. Therefore, the extension of the line is straight.

1.7.2 Midline formation in insects: an inhibition from the dorsal side restricts midline formation ventrally

Experimental observations in insects [5, 31] suggest a totally different mechanism for midline formation [19]. There, a dorsal organizer does not elongate but repels the midline, allowing midline formation only at the largest distance from the organizer, at the opposite ventral side. This occurs more or less simultaneously over the whole AP-extension of, for example, the fruit fly Drosophila, much in contrast to the sequential elongation at the same side in vertebrates. Fig. 1.5 provides a simulation of the proposed mechanism.

1.8 Left-right patterning: the 'left'-signal is induced and repelled by the midline

Considerable progress has been made in the understanding of left-right pattern (for a recent review see [4]). In the following, it will be shown that many observations can be accounted for by the assumption that the midline pattern induces, on a long range, a secondary system that marks the 'left' but represses it at short range (Fig. 1.6). Therefore, the 'left' system depends on the midline system, but becomes shifted to the side. Transiently the 'left' signal may appear on both sides of the midline (notochord). However, these two

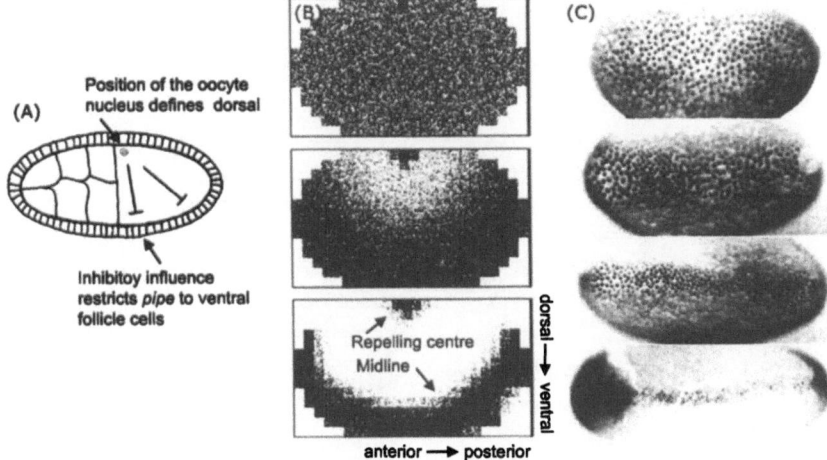

Fig. 1.5. Midline formation in insects. (A) In the *Drosophila* oocyte an interaction of the nucleus with the follicle cells leads to an organizing region that defines dorsal. An inhibitory influence of this centre leads to a restriction of *pipe* expression to the ventral follicle cell along the anteroposterior axis [31]. Via a complex cascade, this leads to a stripe-like expression of *dorsal* in the developing embryo. (B) Model for the formation of midline formation in insects [19]. A spot-like system (top in each figure) has an inhibitory effect on a stripe-forming system. This model predicted that the midline system shows a sequential sharpening (in contrast to the sequential elongation in vertebrates). (C) The *nf-kappa b/dorsal* distribution in the short germ insect *Tribolium*. The protein attains a stripe-like distribution [5]. The expected sequential sharpening is clearly visible (photograph kindly provided by Siegfried Roth, see [5]).

regions will start to compete with each other due to the long-range inhibition inherent in the left-system. As a rule, one side will win this competition. It is a property of such a system that minute asymmetries are sufficient to bias the shift in a predictable way. For mice, evidence exist that an asymmetric flow induced by flagellar movements is decisive [1, 28]. In the absence of such an asymmetry, the decision will be made at random. A good candidate for the 'left'-activator is *Nodal*. It has a positive feedback on its own activation [32, 34]. It also regulates the production of a putative inhibitor, *lefty-2* [1, 6]. The local exclusion of the midline- and 'left'-patterning systems seems to be a late event: *Nodal* is also involved early in mesoderm formation, a precondition for midline formation [6]. Thus, many of the essential ingredients required by the model seem to be present. The correct activation of *Nodal* on the left side requires a long-range communication between the left and right sides [16]. If gap-junctional communication is abolished, *Nodal* activation appears also on the right side, in obvious agreement with the model proposed. Many observations demonstrate an important role of the midline signal in the generation of

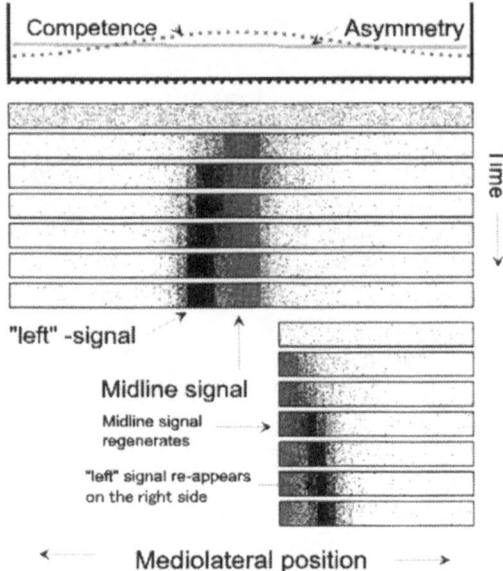

Fig. 1.6. Left-right pattern: simulations in the mediolateral plane [22]. A high concentration of an activator marks the midline (grey). It is formed at the position where the competence for this is highest (upper curve). This induces at longer range a second patterning system that marks 'left' (black). Since the latter is locally suppressed by the midline system, the 'left'-signal is repelled away from the midline. A transient activation occurs on both sides. Due to the long-range inhibition, the peaks compete and a full 'left'-maximum develops only at one side. A minute asymmetry is sufficient to determine which side will "win". After removal of the left side, the midline marker regenerate, triggering again the 'left' signal (black) that normally appears on the left side. It is now shifted to the right since an inhibition from the left side no longer exists, in agreement with the observation [15].

the left-right pattern [3]. In the model, with a reduced activity of the midline system, the squeezing into the neighbouring regions may no longer work and the left signal can remain in the center. The repression of nodal by the midline system, i.e., the notochord, has been shown [17].

The simulation in Fig. 1.6 demonstrates that the model accounts for a crucial observation. Tissue fragments derived from the right side that do not contain notochord and, of course, no activity characteristic of the left side, regenerates first notochordal marker and later *Nodal*, the marker for the left side [15]. In terms of the model, if the left side is removed, the right side no longer has a competitor. Full activation that normally occurs only on the left side will occur in the right side fragment. This outcome is independent of whether or not the left-right asymmetry still exists or not.

1.9 Conclusions

The generation of the primary body axes is certainly a most important first step in the development of higher organisms. Interactions that are based on the interaction of self-enhancing and antagonistic reactions play a crucial role therein. By linking several such systems axes can be formed perpendicular to each other with one extending along the whole anteroposterior axis. Even for the generation of a single axis, several pattern forming systems may be superimposed to cope with the wavelength problem. Thus, the spatial complexity of a developing organism can be understood in terms of an appropriate combination of elements taken from a basic toolbox supplied by the theory.

An essential property of these models is their inherent capability for self-regulation. Normal development can be restored after severe perturbation. The formation of normal animals after dissociation into single cells and re-aggregation (Fig. 1.3) is certainly a most extreme case. By these self-corrections the error propagation into subsequent levels can be avoided, making development a very robust process.

References

1. Adachi, H., Saijoh, Y., Mochida, K., Ohishi, S., Hashiguchi, H., Hirao, A., and Hamada, H. (1999). Determination of left right asymmetric expression of nodal by a left side-specific enhancer with sequence similarity to a lefty-2 enhancer. *Genes Dev.* **13**:1589-1600.
2. Bard, J.B. and Lauder, I. (1974). How well does Turing's theory of morphogenesis work? *J. theor. Biol.* **45**:501-531.
3. Bisgrove, B.W., Essner, J.J. and Yost, H.J. (1999). Regulation of midline development by antagonism of lefty and nodal. *Development* **126**:3253-3262.
4. Capdevila, J., Vogan, K.J., Tabin, C.J. and Izpisua Belmonte, J.C. (2000). Mechanisms of left-right determination in vertebrates. *Cell* **101**:9-21.
5. Chen, G., Handel, K. and Roth, S. (2000). The maternal nf-kappa b/dorsal gradient of tribolium castaneum: dynamics of early dorsoventral patterning in *a short-germ beetle. Development* **127**:5145-5156.
6. Cheng, A.M.S., Thisse, B., Thisse, C. and Wright, C.V.E. (2000). The lefty-related factor xatv acts as a feedback inhibitor of nodal signaling in mesoderm induction and l-r axis development in Xenopus. *Development* **127**:1049-1061.
7. Gierer, A., Berking, S., Bode, H., David, C.N., Flick, K., Hansmann, G., Schaller, H. and Trenkner, E. (1972). Regeneration of hydra from reaggregated cells. *Nature New Biology* **239**:98-101.
8. Gierer, A. and Meinhardt, H. (1972). A theory of biological pattern formation. *Kybernetik* **12**:30-39.
   (available at http://www.eb.tuebingen.mpg.de/abt.4/meinhardt/theory.html)
9. Grens, A., Gee, L., Fisher, D.A. and Bode, H.R. (1996). Cnnk-2, an nk-2 homeobox gene, has a role in patterning the basal end of the axis in hydra. *Dev. Biol.* **180**:473-488.
10. Harland, R. and Gerhart, J. (1997). Formation and function of Spemanns organizer. *Ann, Rev. Cell Dev. Biol.* **13**:611-667

11. Hobmayer, B., Rentzsch, F., Kuhn, K., Happel, C.M., Cramer von Laue,C., Snyder,P., Rothbacher,U. and Holstein,T.W. (2000). Wnt signalling molecules act in axis formation in the diploblastic metazoan hydra. *Nature* **407**:186-189.
12. Hodges, A. (1983). *Alan Turing: the enigma*. Simon and Schuster, New York
13. Keller, R., Danilchik, M. (1988). Regional expression pattern and timing of convergence and extension during gastrulation of Xenopus laevis *Development* **103**:193-209
14. Lacalli, T.C. and Harrison, L.G. (1978). The regulatory capacity of Turing's model for morphogenesis, with application to slime moulds. *J. theor. Biol.* **70**:273-295.
15. Levin, M. and Mercola, M. (1998). Evolutionary conservation of mechanisms upstream of asymmetric Nodal expression: reconciling chick and Xenopus. Dev. *Genetics* **23**:185-193.
16. Levin, M. and Mercola, M. (1999). Gap junction-mediated transfer of left-right patterning signals in the early chick blastoderm is upstream of Shh asymmetry in the node. *Development* **126**:4703-4714.
17. Lohr, J.L., Danos, M.C., Groth, T.W. and Yost, H.J. (1998). Maintenance of asymmetric nodal expression in Xenopus laevis. *Dev. Genetics* **23**:194.
18. Meinhardt, H. (1982). *Models of Biological Pattern Formation*. Academic Press, London
    (available at http://www.eb.tuebingen.mpg.de/abt.4/meinhardt/theory.html)
19. Meinhardt, H. (1989). Models for positional signalling with application to the dorsoventral patterning of insects and segregation into different cell types. *Development* (Supplement):169-180.
20. Meinhardt, H. (1993). A model for pattern-formation of hypostome, tentacles, and foot in hydra: how to form structures close to each other, how to form them at a distance. *Dev. Biol.* **157**:321-333.
21. Meinhardt, H. (1994). Biological pattern-formation - new observations provide support for theoretical predictions. *BioEssays* **16**:627-632.
22. Meinhardt, H. (2001). Organizer and axes formation as a self-organizing process. *Int. J. Dev. Biol.* **45**:177-188.
23. Meinhardt, H. (2002). The radial-symmetric hydra and the evolution of the bilateral body plan: an old body became a young brain. *BioEssays* **24**:185-191.
24. Meinhardt, H. (2003). *The Algorithmic Beauty of Sea Shells* (3rd enlarged edition) Springer, Heidelberg, New York
25. Meinhardt, H. and Gierer, A. (1980). Generation and regeneration of sequences of structures during morphogenesis. *J. theor. Biol.* **85**:429-450.
26. Meinhardt, H. and Gierer, A. (2000). Pattern formation by local self-activation and lateral inhibition. *Bioessays* **22**:753-760.
27. Müller, W. (1990). Ectopic head and foot formation in Hydra: Diacylglycerol-induced increase in positional values and assistance of the head in foot formation. *Differentiation* **42**:131-143.
28. Nonaka, S., Shiratori, H., Saijoh, Y. and Hamada, H. (2002). Determiantion of left-right patterning of the mouse embryo by artificial nodal flow. *Nature* **418**:96-99.
29. Pera, E.M. and Kessel, M. (1998). Demarcation of ventral territories by the homeobox gene nkx2.1 during early chick development. *Dev. Genes Evol.* **208**:168-171.
30. Psychoyos, D. and Stern, C.D. (1996). Restoration of the organizer after radical ablation of Hensen's node and the anterior primitive streak an the chick embryo. *Development* **122**:3263-3273.
31. Roth, S., Neumansilberberg, F.S., Barcelo, G. and Schüpbach, T. (1995). Cornichon and the EGF receptor signaling process are necessary for both anterior-posterior and dorsal-ventral pattern-formation in *Drosophila*. *Cell* **81**:967-978.

32. Saijoh, Y., Adachi, H., Sakuma, R., Yeo, C.Y., Yashiro, K., Watanabe, M., Hashiguchi, H., Mochida, K., Ohishi, S. and Kawabata, M. (2000). Left-right asymmetric expression of lefty2 and nodal is induced by a signaling pathway that includes the transcription factor FAST2. *Molecular Cell* **5**: 35-47.

33. Schnakenberg, J. (1979). Simple chemical reaction system with limit cycle behavior. *J. theor. Biol.* **31**:389-400

34. Schier, A.F. and Shen, M.M. (2000). Nodal signalling in vertebrate development. *Nature* **4**:385-389.

35. Smith, J.C., Conlon, F.L., Saka, Y. and Tada, M. (2000). Xwnt11 and the regulation of gastrulation in Xenopus. *Phil. Trans. R. Soc. Land. B* **355**:923-930.

36. Technau, U. and Bode, H.R. (1999). *HyBra1*, a *Brachyury* homologue, acts during head formation in Hydra. *Development* **126**:999-1010.

37. Turing, A. (1952). The chemical basis of morphogenesis. *Phil. Trans. R. Soc. Land. B.* **237**:37-72.

38. Wolpert, L. (1969). Positional information and the spatial pattern of cellular differentiation. *J. theor. Biol.* **25**:1-47.

39. Wu, L.H. and Lengyel, J.A. (1998). Role of caudal in hindgut specification and gastrulation suggests homology between *Drosophila* amnioproctodeal invagination and vertebrate blastopore. *Development* **125**:2433-2442.

40. Zaffran, S., Das, G. and Frasch, M. (2000). The nk-2 homeobox gene scarecrow (scro) is expressed in pharynx, ventral nerve cord and brain of *Drosophila* embryos. *Mech. Dev.* **94**:237-241.

Spatial Pattern Formation and Morphogenesis in Development: Recent Progress for Two Model Systems

Réka Albert[1], Robert Dillon[2], Chetan Gadgil[3], and Hans G. Othmer[3]

[1] School of Mathematics, University of Minnesota, Minneapolis, MN 55455
[2] Department of Mathematics, Washington State University, Pullman, WA
[3] School of Mathematics, University of Minnesota, Minneapolis, MN 55455

2.1 Introduction

The rapid growth in the number of identified molecular components that comprise signal transduction and gene control networks has focused attention on the difficult problem of understanding how networks of such components function and how they are controlled. In the context of developmental biology, the problem is to understand how the inherited genetic information in a fertilized egg is translated into a normal adult that may comprise several hundred different cell types arranged into numerous specialized tissues and organs. The complexity of the problem stems in part from one of the central facts of developmental biology, which is that the developmental fate of a cell in a multicellular system is determined not only by its genome, but also by its spatial position relative to other cells and the signals it receives from them. Intercellular communication, either through cell-cell contact or through release or presentation of signaling molecules, plays a significant role in determining what part of the genetic code is transcribed at a particular point in space and time in a multicellular system. While gene-sequencing has provided enormous detail concerning which genes are present in the genome of various organisms, much less is understood about the complex networks of regulatory interactions that control how genes are turned on or off at the appropriate time and place in a developing organism.

Recent experimental progress in understanding the qualitative structure or topology of some signal transduction and gene control networks [3] has stimulated new approaches to the modeling of these networks with the goal of understanding their temporal dynamics and control. One property of complex metabolic and regulatory networks which makes the search for general theoretical principles of how they function worthwhile is that they have been very highly conserved in evolution. It is frequently found that similar pathways and molecules are involved in very different stages of development within an organism, or in different organisms, perhaps with very different macroscopic effects, yet the underlying structure of the network remains much the same in very dif-

ferent contexts. This 'conservation of topological connectivity' suggests that it
may be possible to identify regulatory 'modules' within a signal transduction
or gene control network, and that one may be able to understand complex networks by first understanding the behavior of these modules, and then turning
attention to the interaction between modules. In this paper we discuss two examples, the network that controls expression of the segment polarity genes in
Drosophila melanogaster, and the network controlling anterior-posterior (AP)
patterning during limb development in vertebrates. Both examples involve signal transduction by the hedgehog-patched-smoothened pathway. In the former
the protein wingless and its receptor, frizzled, mediate signal transduction,
while in the latter interactions between growth factors and sonic hedgehog
(Shh) influence growth and patterning.

2.2 An outline of development

The adult form of an organism is achieved by numerous repetitions of three
basic types of developmental processes: (i) growth and cell division, (ii) morphogenetic movements of cells and, (iii) determination and molecular differentiation of cells. At the molecular level, determination involves transcription
of a particular battery of genes, whereas molecular differentiation usually involves the production of enzymes or structural proteins that alter the morphology and function of a cell. Both processes are often tightly coupled and
often occur in rapid succession, and therefore we simply speak of differentiation. Differentiation can occur in a single cell such as the *Drosophila* egg, or
in multicellular aggregates wherein several new cell types emerge simultaneously, as in the *Drosophila* embryo. Such space-dependent differentiation or
pattern formation results from the cellular response to an appropriate spatial
pattern of extracellular control signals that must be transduced into an intracellular signal. In a continuous state description, the signal transduction and
gene control networks that govern development can be described by a finite
number of state variables and an evolution equation that determines how the
state changes under prescribed inputs or stimuli. We denote the state vector
by $u(\cdot) \in R^n$ and write the evolution equation in the form

$$\frac{du}{d\tau} = F(u, S), \qquad (2.1)$$

where $S \in R$ represents the stimulus or input to the system. In the context
of gene control the state variables are the levels of inducers or repressors, the
state of promoters, etc. Such a continuous-state description is used later in
the context of patterning in the limb. This type of description also formed the
basis for Turing's [13] analysis of spontaneous pattern formation in reaction-
diffusion systems. More recently this approach has been used to describe the
level of mRNAs, proteins, and other components in *Drosophila* [9, 14].

Turing models involve two or more morphogens that react together and diffuse throughout the system. In Turing's original analysis no cells were distinguished *a priori*; all could serve as sources or sinks of the morphogen. In certain situations a spatially–homogeneous stationary state can, as a result of slow variation in parameters such as kinetic coefficients, become unstable with respect to small non–uniform disturbances, and this can lead to either a spatially non–uniform stationary state or to more complicated dynamical behavior. These nonuniform states may in turn lead, via an unspecified 'interpretation' mechanism, to spatially–ordered differentiation. For mathematical simplicity most analyses of Turing models deal with instabilities of uniform stationary states, but Turing himself recognized the biological unreality of this in stating that 'most of an organism, most of the time is developing from one pattern to another, rather than from homogeneity into a pattern' [13].

The coupling between growth and patterning is important in many developing systems, including the vertebrate limb [4, 8, 12] and in the formation of pigmentation patterns in fish. A Turing model can predict patterns that are qualitatively consistent with experimental observations on angelfish development [7], but finer details of the temporal patterning sequence requires the incorporation of cell movement as well [8]. The developing vertebrate limb bud more than doubles in size during determination of the digits and there are significant changes in shape as well, and therefore growth may be an important determinant of patterning.

An alternate and very different approach to the analysis of regulatory networks quantizes all important variables and describes processes such as gene transcription as either ON or OFF, with no gradation. In the 'discrete-state' approach each network node (mRNA or protein) is assumed to have a small number of discrete states and the regulatory interactions between nodes are described by logical functions similar to those used in programming. Typically time is also quantized, and the network model that describes how gene products interact to determine the state at the next time gives rise to a discrete dynamical system [2, 10, 15]. Such models are more easily analyzed in detail than continuous-state models, and may therefore provide a quick means of testing whether the topology of a proposed network is correct.

In the following section we describe one example in which the Boolean approach has been used to model control of expression of the segment polarity genes in *Drosophila*. This analysis shows that at least in some instances it is the network topology that determines the dynamics, and thus shows an extreme form of robustness with respect to variation of kinetic parameters. In this model spatial interactions are highly-localized and signaling occurs only between nearest neighbors. One outcome of the Boolean analysis is that while specifying only the topology of a control network places few constraints on the dynamics of the network, in the segment polarity network knowledge of the topology of the interactions *together with their signatures* (*i.e.*, whether a species activates or inhibits another species), produces the correct spatial patterns of gene expression in both the wild-type and mutant *Drosophila* embryos.

In the third section we describe a continuous-state model for limb development. This model is very different from the Boolean model of segment polarity genes because spatial interactions are not localized. In fact, the sources of morphogen are localized in specialized regions of the developing limb and they interact only via diffusible intermediates. In this respect it also differs from Turing models, wherein it is assumed that all cells are identical initially. The spatio-temporal distribution of the morphogens in the limb model provides a morphogenetic landscape that is thought to control gene expression in the developing limb.

2.3 Modeling the regulatory network of the segment polarity genes

The genes involved in embryonic pattern formation in the fruit fly *Drosophila melanogaster* are organized in a hierarchical cascade. These genes are expressed in consecutive stages of embryonic development in a spatial pattern that is successively more precisely-defined, the genes at one step initiating or modulating the expression of those involved in the next step of the cascade. The segment polarity genes, initiated by the pair-rule gene products in a periodic spatial pattern, define and maintain the borders between different parasegments and contribute to subsequent developmental processes.

The segment polarity genes encode for the transcription factor engrailed (EN), the cytosolic protein cubitus interruptus (CI) that can be modified into an activating or a repressing transcription factor, the secreted proteins wingless (WG) and hedgehog (HH), and the transmembrane receptor proteins patched (PTC) and smoothened (SMO). The network of interactions between these components is shown in Fig. 2.1. We discuss the hedgehog-related part of this network in detail, since its homolog has a determining role in chick limb development.

The hedgehog protein (HH) binds to its receptor patched (PTC) on the membrane of a neighboring cell. The intracellular domain of PTC forms a complex with smoothened (SMO) in which SMO is inactivated by a post-translational conformation change. Binding of HH to PTC removes the inhibition of SMO, and activates a pathway that results in the conversion of CI into the transcriptional activator CIA, which promotes *wg* and *ptc* transcription. When SMO is inactive, CI is cleaved to form CIR, a transcriptional repressor that represses *wg*, *ptc* and *hh* transcription [6]. This sub-network has a very important regulatory role and appears to be evolutionarily conserved.

The first modeling work focusing on the segment polarity gene network was done by von Dassow and collaborators [14]. They developed a continuous-state model of the core network of five genes (*en*, *wg*, *ptc*, *ci*, and *hh*) and their proteins. This model led to surprisingly robust patterning with respect to

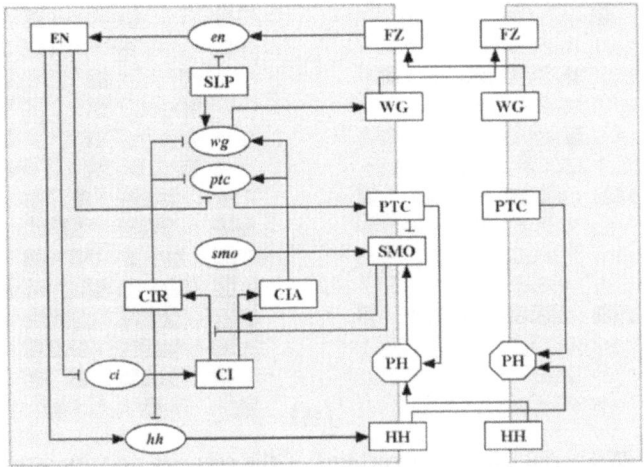

Fig. 2.1. The network of interactions between the segment polarity genes. The shape of the nodes indicates whether the corresponding substances are mRNAs (ellipses), proteins (rectangles) or protein complexes (octagons). The edges of the network signify either biochemical reactions (e.g. translation) or regulatory interactions (e.g. transcriptional activation). The edges are distinguished by their signatures, *i.e.*,whether they are activating or inhibiting. Terminating arrows ($\rightarrow$) indicate translation, post-translational modifications (in the case of CI), transcriptional activation or the promotion of a post-translational modification reaction (e.g., SMO determining the activation of CI). Terminating segments ($\dashv$) indicate transcriptional inhibition or in the case of SMO, the inhibition of the post-translational modification reaction CI$\rightarrow$CIR. (From [1].)

variations in the kinetic constants in the rate laws. This robustness suggests that the essential features involved are the topology of the segment polarity network and the signatures of the interactions in the network (*i.e.*,whether they are activating or inhibiting).

To test this hypothesis, we proposed a Boolean model based not on the biochemical details of the interactions, but rather on their net effect [1]. Motivated by the ON/OFF character of the experimentally-observed gene expression patterns, we admit only two states for each mRNA and protein in the model, corresponding to whether or not they are present. The regulatory interactions between mRNAs and proteins are modeled by Boolean (logical) functions. The state of mRNAs and proteins is updated synchronously at time intervals that correspond to the average duration of a transcription/translation process. We assume that the decay times are of the same order of magnitude as production times, so the state of a protein becomes OFF in one timestep if its mRNA is OFF.

For example, EN is translated from *en*, and therefore $EN_i^{t+1} = 1$ if $en_i^t = 1$. Since EN is a transcription factor, it is assumed that its expression will decay

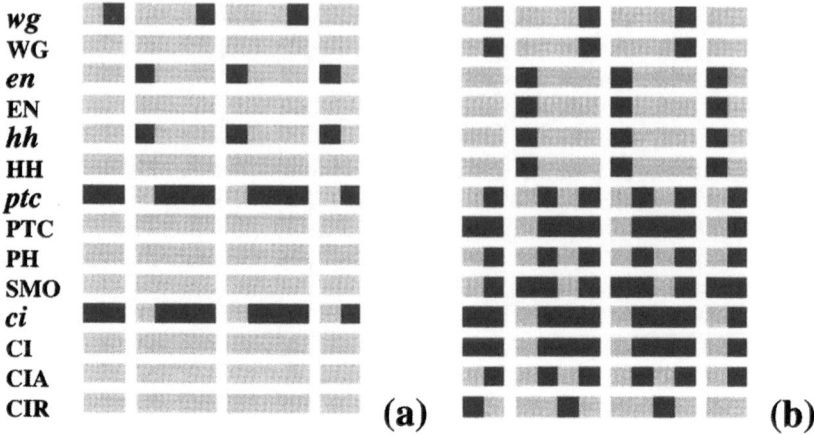

Fig. 2.2. Wild-type expression patterns of the segment polarity genes. Left corresponds to anterior and right to posterior in each parasegment. Horizontal rows correspond to the pattern of individual nodes - specified at the left side of the row - over two full and two partial parasegments. Each parasegment is assumed to be four cells wide. A black (gray) box denotes a node that is ON (OFF). (a) The experimentally-observed initial state before stage 8. *en, wg* and *hh* are expressed in one-cell-wide stripes, while the broad *ptc* and *ci* stripes are complementary to *en*. (b) The steady state of the model when initialized with the pattern in (a). This pattern is in agreement with the observed gene expression patterns during stages 9-11 [1].

sufficiently rapidly that if $en_i^t = 0$, then $EN_i^{t+1} = 0$. These two assumptions mean that EN_i^{t+1} does not depend on EN_i^t, only on the expression of *en*, and therefore $EN_i^{t+1} = en_i^t$.

As an example of transcriptional regulation, *hh* transcription is promoted by EN and inhibited by CIR, and therefore the transcription will proceed only if EN is present and CIR is absent. Since we assume that the mRNA decays in one timestep if not transcribed, *hh* expression only depends on the transcription factors, $hh_i^{t+1} = EN_i^t$ and not CIR_i^t.

We start from the experimentally known initial state of the segment polarity genes, and recover the wild type gene expression pattern as a steady state of the model (see Fig 2.2). Moreover, the model is able to reproduce all the gene patterns observed in null mutants (see Fig 2.3). In agreement with experimental results, the model indicates that the deactivation of either *en, wg* or *hh* leads to a state with no segmentation, while the deactivation of *ptc* causes the broadening of the *wg, en* and *hh* stripes. The success of the Boolean representation suggests that, indeed, the topology and signature of the regulatory interactions does determine the final gene expression pattern. Using the model we gain important insights into the functioning of the segment polarity gene network. We determine all the steady state patterns

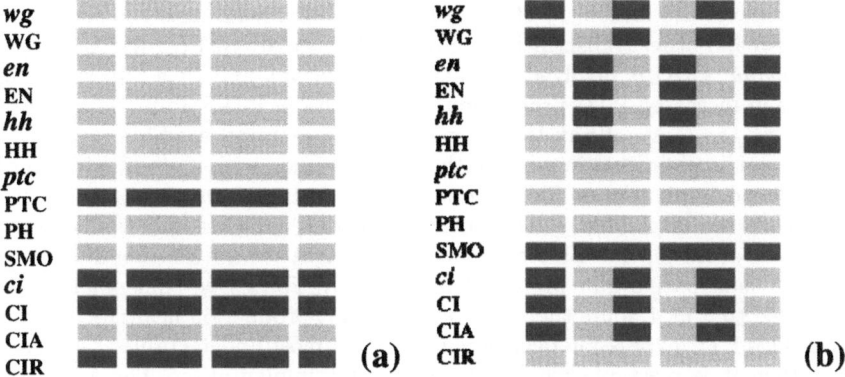

Fig. 2.3. Ectopic expression patterns obtained from the model for certain null mutants. (a) Stable pattern with no stripes for *wg*, *en*, *hh* and *ptc*. This pattern arises in *wg*, *en* or *hh* null mutants. (b) Stable pattern with broadened *wg*, *en* and *hh* stripes obtained for *ptc* null mutants.

and the domains of attraction of these steady states. We find the minimal prepatterning necessary to produce the wild-type spatial expression pattern, and thereby show that the majority of non-initiation errors in the prepattern can be corrected during the temporal evolution from such initial states (see [1]).

2.4 Limb development in vertebrates

A map of the molecular interactions that control patterning in a temporal sequence from limb bud initiation at the embryo flank to digit specification and musculature development has emerged from experimental studies of specific developmental stages [11]. We focus on patterning that leads to digit specification, which occurs after Hamilton Hamburger stage 18. At stage 18 a well-developed zone of polarizing activity (ZPA) comprised of cells that produce and secrete the vertebrate equivalent of hedgehog, Sonic Hedgehog (Shh) is present at the posterior end of the limb bud. A positive feedback interaction, mediated by diffusible intermediates, controls production of Shh and that of one or more fibroblast growth factors (FGFs) produced in the apical ectodermal ridge (AER), a specialized region at the tip of the growing limb. Cell proliferation, and thereby limb outgrowth, is controlled by the FGF concentration.

As in *Drosophila*, transduction of the extracellular Shh signal is mediated by the transmembrane receptor PTC. In the absence of Shh, PTC acts catalytically to inhibit SMOothened (SMO). Inhibited SMO cannot activate downstream transcription factors in the Gli family, but binding of Shh to

PTC relieves the inhibitory effect of PTC on SMO, leading to the activation of Gli1-3 and other transcription factors. In particular, Gli1 (equivalent of the *Drosophila* CIA) is involved in transcription of *ptc*, and thus Shh signaling results in increased PTC production. An increase in PTC production in turn decreases the range of Shh signaling by binding extracellular Shh.

Recent experiments in which either beads soaked with a modified form of Shh ('bead implant'), or Shh-producing cells derived from the ZPA ('tissue implant') are implanted at the anterior end of a growing bud lead to very different results [5]. The Shh in the beads is thought to have a higher diffusivity *in vivo* than native Shh produced by the ZPA. For tissue transplants, it is observed that at 4 hours there is a burst of PTC production near the implant site. This PTC expression decreases until it is almost undetectable at 8 hours, but then it is reestablished by 16 hours after implantation. There is little apparent effect on the PTC expression in the posterior region of the limb bud. For bead implants it was observed that PTC expression in the posterior region increases rapidly, *prior to any PTC production near the implant site*. This region of PTC expression expands anteriorly until the entire AP region shows high PTC expression. Later, PTC expression in the posterior region decreases until the PTC expression domain in the posterior region is reduced to its pre-implant size, whereas the PTC expression near the implant site is maintained.

To explain these observations, we have constructed a new model for the Shh signal transduction pathway in which SMO can exist in two forms: an activated form and an inactive form. Inactivation of SMO is catalyzed by ligand-free PTC, whereas the Shh-PTC complex catalyzes SMO activation. The downstream transcription and translation steps are collapsed into one, in which the rate of PTC production is directly proportional to the fraction of SMO in the active form. This reaction scheme is coupled to a diffusion equation for Shh in a growing domain. The rate of FGF production depends on the local concentration of Shh at the AER, and the rate of Shh production in turn depends on the local FGF concentration in the ZPA region. The limb bud mesoderm is treated as a fluid, and tissue growth is modeled via a source term in the fluid equations that depends on the local FGF concentration. Limb bud growth and shape are determined by solving a coupled system of fluid, chemical and elastic boundary equations [4].

We used this computational model to explain the results of the bead and tissue implants. Comparison of the experimental and computational results in Figures 2.4 and 2.5 shows that the model captures the distinct PTC expression responses to bead and tissue implants. In both types of implant there is an initial gradient of Shh and patched – with high levels of both near the ZPA and low levels in the anterior region near the implants. When the bead or tissue is implanted, there is a diffusive wave of free Shh that sweeps across the limb, providing an additional amount of Shh at each point along the AP axis of the limb. The magnitude of this added Shh concentration is largest near the implant and lowest near the ZPA. In the anterior region with low initial

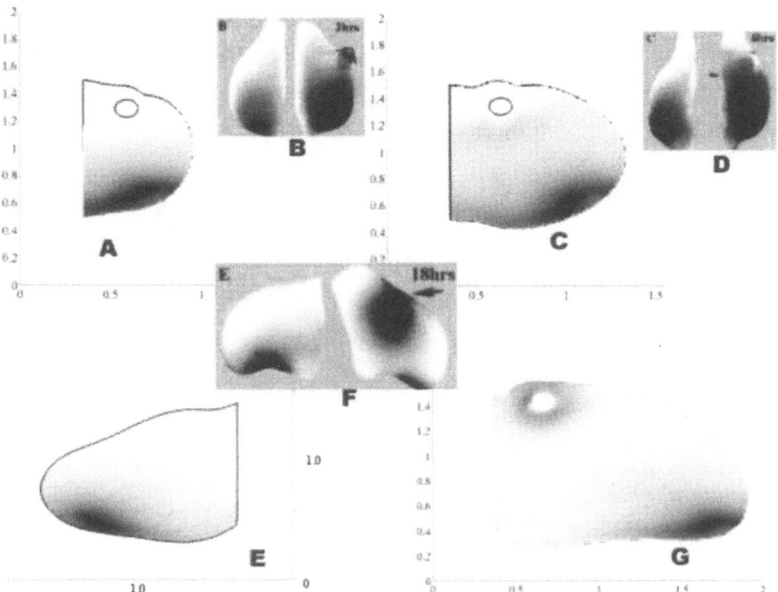

Fig. 2.4. Experimental results and model simulations for a bead implant. Figures show simulated and observed PTC expression levels at 2 (A and B), 6 (C and D) and 18 (F and G) hours after bead implantation. Figure E shows simulated limb growth in the absence of an implant. Figures B, D and F are reproduced from [5] with permission.

concentrations of Shh and patched, the PTC response occurs after a time lag that may be due to the time required for the cells to upregulate PTC and transduce the Shh signal. In the posterior regions of high initial Shh and PTC concentrations, the PTC response can be immediate for even small increases in the concentration of Shh. Thus, if the initial Shh concentration in the bead is large enough, the posterior cells can quickly upregulate PTC. If the initial bead concentration is low, the posterior cells never see enough Shh and thus the only PTC response is the lagged upregulation in the anterior where there is a sufficient level of additional Shh. The tissue implant experimental results can also be interpreted this way. Shh levels in the tissue implant are initially high, but Shh production falls off rapidly since the Fgf concentrations in the anterior are low. If this decline is rapid enough, only the anterior limb sees high enough values for the upregulation of PTC. Shh expression is later upregulated in response to FGF signals from the AER to the transplanted tissue. This in turn leads to a late upregulation of PTC in the limb bud anterior.

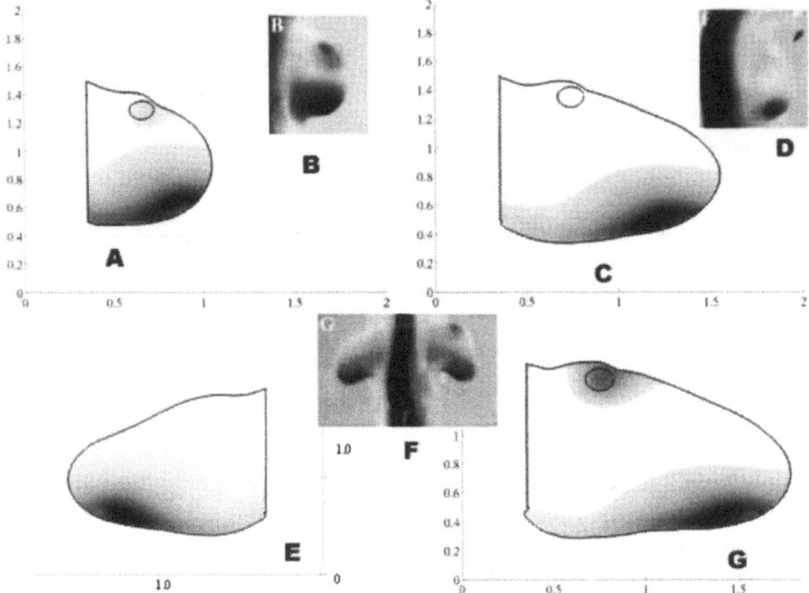

Fig. 2.5. Experimental results and model simulations for tissue implant. Figures show simulated patched expression levels at 6(A), 10(C) and 20h (G) after tissue implantation. Figures B, D and F (reproduced from [5] with permission) show observed PTC expression 4, 8 and 16 hours after tissue implantation.

2.5 Conclusions

We have presented two different modeling approaches for the analysis of patterning in different organisms at different stages of development. The *Drosophila* segment polarity network is an example in which correct patterning at the cellular level is crucial for proper development and survival of the organism. In this system nearest-neighbor interactions are important, but long-range effects are negligible. Our analysis shows that if the topology and signature of the interactions are correct, the output of the network is very robust, in that correct patterning is predicted even when all interactions in the control network are treated as all or none. We speculate that the Boolean approach is so successful in this context because the desired final pattern is also binary in nature, and an 'on-or-off' control scheme produces this robustly.

In the case of limb development in chick the morphogens are produced in specialized regions on the boundary of the limb, but diffuse and interact with receptors throughout the growing limb. Here long-range interactions are clearly important, and the feedback loop in which a graded Shh distribution leads to a graded PTC expression further shapes the Shh distribution. Of course the response is binary at the level of gene transcription, in that it is either on or off, but in the limb long-range signaling produces smooth morphogen distributions that trigger on-off responses, presumably via a set of thresholds, and a Boolean model for the PTC-SMO-Shh interactions would probably be inappropriate. Moreover, the signal transduction network is also relatively robust in the sense that it was not necessary to fine-tune parameters to obtain the desired response, but we have not made a detailed study of this aspect. Thus there is no single type of model that is appropriate in all contexts; the nature of the mathematical model chosen to represent the system must reflect the nature of the response to be predicted.

Acknowledgments

This work was supported in part by NIH Grant #GM 29123 and NSF grant DMS-9805494. Robert Dillon was supported in part by NSF grant DMS-989051 and DMS-0109957.

References

1. Albert,R.,and Othmer, H.G. (2002). The topology of the regulatory interactions predicts the expression pattern of the segment polarity genes in *Drosophila melanogaster*. Submitted to *J. Theor.Biol.*
2. Boder, J.W.,and Bradley, M.K. (2001). Programming the Drosophila Embryo. *Cell Biochem. and Biophys,* **34**:153-189.
3. Davidson, E.H., Rast, J.P., Oliveri, P., Ransick, A., Calestani, C., Yuh, C.H., Minokawa, T., Amore, G., Hinman, V., Arenasmena, C., Otim, O., Brown, C.T., Livi, C.B., Lee, P.Y., Revilla, R., Rust, A.G., Pan, Z.J., Schilstra, M.J., Clarke, P.J.C., Arnone,M.I., Rowe,L., Gameron,R.A., McClay,D.R., Hood,L., Bolour,H., Rowen,L., Cameron,R.A.,Mc,Clay.D.R., Hood,L. and Bolouri,H. (2002). A genomic regulatory network for development. *Science* **295**:1669-1678.
4. Dillon, R.and Othmer, H.G. (1999). A mathematical model for outgrowth and spatial patterning of the vertebrate limb bud. *J of Theor. Biol.*
5. Drossopoulou, G., Lewis, K.E., Sanz-Ezquerro, J.J., Nikbakht, N., McMahon, A.P., Hoffman, C.,and Tickle, C. (2000). A model for anteroposterior patterning of the vertebrate limb based on sequential long- and short-range Shh signalling and Bmp signalling. *Development* **127**:1337-1348.
6. Ingham, P.W. and McMahon, A.P. (2001). Hedgehog signaling in animal development: paradigms and principles. *Genes Dev.* **15**:3054-3087.
7. Kondo, S.,and Asai, R.(1995). A reaction-diffusion wave on the skin of the marine angelfish Pomacanthus. *Nature* **376**:(6543):765-768.
8. Painter, K.J., Maini, P.K., and Othmer, H.G. (1999). Stripe formation in juvenile Pomacanthus explained by a generalized Turing mechansim with chemotaxis. *Proc. Net'l.Acad. Sci.* **96**(May):5549-5554.
9. Reinitz, J., and Sharp, D.H. (1995). Mechanism of eve stripe formation.*Mechanisms of Development* **49**:133-158.
10. Sánchez, L., and Thieffry, D. (2001). A logical analysis of the *Drosophila* gap-gene system. *J.theor.Biol.* **211**:115-141.

11. Schaller, S.A., Li, S., Ngo-Muller, V., Han, M.J., Omi, M., Anderson, R. and Muneoka, K. (2001). Cell biology of limb patterning. *Int Rev Cytol* **203**:483-517.
12. Schnell, S., Painter, K,J., Maini, P.K., and Othmer, H.G. (2001). *Mathmatical Models for Biological Pattern Formation. Maini, P.K. and Othmer, H.G. eds.* New York:Springer-Verlag.Chap.Spatiotemporal pattern formation in early development: A review of primitive streak formation and somitogenesis.
13. Turing,A.M. (1952). The chemical basis of morphogenesis. *Phil. Trans.* **237**:37-72.
14. von Dassow,G., Meir,E., Munro,E.M., and Odell,G.M. (2000). The segment polarity network is a robust developmental module. *Nature* **406**:188-192.
15. Yuh,C.-H., Bolouri,H., and Davidson,E.H. (2001). Cis-regulatory logic in the endo16 gene: switching from a specification to a differentiation mode of control. *Development* **128**:617-629.

3

Alan Turing and "The Chemical Basis of Morphogenesis"

Vidyanand Nanjundiah

Indian Institute of Science and Jawaharlal Nehru Centre for Advanced Scientific
Research Bangalore 560012 India
E-mail: vidya@ces.iisc.ernet.in

3.1 Introduction

A. M. Turing (1912-1954) was one of the foremost thinkers of the 20th century. He contributed at least three original ideas in his life, a claim that not
many can make. They were the 'Turing machine', the 'Turing test' and the
'Turing instability'. The aim of this article is to make some comments on the
last of these. It forms part of the only biological work by him published in
his lifetime, entitled "The Chemical Basis of Morphogenesis" [24]. My comments begin as questions pertaining to its background, immediate reception
and subsequent impact from the viewpoint of biology. No attempt is made
to discuss the mathematical aspects of Turing's analysis, interesting as they
are, because to do so would be redundant for the readers of this volume. This
being Turing's sole publication dealing with a conventional biological problem
(i.e., ignoring for the moment what he had to say on the question of Mind),
I will use "Turing's paper" and "Turing's contribution to biology" as if the
two terms were interchangeable. Strictly speaking, this is not correct. Wardlaw [26] finished and wrote up on his own the ideas about phyllotaxis that he
had been discussing with Turing. There is also a volume of Turing's Collected
Works that includes later, incomplete developments on morphogenesis which
appeared after his death [20]. More pertinently, neither contains essential information bearing on our assessment of *The Chemical Basis of Morphogenesis*
– which, after all, is the only thing that Turing did that many biologists know,
or know about.

The questions that I pose are: (a) What are the essential features of this
work? (b) How did biologists react to its publication at the time or soon
thereafter? (c) How are "Turing models" for biological patterning seen today?
(d) Who were Turing's predecessors; in particular, was D'Arcy Thompson one?
(e) Finally, how should one assess the significance of Turing's contribution?

In brief, my answers are as follows. (a) The central feature of Turing's
approach is its originality; (b) in the beginning, biologists responded either
coolly or not at all; (c) the paper has acquired an iconic status today, which

has made it difficult to assess its relevance; (d) the one person with a possible claim to have anticipated Turing's approach technically, N. N. Rashevsky, had hardly any impact, either on Turing or since, whereas D'Arcy Thompson was a genuine predecessor in that he too looked for solutions to the problems of form and, as with Turing, his attitude was anti-evolutionary (in D'Arcy Thompson's case, explicitly anti-Darwinian); and (e) if we wish to assess the relevance of Turing's ideas for biological patterning, the appropriate way to do so is – paradoxically – from an evolutionary perspective.

Before expanding on these points, I should like to indicate a few general references for material pertaining to Turing. The biography by Hodges [7] is readable and extremely informative. Its contents keep getting updated on the web site maintained by Hodges (for a guide see http://www.turing.org.uk/ turing). A second web site, maintained by Swinton (http://www.swintons. net/jonathan/turing.htm), contains a higher proportion of science and is almost as indispensable. Saunders [20] provides a succinct and penetrating assessment of Turing's contributions to biology. He touches on some issues that I raise here, especially the importance of asking how Turing's views fit, or do not fit, the conventional picture of how evolution operates. A recent book by Keller [9] looks at Turing's paper as part of the broader enterprise of mathematical biology; she raises questions very similar to mine (see especially chapter 3). Those who take a look at these and other writings on Turing will see how much of what follows has been borrowed from others.

3.2 Features of the paper

In "The Chemical Basis of Morphogenesis" Turing tried to demonstrate that systems of mutually reacting and diffusing chemicals could be used to illustrate the mysterious origin of biological form within a previously formless structure – for example within a developing embryo. His motivation for doing so seems to have been to refute the 'argument from design' once and for all, eventually by way of computer simulations of the brain. As is well known, the argument from design refers to William Paley's metaphor of watch and watchmaker. The intricacy of construction of a watch, the interdependence of its component parts and the evident function it possesses – that of recording the passage of time - means that it cannot have come together by chance; a watchmaker must have constructed it according to a plan. How much less likely, then, that a living organism, so much more complex in its construction and working than a mere watch, could have come into being without a designer. One would have thought that the explanatory power of natural selection – the 'Blind Watchmaker' [5] - would have dealt the argument from design a fatal blow. Nevertheless, variations on Paley's argument continue to survive in different guises. The development of form during embryogenesis and the abilities of the human mind provided, and in the latter case still provide today, a situation with just sufficient ignorance of mechanism for mystical and vitalist thinking to thrive. One should note that Turing's attempt to counter the argument from design was based on natural law, not on evolution, and the distinction is important (see below).

Turing's focus of attention was the dramatic breakdown of symmetry that occurs, apparently in a homogeneous and isotropic structure (the fertilized egg) and *without* any external cues. This was something that happened time after time during embryogenesis. He proved that it was possible to rig up plausible chemical reactions that, when combined with diffusion, did the job. At least two chemical species were needed and they had to be capable of stimulating or inhibiting each other's synthesis. Overall, the chemical reactions had to be non-linear. In mathematical language, Turing demonstrated that given the right set of parameters (rates of reaction, feedbacks, diffusion coefficients), the starting uniform state was unstable with respect to spontaneously occurring inhomogeneous perturbations. The instability, whose existence could be inferred by studying the behaviour of the system at short times following its onset (i.e., in a "linearised" regime), was ultimately checked by non-linear effects. This is known today as the Turing instability. The instability could occur in one or more dimensions. The outcome could be a stable, time-independent, pattern of concentrations; under certain conditions it could also be a pattern that exhibited temporal oscillations. Remarkably, in the abstract of the paper Turing says that his purpose is "to discuss a possible mechanism by which the genes of a zygote may determine the anatomical structure of the resulting organism", but what he actually deals with is many steps removed from genes.

As many have pointed out, *Morphogenesis* does not deal with morphogenesis – the generation of form – at all, but with what we have come to call pattern formation. The sorts of structures that are sought to be explained are the distribution of spots on a leopard, stripes on a zebra, arms on a starfish or tentacles on a *Hydra*, not the shape of a fish or an elephant. There is no role for growth in Turing's published model. The solution of his equations are prepatterns [22], variations in time and space that foreshadow the pattern that the relevant set of tissues will adopt finally. In order to go on and give rise to a new form they need to pass through many intervening internal states including those related to differential adhesiveness, relative movement and differential gene expression. But however inappropriate the term 'morphogenesis', it helped Turing to coin a new word, 'morphogen', for 'form producer'. It is universally understood today to mean a chemical species involved in pattern formation. Turing's own analogy was to 'evocator', the word invented by Waddington [25] to stand for something that directed a responsive tissue to react to its presence in a particular way. Turing raises the question whether the morphogens that he is speaking of are enzymes. He concludes that they are more likely to be small molecules such as hormones that are capable of diffusing in a tissue without much hindrance. Conceivably they are the products of chemical reactions catalysed by other morphogens whose synthesis is catalysed by other morphogens, ..., with the causal series ending with an enzyme whose formation is catalysed by genes. It is interesting that Turing's paper appeared one year before the double helix structure of DNA was announced (also from Cambridge). So there can be no question of his having transcriptional cascades in mind, but the analogy is suggestive.

A striking feature of *Morphogenesis* is that it seems to lack antecedents [20]. The text contains only six references. Three are to biological works; these are by W. D'Arcy Thompson, C. M. Child and C. H. Waddington. In no case does Turing make use of one or the other of these as a point of departure for his own thinking. Thompson's name would seem to be mentioned in the references because he too tried to explain biological form on the basis of physical principles; Child is quoted for his description of tentacle regeneration in *Hydra*; and Waddington for his concept of evocators. The remaining three citations are to technical points – the mathematical form of the simplest enzyme-catalysed chemical reaction as derived from the law of mass action (Michaelis and Menten), getting a value for membrane permeability (Davson and Danielli) and making use of a formula for spherical harmonics (Jeans).

As Saunders [20] says, Turing "believed that the most important problem of contemporary biology was to account for pattern and form, and, as in everything that he did, he was not to be diverted from his view by what others were doing. He also believed that the solution was to be found in physics and chemistry, and so it was to these subjects and the sort of mathematics that could be applied to them that he turned". In the event the only physics and chemistry that he made use of was Fick's law of diffusion and the law of mass action. The biological phenomena which were at the back of his mind were the tentacle pattern in *Hydra*, whorled leaf patterns and gastrulation; he was also able to show that his model could give rise to dappled patterns such as seen in animal coats. Ironically, considering his other achievements, a serious obstacle that he faced in all this was that the computational facilities available were (by today's standards) terrible.

Turing recognized that there was one aspect of symmetry that his theory could not address, namely handedness: there was no escaping from the result that for every 'left-handed' Turing instability there had to be a corresponding 'right-handed' instability. He put this aside as a problem for the future while raising the possibility that there might be inhomogeneities in the starting material (the fertilized egg) that would help to select solutions of one handedness over another – a fact that we might take for granted now.

3.3 Turing's predecessors

A number of people had raised the possibility that spatial differences in concentrations of a chemical, or in physiological states, could lead to overt differentiation in a tissue. None of them seem to have had any influence on Turing's thinking; none were cited by him.

Morgan [14] postulated the existence of an anterior-posterior material gradient along the body axis of the planarian as being responsible for the observed gradient in the time needed for the regeneration of a new head following amputation. Kolmogorov, Piskunov and Petrovsky [11] studied a one-component, one-dimensional diffusion equation with a 'source' term and showed that in an infinite medium traveling wave solutions were possible. (One should note here that the Turing instability requires at least two components.) Rashevsky [18] and Weinberg (cited by Rashevsky) too discussed the possibility of a reaction-diffusion system with spatially inhomogeneous solutions. However, unlike Turing's later approach, both Rashevsky and Weinberg tried to model the system in terms of the manner in which it developed starting from initial conditions. The generality of their approach was limited by the fact that explicit solutions could be worked out only for the simplest cases of chemical reaction. Besides, the problem that Rashevsky addressed was that of cell division, and the way he went about it neither caught on then nor looks interesting now (see [9] for more on the reactions to Rashevsky and why his enterprise of establishing a research program of mathematical biology foundered). Spiegelman [21] introduced two concepts that were to become important later, physiological competition leading to a hierarchy of tissue types and autocatalysis as a means whereby concentrations in a tissue could rise spontaneously. Rose [19] (which may have been too late anyway for Turing), following Child [3] and harking back to Morgan [14], stated that qualitative differences in a tissue might be preceded by quantitative differences in levels of regulatory substances. For example, a combination of initial differences in the rates of a reversible physiological reaction that was self-limiting, followed by differential competition in the tissue for binding a common product, could lead to binding sites of relatively low affinity being unoccupied as one moved away from a local region of high concentration of the substrate.

With the exception of Rashevsky, none of those listed here tried to tackle the question of pattern arising in a tissue that was previously uniform. Being biologists, they probably thought that complete uniformity was an idealization with no counterpart in any real embryo or tissue. Rashevsky on the other hand did consider how spatial and/or temporal differentiation might occur within a homogeneous mass, but his approach was too far removed from that of biologists for there to be any useful response from them – the dialogue that might have been never took off.

3.4 D'Arcy Thompson

Turing had a predecessor, D'Arcy Thompson, who influenced his broad philosophy if not specific approach. At first sight the two approaches seem not to resemble one another. Where D'Arcy Thompson's arguments leant heavily on geometrical reasoning, Turing depended on algebra and calculus. D'Arcy Thompson is best known for *On Growth and Form* [4], a classic in which he tried to tackle the problems of biological form by using concepts from physics and engineering and the language of geometry. He began by observing that

cells and tissues were ultimately forms of ordinary physical matter. Therefore their shapes were subject to the same laws as any other form of matter,
in particular inanimate matter. His approach was materialistic and strongly
anti-vitalist; he complains that zoologists are "deeply reluctant to compare the
living with the dead". For D'Arcy Thompson the laws of growth combined
with the principles of mechanics would yield minimum-energy configurations
that could describe form. As an attitude this was, and remains, quite the opposite of the usual approach in biology. As Bonner [4] remarks of *On Growth
and Form*, "... the reader must often be prepared for disappointment if he
expects to find immediate causes. Most experimental scientists are only men-
tally satisfied if they can understand a particular form by the configuration of
its immediate precedents, and the precedents in turn are analysed in the same
way so that an epigenetic chain is exposed; this is the basis, for instance, of
'causal embryology'. D'Arcy Thompson, on the other hand, was quite satisfied
with a mathematical description or a physical analogy." Not only was Turing's
mathematical style different, his analysis, unlike D'Arcy Thompson's, invoked
a causal chain – of a sort. In his model, once an instability got going, different modes would begin to compete with each other and lead to the selection
of the fastest-growing mode (albeit non-linearities in the chemical reactions
would make it difficult if not impossible to predict the pattern that would be
stabilized finally).

The differences hide a deeper similarity. The approaches of both Turing
and D'Arcy Thompson were at heart non-evolutionary; with D'Arcy Thompson it was explicitly anti-Darwinian. His deepest objection to Darwinian thinking seems to have been that for him the organism was a whole, an entirety,
whose physical body was marked by an attention to detail that was inevitable
for any complex physical structure whose stability had to be assured – for example a bridge. The relative proportions of such a structure had to conform
strictly to the rules imposed by mechanical engineering criteria on physical
bodies. In particular, one could not tinker with a part of the structure without
affecting the functional efficiency of the whole. In the language of physics, the
interactions between the parts would be strong. Thus, for him, the supposedly
Darwinian assertion - that evolutionary modifications could result in gradual
changes in the shape and conformation of one body part without affecting all
others, or at most affecting them only slightly - was in contradiction to the
laws of equilibrium that must govern all configurations of matter, therefore
also living matter. Notions such as natural selection modifying the shape of
a jaw or angle of a fin but doing nothing else were absurd, because these
were parts of a body that made no sense except in the context of the whole.
Besides, if mechanical considerations sufficed (as he thought they did) to explain the dramatic changes in form in an embryo as it developed into an adult,
why might similar considerations not be useful for explaining the more slight
differences in forms between individuals of one species and another?

It is true that Turing does not say anything explicitly against natural
selection. All the same, an approach that is based on physics and chemistry

as applied 'globally' - to the tissue in its entirety - is rather different from one that is based on assuming that genes and gene products are the basic elements that go into building form and pattern. The latter, local, approach, typical in contemporary developmental biology, demands a research program that is devoted to asking what makes a particular gene active in a particular cell at a particular time. That is not the same as asking what global tissue patterns are consistent with a particular combination of gene products. The local view is a static view, or a succession of static views; it forms the staple of most experimental research. The global view is inherently dynamic. The gene-based way of studying patterns has an implicit assumption embedded in it: namely, that the genes whose activities underlie the development of a particular pattern also testify, by virtue of the fact that they are genes, that the pattern must be a product of evolution.

In saying this I have in mind the commonly expressed picture of evolution as something that "ultimately" takes place in gene space. As a corollary, if a biological pattern whose genetic correlates are reasonably well-understood exhibits variations within the same or different species, there is the hope that DNA sequencing studies (say) might throw light on the genetic changes that have accompanied its evolution. On the face of it, no such hope exists in the case of a pattern that is understood as the consequence of a Turing instability (except in the unlikely case, wherein the reacting chemicals are primary gene products). Further, it is common to think of the physical and chemical properties of biological tissue as constituting constraints to evolutionary change, not as facilitators of such change. A purely physical explanation for pattern, if successful, might even raise the question whether it is meaningful to think about it in evolutionary terms at all. Thus Turing's attitude to the formation of pattern exemplified a non-evolutionary approach. In this sense it was similar in its underlying philosophy to that of D'Arcy Thompson. (An essay on Turing by J. M. Kowalik (http://ei.cs.vt.edu/~history/Turing.html) supports this. To quote: "Instead of asking why a certain arrangement of leaves is especially advantageous to a plant, he tried to show that it was a natural consequence of the process by which the leaves are produced".)

It may be worth reiterating briefly the two respects in which evolutionary explanations differ radically from non-evolutionary physical explanations. Firstly, a physical explanation of an entity implies an understanding (to whatever level is demanded) of that individual entity and that entity alone. In the real world, the rest of the universe is thought of as a nuisance; its interfering effects are sought to be minimized in the laboratory. This is reflected in the experimental physicist's attempt to 'prepare' and 'isolate' a 'system'. By doing so, and with skill and luck, the hope is to understand everything possible about the entity of interest. The evolutionary biologist's way of explaining a living organism could not be more different. To begin with, the environment of the organism is far from being something whose influence has to be minimized on the grounds that it masks the essence of its being. In fact the environment is a central element in defining what the organism is. More than that, for an evolutionist the 'minimal' object that is capable of being understood is not the individual organism at all, but rather a whole population of organisms, together constituting a recognizable class (the species) to which the starting organism belongs. Secondly, because of evolution, living creatures are products of history. They make sense - are capable of being understood - only *in the context* of their history. As with historical explanations in general, accidents and vagaries of chance, not to mention other histories, need to be taken into account. On the other hand, striving for an evolutionary explanation of a physical object would be like saying that to understand a hydrogen atom one had to examine all the hydrogen atoms in the universe, and besides, to ask how they got that way. It would be a foolish exercise.

3.5 Reactions to Turing's paper then and now

Hardly any biologist seems to have read *Morphogenesis* in the early years, or if they read it, they did not cite it. Considering that it is often referred to as a seminal piece of work, the number of citations that it received in the ten years following publication averaged less than two per year (`http://www-xdiv.lanl.gov/XCM/pearson/t_hist.gif`). One year after its publication, biology was convulsed by the implications of the paired structure of DNA; and that was followed by a dizzying pace of work on unraveling the genetic code and the various mechanisms involved in regulating gene expression. Turing's thoughts on pattern formation were far from the mainstream of biological thought.

C. H. Waddington [25] was a prominent exception to the rule. But his reaction was mixed. While welcoming the fact that Turing had attacked the problem of pattern formation at all, Waddington noted that there were unresolved issues. For one thing, one of the pleasing outcomes of the model was a 'natural' scale of length composed of a chemical rate and diffusion coefficent. Therefore the number of pattern elements within a given region of physical space must vary with the size of the region. But, said Waddington, development is a dynamic process. Any pattern that is predicted by theory must therefore also be dynamic; it too must keep changing as the system develops. How can this happen with a fixed length scale unless the pattern has some other way (not provided for in the model) of accommodating itself to changing tissue dimensions? In biological language, what Waddington was saying was that Turing patterns were qualitatively imprecise; could not regulate. The same point has been made by later workers [2]; some have tried to get around this limitation with the help of *ad hoc* measures [8, 17].

Waddington's other criticism was couched in a form that will appear familiar to theorists. He said Turing had oversimplified things, but in a way that – counter-intuitively - increased, rather than decreased, the difficulty of the problem. "He seems to have a somewhat scanty knowledge of biology and, like many of those unfamiliar with the facts of embryology, he thought that its problems, which are difficult enough in all conscience, were even harder than they actually are". In particular, Waddington questioned the need for assuming an initial homogeneous and isotropic condition: "We probably do not in practice often have to deal with completely homogeneous systems in which one can appeal only to stochastic processes to generate form as Turing did". Waddington read and commented on what Turing did because he was exceptional in the lively interest that he took in theory and mathematical modeling. For most developmental (experimental) biologists, however, Turing's paper meant little. My guess is that that remains the case today.

I know of only one early technical exposition of Turing's ideas for a biological readership, and that is in the small book *Mathematical Ideas in Biology* by Maynard Smith [12]. Maynard Smith focused on what, thanks to Gierer and Meinhardt [6], subsequently became the most common way of looking at Turing models, typified in the phrase 'short-range activation, long-range inhibition'.

A surge in theoretical and mathematical models of development began in the late 1960s and continues till today. Turing's paper soon acquired iconic status in the field and began to be extensively cited, a state of affairs that continues. On occasion the citations appear in articles concerned with phenomena that plausibly depend on an underlying Turing mechanism (e.g., [10, 13]). In other cases the link between the observed phenomenon and a Turing instability appears tenuous except to the extent that one can draw a formal analogy to 'short-range activation and long-range inhibition' [23]. The point to be noted in all this is that we do not have as yet a single case of pattern formation in a biological system which can be described by a Turing model and in which the relevant molecular species are known. This could be one reason why Turing's name appears to be widely known among theoretical biologists (and not just for *Morphogenesis*), somewhat less so among developmental biologists and hardly at all among biologists in general.

The other reason is that the tremendous advances in developmental biology in recent years, especially advances in our understanding of how spatial patterns develop, have all been by way of insights derived from cleverly designed experiments. One would be hard put to come up with an experiment in which the motivation was provided by a mathematical model, the Turing model in particular. Not only that, the findings from the experiments have been by and large accounted for – to general satisfaction - in terms of specifics involving the regulation of gene transcription, local cell-cell interactions and so on, in other words in a manner quite removed from a global picture as sketched by Turing. Let us consider a specific example. The pair-rule genes of *Drosophila melanogaster* are expressed along the body of the developing fly embryo in seven "zebra" stripes separated by gaps. This pattern would seem to be tailor-made for an explanation based on a Turing instability with seven wave-lengths. But detailed study of the genetic basis of the pattern shows that things are far more complicated and, in a sense, far more arbitrary. The spatial pattern of transcription of these genes is regulated independently, meaning in a stripe-by-stripe fashion, with the help of proteins encoded by many other genes. As has been pointed out, aptly, the stripes are made inelegantly [1]. It should be emphasized that one cannot get around the difficulty of reconciling this explanatory framework with a framework based on a Turing mechanism by simply saying that both could reflect steps within the same causal chain. After all (one might continue), morphogens must either be gene products - say proteins - or smaller molecules whose production is catalysed by gene products. This seemingly plausible argument is in fact incorrect. A gene-based, 'stripe-by-stripe' explanation of the spatial pattern of expression of pair rule genes differs fundamentally from an explanation based on a Turing instability. The former depends on local factors: the factors that regulate gene expression within a cell plus the interactions between a cell and its neighbours. In short, the individual cell is the ultimate source of the pattern. In contrast, the latter is an explanation based on global interactions; the cell plays the role of a reporter or indicator of what is essentially a systemic phenomenon.

3.6 Turing models and contemporary biology

What might be the reason why the reaction to Turing among experimental biologists has been as cool as it is? Keller [9] poses the larger question of why mathematical biology as a whole has not caught on among biologists and brings up the 'two cultures' of theoretical and experimental science as a possible reason. One can make a long list of ways in which the two approaches differ. Experimental biology is still largely descriptive - or, in speaking of cause and effect, it tends merely to push back the problem one stage, whereas theoreticians strive for 'deep' explanations. Experimentalists look for immediate causes of what they observe in terms of a particular genetic state or cell state; unlike theorists, by training they are uncomfortable with the notion of randomness or stochasticity as causes. Theorists are interested in stripping a phenomenon to its bare essentials; they strive for minimal models. On the other hand, experimenters often want to pin down all possible factors that bear on a situation; one might say that they are interested in 'maximal' models. The one explanatory framework in biology, evolution by natural selection, is too far removed from the minutae of experimental detail for it to explain why, as an explanation of a phenomenon, one proximal cause should serve better than another.

How then should one view Turing's model in the light of contemporary biology? I wish to suggest that the correct thing to do is to look at Turing's model in an evolutionary framework. This may seem odd in the light of the intrinsically anti-evolutionary methodology of global models for patterning. But global models may tell us something about the evolutionary antecedents of present-day patterns, antecedents dating from a period when the link between genes and development was not as intimate as it is today. This suggestion is based on what Newman [15] has said in the context of the very same stripe-forming mechanism in *Drosophila* that was mentioned earlier: "The complex, multicomponent segment-forming systems found in contemporary organisms (e.g., *Drosophila*) are the products of evolutionary recruitment of molecular cues .. that increase reliability and stability ..". In short, if one adopts the view that primary physical and chemical processes were far more important for building bodies during the past than they are now, and that genetic regulatory mechanisms were themselves selected because they made development increasingly reliable, it may be that global systems for patterning (based on physics and chemistry) became gradually overlaid by detailed gene-based regulatory mechanisms that ensured that variations were kept to a minimum [16]. Seen from this perspective, it is noteworthy that Turing models are being actively explored in palaeobiological studies.

Acknowledgements

I am grateful to Drs. T. Sekimura and K. Inouye for giving me the opportunity to think about Turing's paper. To Dr. E. F. Keller I express my thanks for sending me chapter 3 of her book in advance of its publication.

References

1. Akam, M. (1989). Making stripes inelegantly. *Nature* **341**: 282-283.
2. Bard, J. and Lauder, I. (1974). How well does Turing's theory of morphogenesis work? *J. theor. Biol.* **45**: 501-531.
3. Child, C. M. (1941). *Patterns and Problems in Development.* University of Chicago press.
4. D'Arcy Thompson, W. (1917; 1942; 1971). *On Growth and Form.* (Abridged edition) J. T. Bonner, ed., Cambridge University Press.
5. Dawkins, R. (1986). *The Blind Watchmaker.* New York: Norton.
6. Gierer, A. and Meinhardt, H. (1972). A theory of biological pattern formation. *Kybernetik* **12**: 30-39.
7. Hodges, A. (1983). *Alan Turing: The Enigma of Intelligence.* Burnett Books, London.
8. Hunding, A. and Sørensen, P. B. (1988). Size adaptation of Turing prepatterns. *Bull. Math. Biol.* **26**: 27-39.
9. Keller, E. F. (2002). *Making Sense of Life.* Harvard University Press.
10. Kondo, S. and Asai, R. (1995). A reaction-diffusion wave on the skin of the marine angelfish Pomacanthus. *Nature* **376**: 765-768.
11. Kolmogorov A. N., Petrovski, I. G. and Piskunov, N. S. (1937). Study of the diffusion equation with a concentration-dependent source term and an application to a biological problem. *Moscow Univ. Bull. Ser. Internat. Sect.* A 1: 1-25. (in Russian)
12. Maynard Smith, J. (1968). *Mathematical Ideas in Biology.* Cambridge University Press.
13. McNally, J. G. and Cox, E. C. (1989). Spots and stripes: the patterning spectrum in the cellular slime mould Polysphondylium pallidum. *Development* **105**: 323-333.

14. Morgan, T. H. (1905). "Polarity" considered as a phenomenon of gradation of materials. *J. exp. Zool.* **2**: 495-506.
15. Newman, S. A. (1993). Is segmentation generic? *BioEssays* **15**: 277-283.
16. Newman, S. A. (2002). Developmental mechanisms: Putting genes in their place. *J. Biosciences* **27**: 97-104.
17. Othmer, H. G. and Pate, E. (1980). Scale invariance in reaction-diffusion models of spatial pattern formation. *Proc. Natl. Acad. Sci. USA* **77**: 4180-4184.
18. Rashevsky, N. (1960). *Mathematical Biophysics: Physico-Mathematical Foundations of Biology*, Vols. 1, 2. New York: Dover, 1960.
19. Rose, S. M. (1952). A hierarchy of self-limiting reactions as the basis of cellular differentiation and growth control. *Am. Nat.* **LXXXVI**: 337-354.
20. Saunders, P. T. (1993). Alan Turing and biology. *IEEE Annals of the History of Computing* **15**(3): 33-36.
21. Spiegelman, S. (1945). Physiological competition as a regulatory mechanism in morphogenesis. *Quart. Rev. Biol.* **20**(2): 121-146.
22. Stern, C. (1968). *Genetic Mosaics and Other Essays*. Harvard University Press.
23. Theraulaz, G., Bonabeau, E., Nicolis, S. C., Solé, R. V., Fourcassié, V., Blanco, S., Fournier, R., Joly, J -L., Fernández, P., Grimal, A., Dalle, P. and Deneubourg, J-L. (2002). Spatial patterns in ant colonies. *Proc. Natl. Acad. Sci. USA* **98**: 9645-9649.
24. Turing, A. M. (1952). The chemical basis of morphogensis, *Phil. Trans. Roy. Soc.* B, **237**, 37-72.
25. Waddington, C. H. (1962). *Patterns and Problems in Development*. Columbia University Press, New York, USA.
26. Wardlaw, C. W. (1953). A commentary on Turing's diffusion-reaction theory of morphogenesis. *New Phytol.* **52**: 40-47.

4

On the Stochastic Geometry of Growth

Vincenzo Capasso

MIRIAM and Dept. of Mathematics, University of Milan
via Saldini 50, 20133 Milano, Italy.
email: vincenzo.capasso@mat.unimi.it

4.1 Introduction

The pioneering book by D'Arcy Thompson, entitled "On Growth and Form" [13], was perhaps the first to consider applying (deterministic) mathematics to problems in biology, in particular those problems associated with the growth of biological objects. However most people nowadays are aware of the fact that we cannot ignore stochasticity in real biological phenomena. The scope of this chapter is to introduce relevant nomenclature and mathematical methods for the analysis of geometries related to stochastic birth-and-growth processes, thus providing a guided tour in a selected bibliography. First of all let us introduce the current terminology in this framework. The most important chapters of mathematical interest in this context are the following, for which we refer to the relevant literature: While *MORPHOGENESIS (PATTERN FORMATION)* deals with the mathematical modelling of the causal description of a pattern (a direct problem) [11, 14] (see Fig. 4.1), *STOCHASTIC GEOMETRY* deals with the analysis of geometric aspects of "patterns" subject to stochastic fluctuations (direct and inverse problems) [1, 12] (see Figs. 4.2, 4.3).

4.2 Elements of stochastic geometry

The scope of stochastic geometry is the mathematical and statistical analysis of the spatial structure of patterns which are random in location and shape. In this context the mathematical interest is in spatial *occupation*, so that geometric measure theory is involved in the presence of stochastic fluctuations.

Examples are provided by forest growth, tumor growth, crystallization processes, etc. (see also various contributions to this proceedings, and in particular the paper by T. Ubukata, regarding sea shells).

In a more detailed description, all these processes are birth-and-growth processes. In forest growth births start from seeds randomly dispersed in a

Fig. 4.1. A picture of a butterfly showing the complex pattern (stripes and spots) on the wing.

region of interest, and grow due to nutrients in the soil that may be randomly distributed themselves or driven by a fertilization procedure; in tumor growth abnormal cells are randomly activated and develop thanks to an underlying nutrition field driven by blood circulation (angiogenesis); in crystallization processes such as sea shells, polymer solidification, nucleation and growth may be due to an underlying biochemical field, to temperature cooling, etc. All these kinds of phenomena are subject to random fluctuations, with the same underlying field, because of intrinsic reasons or because of the coupling with the growth process.

In all the above examples a bounded region of interest $E \subset \mathbf{R}^d$ in a space of dimension $d \geq 2$ is decomposed as a random tessellation (see Figs. 4.8, 4.9).

Given a random object $\Sigma \in \mathbf{R}^d$, a first question that arises is *WHERE is this object located ?* Being a random object, the only answer we may provide concerns the probability that a point x belongs to Σ, or else the probability that a compact set K intersects Σ.

In the presence of stochastic fluctuations, we need to refer to an underlying probability space $(\Omega, \mathcal{A}, P)$. The theory of Choquet-Matheron [9], shows that it is possible to assign a unique probability law associated with a *RACS* (**random closed set**) $\Sigma \in \mathbf{R}^d$ on the measurable space $(\mathcal{F}, \sigma_{\mathcal{F}})$ of the family of closed sets in $\mathbf{R}^d$ endowed with the σ-algebra generated by the hit-or-miss topology, by assigning its **hitting function** T_Σ. That is we define a RACS Σ as a random object

$$\Sigma : (\Omega, \mathcal{A}, P) \longrightarrow (\mathcal{F}, \sigma_{\mathcal{F}}).$$

Then we denote by $\mathcal{K}$ the family of compact sets in $\mathbf{R}^d$. The hitting function of Σ is defined as

$$T_\Sigma : K \in \mathcal{K} \longmapsto P(\Sigma \cap K \neq \emptyset).$$

Actually we may consider the restriction of T_Σ to the family of closed balls $\{B_\varepsilon(x); x \in \mathbf{R}^d, \varepsilon \in \mathbf{R}_+ - \{0\}\}$.

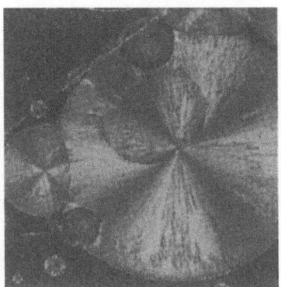

Fig. 4.2. Sweets or phtalate crystals?

Example: Homogeneous Boolean model [1, 12]

Let $N = \{X_1, X_2, \cdots, X_n, \cdots\}$ be a homogeneous spatial Poisson process in $\mathbf{R}^2$, with intensity $\lambda > 0$:

$$\forall A \in \mathcal{B}_{\mathbf{R}^2} : P[N(A) = n] = exp(-\lambda \nu^2(A))\frac{\lambda \nu^2(A)}{n!}, n \in \mathbf{N}.$$

Let $\{\Sigma_1, \Sigma_2, \cdots, \Sigma_n, \cdots\}$ be a sequence of i.i.d. RACS all having the same distribution as a *primary grain* Σ_0, (e.g. a ball with a random radius R_0)

$$\Sigma := \bigcup_n \{X_n \oplus \Sigma_n\}.$$

In this case the hitting function is given by

$$T_\Sigma(K) = 1 - exp\{-\lambda \mathbf{E}[\nu^2(\Sigma_0 \oplus \check{K})]\}, K \in \mathcal{K},$$

where

$$\check{K} = \{-x | x \in K\},$$

and

$$A \oplus B = \{x + y | x \in A, y \in B\}.$$

Example: Birth-and-growth model; an inhomogeneous Boolean model [4, 2](see Fig. 4.3).

Consider the marked point process (MPP) N defined as a random measure given by

$$N = \sum_{n=1}^{\infty} \epsilon_{T_n, X_n}$$

where

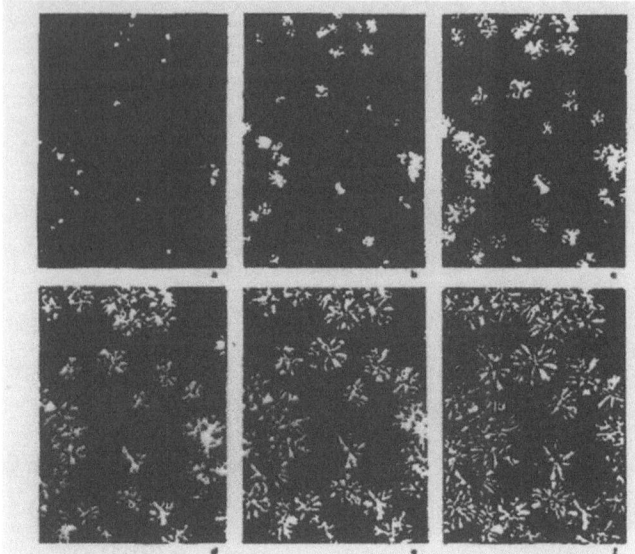

Fig. 4.3. Forest growth or crystallization process?

- T_n is an $\mathbf{R}_+$-valued random variable representing the time of birth of the $n-$th nucleus,
- X_n is an E-valued random variable representing the spatial location of the nucleus born at time T_n,
- $\epsilon_{t,x}$ is the Dirac measure on $\mathcal{B}_{\mathbf{R}_+} \times \mathcal{E}$ such that for any $t_1 < t_2$ and $B \in \mathcal{E}$,

$$\epsilon_{t,x}([t_1, t_2] \times B) = \begin{cases} 1 & \text{if } t \in [t_1, t_2], x \in B, \\ 0 & \text{otherwise} . \end{cases}$$

We have that

$$N(A \times B) = \sharp\{T_n \in A, X_n \in B\}, A \in \mathcal{B}_{\mathbf{R}_+}, B \in \mathcal{E}$$

is the (random) number of nuclei born during A, in the region B.

Let $\Theta^t_{T_n}(X_n)$ be the RACS obtained as the evolution up to time $t > T_n$ of the nucleus born at time T_n in X_n, according to some growth model. The *germ-grain* model is given by

$$\Theta^t = \bigcup_{T_n < t} [\Theta^t_{T_n}(X_n) \oplus X_n].$$

In this case, if the nucleation rate in the free space is $\lambda(t), t \in \mathbf{R}_+$, let S be the r.v. in $\mathbf{R}_+$ with probability density

$$f(s) = \frac{\lambda(s)}{\lambda(0,t)} I_{(0,t)}(s), s \in \mathbf{R}_+,$$

with $\lambda(0,t) = \int_0^t \lambda(s)ds$ and let $\Theta_S^t = \Theta_\bullet^t \circ S$. The hitting function of Θ^t is now given by

$$T_{\Theta^t}(K) = 1 - exp\{-\lambda(0,t)\mathbf{E}_S[\nu^d(\Theta_S^t \oplus \check{K})]\}, K \in \mathcal{K}.$$

Example: A tumor growth model based on an inhomogeneous Boolean model [6]

A model for tumor growth has been proposed in [6] based on a discrete time birth-and-growth model in $\mathbf{R}^d, d = 2, 3$. Given a RACS X_i representing the region occupied by the tumor mass at time $i \in \mathbf{N}$, a spatial point process $\Phi_{i+1} = \{s_k\}_{k \in \mathbf{N}}$ is generated with an intensity

$$\lambda_{i+1}(s) = \begin{cases} \lambda & if \ s \in X_i \\ 0 & if \ s \notin X_i \end{cases}$$

where λ is a given positive real number.

The region occupied by the tumor mass at time $i + 1$ is modelled as a Boolean model

$$X_{i+1} = \bigcup_k \{Z_{i+1}(s_k) | s_k \in \Phi_{i+1} \cap X_i\}$$

where the $Z_{i+1}(s_k)$ are i.i.d. balls (disks) with a random radius R_{i+1} centered at s_k (see Fig. 4.4).

The hitting function of the model is now

$$T_{X_{i+1}}(K) = 1 - exp\{-\lambda\mathbf{E}[\nu^d((Z_{i+1} \oplus K) \cap X_i)]\}, K \in \mathcal{K}, i \in \mathbf{N}.$$

Based on this model, procedures have been proposed for the estimation of the relevant parameters of birth λ and growth R_i [6].

4.2.1 Stochastic geometric measures

Consider the measure spaces $(\mathbf{R}^d, \mathcal{B}_{\mathbf{R}^d}, \nu^d)$, and $(\mathcal{F}, \sigma_{\mathcal{F}}, P_\Xi)$ where $\mathcal{F}$ is the family of closed subsets of $\mathbf{R}^d$, $\sigma_{\mathcal{F}}$ is the σ-algebra generated in $\mathcal{F}$ by the hit-or-miss topology [9], and P_Ξ is the probability measure induced by a RACS Ξ on $(\mathcal{F}, \sigma_{\mathcal{F}})$. Correspondingly we shall denote by E_Ξ expected values computed with respect to this law.

A quantitative description of the RACS Ξ can be obtained in terms of mean densities of volumes, surfaces, edges, vertices, etc., at the various Hausdorff dimensions, in the following way.

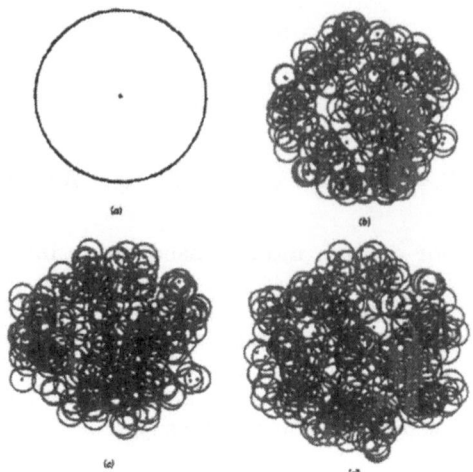

Fig. 4.4. Simulation of the growth of a tumor via a Boolean model (Picture taken from N. Cressie, Statistic for Spatial Data, Wiley, New York[6], 1993. This material is used by permission of John Wiley & Sons Inc.)

Proposition 1 *Let Ξ be a d–dimensional RACS in $\mathbf{R}^d$, having boundary of Hausdorff dimension $d-1$. The* **mean local volume density** *and* **mean local surface density**, *respectively, of the RACS Ξ at point $x \in \mathbf{R}^d$ are given by*

$$V_V(x) = \lim_{r \to 0} \frac{\mathbf{E}_\Xi[\nu^d(\Xi \cap B(x,r))]}{\nu^d(B(x,r))}$$

$$S_V(x) = \lim_{r \to 0} \frac{\mathbf{E}_\Xi[\nu^{d-1}(\partial\Xi \cap B(x,r))]}{\nu^d(B(x,r))},$$

provided that the limits exist and are a.e. finite.

It can be shown [8, 3] that

$$V_V(x) = P(x \in \Xi).$$

(see Figs. 4.5 and 4.6).

4.2.2 The dynamical case

In the dynamical case, such as a birth-and-growth process, the RACS $\Xi(t)$ may depend upon time so that a second question arises, i.e. *"WHEN"* is a point $x \in E$ reached (*captured*) by a growing stochastic region $\Xi(t)$; or vice versa up until when does a point $x \in E$ survive capture?

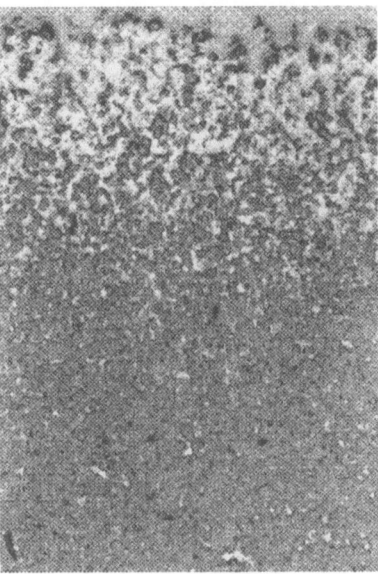

Fig. 4.5. A copper-tungsten alloy

To this aim a relevant quantity is the so-called *survival function* or *porosity* of a random set:

$$p_x(t) = P(x \notin \Xi(t)) = 1 - V_V(x,t).$$

We may introduce the (random) time $T(x)$ of survival of a point $x \in E$ with respect to its capture by the growing region $\Xi(t)$, so that

$$p_x(t) = P(T(x) > t).$$

Correspondingly the **hazard function** $h(x,t)$ can be defined as the rate of capture by the crystallization process (see Fig. 4.7, i.e.

$$h(x,t) = \lim_{\Delta t \to 0} \frac{P(x \in (\Xi(t+\Delta t)|x \notin \Xi(t))}{\Delta t}.$$

The probability density function of $T(x)$ is such that

$$f_x(t) = \frac{d}{dt}(1 - p_x(t)) = \frac{\partial V_V(x,t)}{\partial t}$$

so that

$$h(x,t) = \frac{f_x(t)}{p_x(t)}$$

52   Vincenzo Capasso

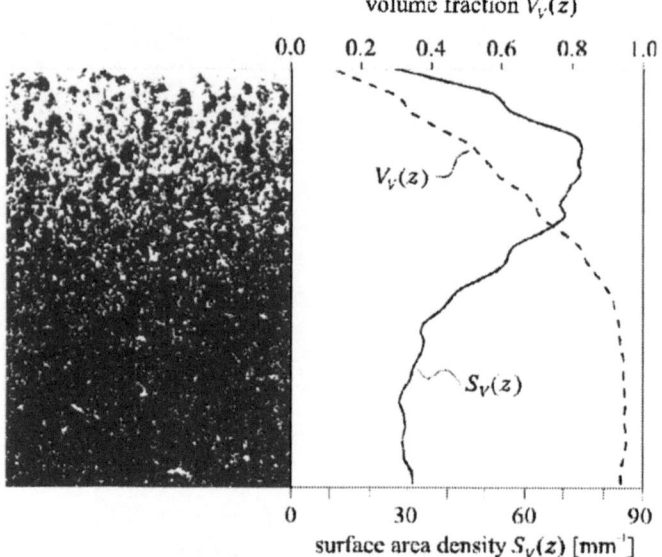

Fig. 4.6. An estimate of S_V and V_V from Fig. 4.5.

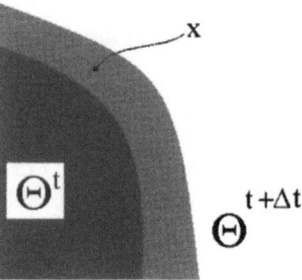

Fig. 4.7. Geometrical representation of the hazard function.

and

$$f_x(t) = p_x(t)h(x,t)$$

from which we immediately obtain

$$\frac{\partial V_V(x,t)}{\partial t} = (1 - V_V(x,t))h(x,t).$$

4.2.3 Stochastic tessellations

A random subdivision of space may need further information to be characterized. For example in the birth-and-growth process described above we may include an additional feature known as impingement, by assuming that at points
of contact of their growth front grains stop growing. In this case the spatial
region in $\mathbf{R}^d$ in which the process occurs is divided into cells (random Johnson-
Mehl tessellation [7, 10]), and interfaces (n-facets, $n = 0, 1, 2, \cdots, d$) at different Hausdorff dimensions (cells, faces, edges, vertices) appear. As above, we
may describe quantitatively the tessellation by means of mean densities of the
n-facets with respect to the d-dimensional Lebesgue measure [5].

By referring to the birth-and-growth process described in Section 4.2 we
introduce the following definitions [5]:

Let us denote by $\tau_i(y)$ the random time at which a point $y \in E$ is reached
by a grain freely grown (disregarding impingement) from a germ $a_i = (X_i, T_i)$,
where X_i is its random spatial location and T_i its random time of birth. We
admit $\tau_i(y) = \infty$, which corresponds to the case in which the point y is never
covered by the i-th grain.

Definition 1. *A* **cell** *$C_i(t)$ of the (incomplete) tessellation of the region $E \subset$
$\mathbf{R}^d$, generated by the nucleus $a_i = (X_i, T_i)$, at the time of observation t is the
non-empty set*

$$C_i(t) = \{ y \in E | \tau_i(y) \leq t \text{ and } \tau_i(y) \leq \tau_j(y) \; \forall j \neq i \}.$$

We denote by $C_e(t)$ the **empty region** at time t, i.e.

$$C_e(t) := \{ y \in E | \tau_i(y) > t, \; \forall i \in \mathbf{N} \}.$$

More generally we may call the **cell** of a random tessellation any element
of a family of RACS's partitioning the region E in such a way that any two
distinct elements of the family have empty intersection of their interiors. It is
clear that this last definition may also be used in the static (time independent)
case.

Let us now introduce a rigorous concept of "interface" at different Hausdorff dimensions.

Definition 2. *An* **n-facet** *at time t ($0 \leq n \leq d$) is the non-empty intersection
between $m + 1$ cells, with $m = d - n$ and $k_0, \ldots, k_m \in \mathbf{N}$.*

Note that in the previous definition:

- $d =$ dimension of the space in which the tessellation takes place
- $n =$ Hausdorff dimension of the interface under consideration
- $m + 1 =$ number of cells that form such an interface

	$d = 1$	$d = 2$	$d = 3$
$n = 0$	**vertex**=intersection of 2 cells	**vertex**=intersection of 3 cells	**vertex**=intersection of 4 cells
$n = 1$	**cell**	**edge**=intersection of 2 cells	**edge**=intersection of 3 cells
$n = 2$		**cell**	**face**=intersection of 2 cells
$n = 3$			**cell**

Table 4.1. Specific meaning of "n-facets" in spaces of dimension d

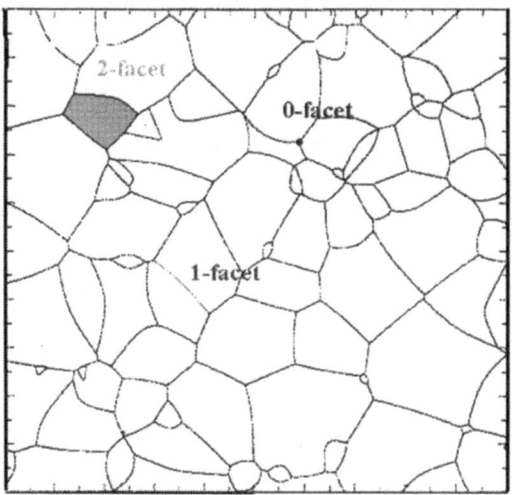

Fig. 4.8. $n-$facets for a tessellation in $\mathbf{R}^2$

(see also Table 4.1, and Fig. 4.8).

Consider now the union of all n-facets at time t, $\Xi_n(t)$.

For any Borel set B in $\mathbf{R}^d$ one can define the *mean n-facet content* of B at time t as the measure

$$\mathcal{M}_{d,n}(t, B) = E_{\Xi(t)}\left[\lambda_n(B \cap \Xi_n(t))\right] \tag{4.1}$$

where λ_n is the n-dimensional Hausdorff measure (coinciding with the n-dimensional Lebesgue measure ν_n for integer and positive values of n). Note that, with the previous definitions, $\Xi_d(t) \equiv \Xi(t)$, so that $\mathcal{M}_{d,d}(t, B)$ is the d-dimensional volume of the portion of the set B occupied by cells at time t.

If the probability distribution of the germs a_i which generate the tessellation is non-atomic, then $\mathcal{M}_{d,n}(t, \cdot) \ll \nu_d$ where ν_d is the d-dimensional

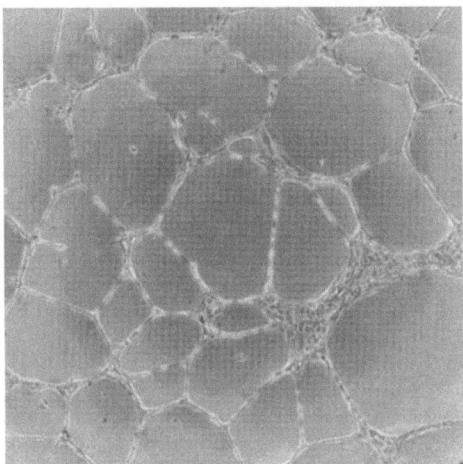

Fig. 4.9. A real picture showing a spatial tessellation due to the vascularization of biological tissue: endothelial cells form a network of blood vessels. This picture is used by permission of Dr. Federico Bussolino.

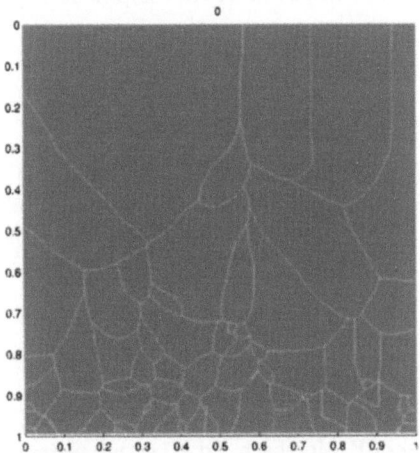

Fig. 4.10. An inhomogeneous Johnson-Mehl tessellation.

Lebesgue measure [3]. Whenever this is true, a density $\mu_{d,n}(t,x)$ exists such that for all Borel sets B in $\mathbf{R}^d$

$$\mathcal{M}_{d,n}(t,B) = \int_B \mu_{d,n}(t,x)dx. \tag{4.2}$$

Definition 3. *The function* $\mu_{d,n}(t,x)$ *defined by (4.2) is called the local mean n-facet density of the (incomplete) tessellation at time t.*

In particular $\mu_{d,d}(t,x)$ is the mean local volume density of the occupied region at time t, and $\mu_{d,d-1}(t,x)$ is the surface density of the cells. In order to obtain evolution equations for these densities additional assumptions are needed about the growth process of cells. For example for a growth along the local normal at each point x of the outer surface with a speed $G(x,t)$ and a free space nucleation rate $\lambda(x,t)$, depending both on time and space, the following evolution equations have been obtained [5]

$$\frac{\partial}{\partial t}\mu_{d,n}(x,t) = c_{d,n}\frac{(h_{m+1}(x,t))}{(m+1)!}(1 - V_V(t,x))(G(x,t))^{-m}$$

where $c_{d,n}$ is a constant depending only on the space dimension, and

$$h_k(x,t) = (h(x,t))^k, \quad k = 2,3,\cdots. \tag{4.3}$$

Acknowledgements
It is a pleasure to acknowledge the important contribution of M. Burger in Linz and A. Micheletti in Milan in the development of joint research projects relevant for this presentation.

References

1. Barndorff-Nielsen, O.E., Kendall, W.S., van Lieshout, M. N. M, Eds. *Stochastic Geometry. Likelihood and Computation*, Chapman & Hall-CRC, Boca Raton,.
2. Capasso, V. Ed. (2003). *Mathematical Modelling for Polymer Processing. Polymerization, Crystallization, Manufacturing*, Springer-Verlag, Heidelberg. In press.
3. Capasso, V., Micheletti, A. (2000). Local spherical contact distribution function and local mean densities for inhomogeneous random sets, *Stochastics and Stoch. Rep.*, **71**, 51-67.
4. Capasso, V., Micheletti, A. (2003). Stochastic geometry of spatially structured birth-and-growth processes. Application to crystallization processes. In *Spatial Stochastic Processes* (E. Merzbach, Ed.). Lecture Notes in Mathematics - CIME Subseries, Springer-Verlag, Heidelberg. In press.
5. Capasso, V., Micheletti, A., Burger, M. (2001). Densities of n-facets of incomplete Johnson-Mehl tessellations generated by inhomogeneous birth-and-growth processes. Preprint,.
6. Cressie, N. (1993). *Statistics for Spatial Data*, Wiley, New York.
7. Johnson, W.A., Mehl, R.F. (1939). Reaction kinetics in processes of nucleation and growth, *Trans. A.I.M.M.E.*, **135**, 416-458.
8. Kolmogorov, A.N. (1956). *Foundations of the Theory of Probability*, Chelsea Pub. Co., New York.
9. Matheron, G. (1975). *Random Sets and Integral Geometry*, Wiley, New York,.

10. Møller, J. (1992). Random Johnson-Mehl tessellations, *Adv. Appl. Prob.*, **24**, 814-844.
11. Murray, J.D. (1989). *Mathematical Biology*, Springer-Verlag, Heidelberg,.
12. Stoyan, D., Kendall, W.S., Mecke, J. (1995). *Stochastic Geometry and its Application*, John Wiley & Sons, New York,.
13. Thompson, D.W. (1970). *On Growth and Form (1917)*, Cambridge University Press, Cambridge,.
14. Turing, A.M. (1952). The mechanical basis of morphogenesis. *Phil. Trans. Roy. Soc. Lond.*, B **237**, 37-72.

5

The Moving Grid Finite Element Method Applied to Biological Problems

Anotida Madzvamuse[1], Roger D.K. Thomas[2], Toshio Sekimura[3], Andrew J. Wathen[1], and Philip K. Maini[4]

[1] Oxford University Computing Laboratory, Oxford OX1 3QD, U.K.
[2] Department of Geosciences, Franklin & Marshall College, Lancaster, Pennsylvania, 17604-3003, U.S.A.
[3] Department of Biological Chemistry, College of Bioscience and Biotechnology, Chubu University, Kasugai, Aichi 487-8501, Japan.
[4] Centre for Mathematical Biology, Mathematical Institute, 24-29 St Giles. Oxford OX1 3LB, U.K

Summary. This paper presents a novel numerical technique, the moving grid finite element method, to solve generalised Turing [20] reaction-diffusion type models on continuously deforming growing domains. Applications to the development of bivalve ligaments and pigmentation colour patterns in the wing of the butterfly *Papilio dardanus* will be considered, by way of examples.

5.1 Introduction

It is half a century since the appearance of Turing's seminal paper [20] on the chemical basis of morphogenesis which gave rise to the emergence of reaction-diffusion theory in developmental biology. He considered a system of two reacting and diffusing chemicals (which he termed morphogens) and demonstrated the possibility that, although in the absence of diffusion, the system tends to a linearly stable uniform steady state, in the presence of diffusion, the system evolves, due to diffusion driven instability, to a spatially non-uniform pattern. Since then, many nonlinear reaction-diffusion models have been proposed [10] and analysed, mainly on geometrically simple fixed domains.

Nature is more complicated, however. For example, butterfly wing pigmentation patterns, animal coat markings [10] and shell pigmentation patterns [8] occur on geometrically complex growing surfaces. Kondo and Asai [6] have shown that domain growth can play an important role in pattern formation. To compute the outcome of a pattern generator operating on a continuously deforming and growing domain requires novel applications of numerical computational methods. Of particular interest is the moving grid finite element method [1]. We employ this method to compute solutions of a general Turing system of two chemical morphogens on fixed, complicated and growing domains in one and two dimensions.

In section 5.2 we present the theory behind the moving grid finite element method applied to a generalised Turing reaction-diffusion model on a continuously deforming domain. Two biological applications are considered. Section 5.3 analyses the essentially one-dimensional growth patterns of certain bivalve ligaments. Section 5.4 considers the two-dimensional wing colour patterning in the butterfly *Papilio darndanus*. Finally, in section 5.5 we discuss future research.

5.2 Moving grid finite element method

We write the non-dimensional form of the two species Turing reaction-diffusion model on a continuously deforming domain $\Omega(t)$ in the form [3]:

$$\frac{\partial u}{\partial t} + \nabla \cdot (\mathbf{a}\,u) = \gamma\,f(u,v) + p_3(u,v) + \nabla^2 u, \qquad (5.1)$$

$$\frac{\partial v}{\partial t} + \nabla \cdot (\mathbf{a}\,v) = \gamma\,g(u,v) + q_3(u,v) + d\,\nabla^2 v, \qquad (5.2)$$

where $u(\mathbf{x}, t)$ and $v(\mathbf{x}, t)$ are chemical concentrations at spatial position $\mathbf{x}$ and time t. We define $\mathbf{a}(\mathbf{x}, t)$ as the velocity field. The kinetic functions f, p_3, g and q_3 describe the nonlinear reaction between the chemicals, with p_3 and q_3 bivariate cubic polynomials. Here f and g encode some of the familiar and often used reaction schemes:
(see website [1] for more details and a freely downloadable software). The parameter values γ and d represent the reaction timescale and ratio of diffusion coefficients respectively. Typically, boundary conditions on the spatial domain are either zero flux (Neumann) or fixed (Dirichlet) or both.

The model equations (5.1) and (5.2) assume that the domain $\Omega(t)$ deforms continuously and uniformly in time. This enables us to solve the system numerically by use of moving grid finite elements [1]. We first derive an equivalent weak form over a space V. Multiplying (5.1) and (5.2) by $w \in V$ and applying Green's theorem we seek to find $u, v \in V$ such that:

$$\left(\frac{\partial u}{\partial t}, w\right) + (\nabla \cdot (\mathbf{a}\,u), w) = (\gamma\,f(u,v) + p_3(u,v), w) + \left(\nabla^2 u, w\right), \qquad (5.3)$$

$$\left(\frac{\partial v}{\partial t}, w\right) + (\nabla \cdot (\mathbf{a}\,v), w) = (\gamma\,g(u,v) + q_3(u,v), w) + d\left(\nabla^2 v, w\right), \qquad (5.4)$$

where $(u, w) = \int_{\Omega(t)} u\,w\ d\Omega(t)$ is the L_2-inner product. Let $V^h \subset V$ be a finite-dimensional space consisting only of simple functions depending only

---

[1] http://web.comlab.ox.ac.uk/oucl/work/andy.wathen/software.html

on finitely many parameters. The Galerkin Formulation [12] seeks to find u^h, $v^h \in V^h$ such that

$$\left(\frac{\partial u^h}{\partial t}, w^h\right) + \left(\nabla \cdot \left(\mathbf{a}\, u^h\right), w^h\right) = \left(\gamma\, f(u^h, v^h) + p_3(u^h, v^h), w^h\right) - \left(\nabla u^h, \nabla w^h\right),$$

$$\left(\frac{\partial v^h}{\partial t}, w^h\right) + \left(\nabla \cdot \left(\mathbf{a}\, v^h\right), w^h\right) = \left(\gamma\, g(u^h, v^h) + q_3(u^h, v^h), w^h\right) - d\left(\nabla v^h, \nabla w^h\right),$$

for all $w^h \in V^h$ where zero-flux boundary conditions have been applied. Here u^h and v^h are the finite element approximations to u and v respectively, defined as

$$u^h(\mathbf{x}, t) = \sum_{i=0}^{N+1} u_i^h(t)\, \alpha_i(\mathbf{x}, \mathbf{s}(t)) \quad \text{and} \quad v^h(\mathbf{x}, t) = \sum_{i=0}^{N+1} v_i^h(t)\, \alpha_i(\mathbf{x}, \mathbf{s}(t))$$

where $\mathbf{x} \in \mathbb{R}^m$ indicates the spatial coordinates and $\mathbf{s}(t)$ represents the moving grid in time. The time derivative of u^h (or similarly v^h) is given by ([2, 5])

$$\frac{\partial u^h}{\partial t} = \sum_{i=0}^{N+1} \left[\dot{u_i}^h - \dot{x}_i\, u_x^h\right] \alpha_i(\mathbf{x}, \mathbf{s}(t)) \tag{5.5}$$

in one dimension and

$$\frac{\partial u^h}{\partial t} = \sum_{i=0}^{N+1} \left[\dot{u_i}^h - \left(\dot{x}_i\, u_x^h + \dot{y}_i\, u_y^h\right)\right] \alpha_i(\mathbf{x}, \mathbf{s}(t)) \tag{5.6}$$

in two dimensions. The effect of domain growth on the finite element formulation is to add extra terms as illustrated in (5.5) and (5.6). The spatial discretisation gives rise to a semi–discrete system of nonlinear ordinary differential equations. We use the Backward Euler finite difference scheme to discretise the ordinary differential equations in time. In one dimension the discretisation gives rise to symmetric, tridiagonal and diagonally dominant systems which can be solved using the Thomas algorithm [9]. In two dimensions we use a preconditioned Conjugate Gradient method [13].

5.3 Growth patterns in bivalve ligaments

The bivalve ligament is the uncalcified, elastic part of the bivalve shell which joins the two valves dorsally (Fig. 5.1, top panel). In the family Arcidae, these ligaments typically consist of oblique lamellar and fibrous sheets, alternating along the hinge so that their attachments on the two valves form characteristic chevron patterns. New elements are added at or near the middle of the growth zone as the ligament expands ventrally (see [7, 18] for more details). In the family Noetiidae, new elements are added to each end of the ligament,

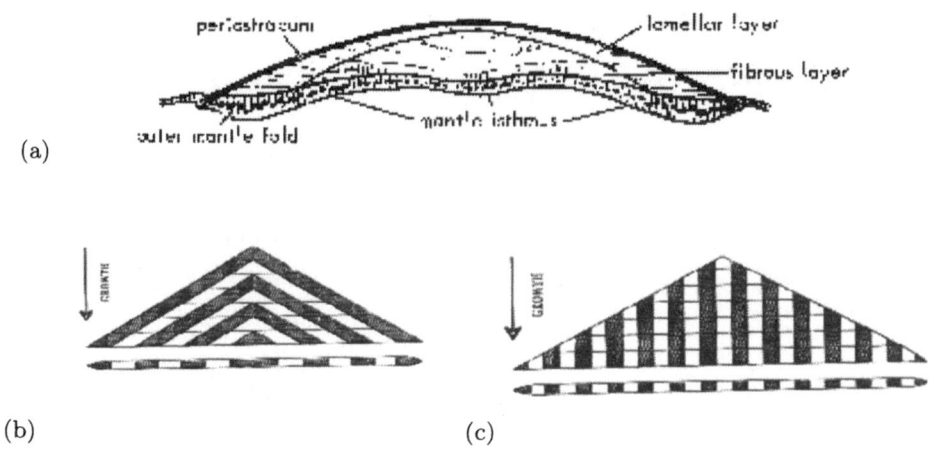

(a)

(b)   (c)

Fig. 5.1. (a): Bivalve ligament in longitudinal section (Adapted from Fig. 51, Trueman [19]). Schematic showing growth patterns of the duplivincular ligament typical of (b) *Glycymeris* (arcoid) and (c) the noetiid ligament showing a cross–sectional view of the ligament as it is inserted on the attachment area of each valve [18].

anteriorly and posteriorly (see Fig. 5.1). By solving numerically the Schnakenberg [14] reaction-diffusion model on a one-dimensional growing domain with fixed parameter values we generate a variety of patterns consistent with those observed in nature (see Fig. 5.2 top panel). Similar results can be obtained in two dimensions as illustrated in Fig. 5.2 (bottom panel). This investigation shows that the noetiid growth pattern can be derived from the chevron pattern by simply fixing the value of one of the morphogens at the centre of the domain. These results suggest that the noetiid growth pattern could have evolved independently more than once, and not necessarily from a single common ancestor as the existing classification implies.

5.4 Colour patterning in *Papilio dardanus*

For many decades scientists have been fascinated by the spectacular colour patterns of butterfly wings. On the basis of the pioneering work of Schwanwitsch [15] and Süffert [17] on the nymphalid ground plan, the seemingly complicated colour patterns on butterfly wings can now be understood as a composite of a relatively small number of pattern elements. A number of mathematical models have been put forward to account for the diversity of colour patterning (see [10, 11] for review). Nijhout [11] proposed a specific ground plan for *Papilio dardanus*, a species known for its spectacular phenotypic polymorphism in females. Recently, Sekimura *et al.* [16] proposed a global wing

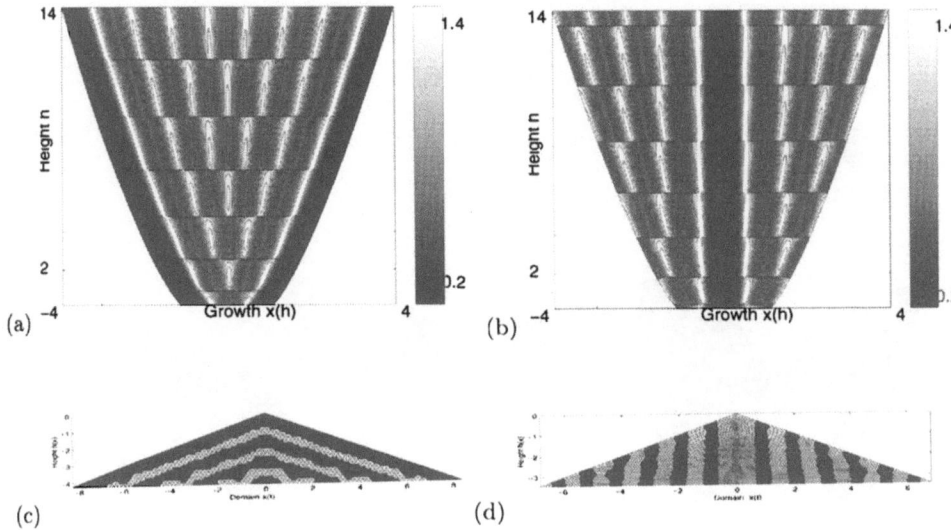

Fig. 5.2. (a), (b): Numerically computed results for u corresponding to the Schnakenberg reaction scheme with the boundaries (a) $v = 0.7$, $u_n = 0$ along $x = -1$ and $x = 1$. (b) $v_n = 0$, $u_n = 0$ along $x = -1$ and $x = 1$ with $v = 0$ along $x = 0$. (a) is consistent with arcoid patterning, while (b) is similar to the patterning observed in the noetiids (compare with Fig 5.1 (b) and (c)). (c), (d): Two-dimensional results consistent with the patterns observed in one dimension.

colouration hypothesis due to stripe-like patterns of some pigment inducing morphogens. By solving the Gierer-Meinhardt [4] reaction diffusion model on a geometrically accurate (fixed) wing shape, we can capture the details of the diverse colour patterns exhibited in the wing of *Papilio dardanus* (Fig. 5.3) by simply modifying the gradient threshold above which pigmentation occurs.

5.5 Discussion

Although we have illustrated results for the Schnakenberg and Gierer-Meinhardt reaction models, similar results can be obtained for other common kinetic models. The power of the moving grid finite element method applied to biological problems is that dynamically deforming complex geometries can be dealt with easily and efficiently without many changes in the numerical code. By solving the paradigm Turing model on continuously deforming domains we have shown that when a particular species exhibits morphological diversity, such diversity may arise through a simple modification of a basic ground plan. The generality of our numerical method allows us to study other biological patterns observed experimentally. We are currently carrying out computational

Fig. 5.3. Top panel: Some mimetic forms of females of *Papilio dardanus*. From left to right: cenea, hippocoon, and planemoides. (Courtesy of Dr. A. P. Volger and Dr. A. Cieslak of the Natural History Museum, London and Imperial College, Silwood Park). Bottom panel: Numerical results of the Gierer-Meinhardt model [4], showing patterns corresponding to hippocoon, cenea and planemoides respectively (see [16] for full details).

studies of wing development in *Papilio dardnus* from the larval imaginal disc to the adult wing, and will use our model and numerical scheme to make predictions on the effects of experimental manipulation.

Acknowledgements

This work (AJW & AM) was supported by the EPSRC Life Sciences Initiative grant (GR/R03914). PKM acknowledges support from a Royal Society Leverhulme Trust Senior Research Fellowship. This paper was written when PKM was a foreign Visiting Fellow at the Laboratory of Nonlinear Studies and Computation, University of Hokkaido, Sapporo, Japan. TS was supported in part by grant from the Human Frontier Science Program (GR0323/1999-M) and Grant-in-Aid for Scientific Research from the Ministry of Education, Science, Sports and Culture of Japan (No.14034258).

References

1. Baines, M.J. (1994). *Moving Finite Elements*. Monographs on Numerical Analysis, Oxford: Clarendon Press.

2. Baines, M.J. and Wathen, A.J. (1988). Moving finite element method for evolutionary problems. *I. Theory J. Comp. Phys.* , **79**, 245-269.
3. Crampin, E.J., Gaffney, E.A., and Maini, P.K. (1999). Pattern formation through reaction and diffusion on growing domains: Scenarios for robust pattern formation, *Bull. Math. Biol.*, **61**, 1093-1120.
4. Gierer, A. and Meinhardt, H. (1972). A theory of biological pattern formation. *Kybernetik.*, **12**, 30-39.
5. Jimack, P.K. and Wathen, A.J. (1991). Temporal derivatives in the finite element method on continuously deforming grids. *SIAM J. Numer. Anal.*, **28**, 990-1003.
6. Kondo, S. and Asai, R. (1995). A reaction-diffusion wave on the skin of the marine anglefish, Pomacanthus. *Nature*, **376**, 765-768.
7. Madzvamuse, A., Thomas, R.D.K., Maini, P.K., and Wathen, A.J. (2002). A numerical approach to the study of spatial pattern formation in the ligaments of arcoid bivalves. *Bull. Math. Biol.*, **64**, 501-530.
8. Meinhardt, H. (1995). *The Algorithmic Beauty of Sea Shells*, Heidelberg, New York, Springer-Verlag.
9. Morton, K.W. and Mayers, D.F. (1994). *Numerical Solution of Partial Differential Equations*. Cambridge University Press.
10. Murray, J.D. (1993). *Mathematical Biology*. Springer-Verlag, Berlin, 2nd edition.
11. Nijhout, H.F. (1991). *The Development and Evolution of Butterfly Wing Patterns*. Washington, DC. Smithsonian Institution Press.
12. Reddy, T.N. (1984). *An Introduction to the Finite Element Method*, McGraw-HIll.
13. Saad, Y. (1986).*Iterative Methods for Sparse Linear Systems*. PWS Publishing Co.
14. Schnakenberg, J. (1979). Simple chemical reaction systems with limit cycle behaviour. *J. Theor. Biol.*, **81**, 389-400.
15. Schwanwitsch, B.N. (1924). On the ground plan of wing-pattern in the nymphalids and certain other families of rhopalocerous Lepidoptera. *Proc. Zoo. Soc. Lond.* B, **34**, 509-528.
16. Sekimura, T., Madzvamuse, A., Wathen, A.J., and Maini, P.K. (2000). A model for colour pattern formation in the butterfly wing of *Papilio dardanus*. *Proc. Roy. Soc. Lond. B*, **267**, 851-859.
17. Süffert, F. (1927). Zur vergleichende analyse der schmetterlingszeichung. *Biol. Zbl.*, **47**, 385-413.
18. Thomas, R.D.K., Madzvamuse, A., Maini, P.K., and Wathen, A.J. (2000). Growth patterns of noetiid ligaments: Implications of developmental models for the origin of an evolutionary novelty among arcoid bivalves. *The Evol. Biol. of the Biv. Geological Soc. Lond.*, **177**, 279-289.
19. Trueman, E.R. *Ligament.* In: Moore R.C. (ed.) (1969) *Treatise on Invertebrate Paleontology.* Geo. Soc. Amer. and Univ. of Kansas. Part N. Mollusca, **6**, 58-64.
20. Turing, A. (1952). The chemical basis of morphogenesis. *Phil. Trans. R. Soc. Lond. B*, **237**, 37-72.

Part II

Morphogenesis and Pattern Formation in Animals

Regulation of Pattern Formation by the Interaction between Growth Factors and Proteoglycans

Naoto Ueno and Bisei Ohkawara

Department of Developmental Biology, National Institute for Basic Biology, 38 Nishigonaka, Myodaiji-cho, Okzaki 444-8585, Japan

6.1 Introduction

During early development, a series of cell-to-cell interactions take place in the patterning germ layers by "morphogens". Morphogens are emanated from one population of cells and act on the other population of cells and believed to direct cells to give rise to a variety of cell types. It is hypothesized that a gradient of morphogen activity induces target genes in responding cells with different thresholds of response to the morphogen, which eventually leads to the patterning of tissues. That is, extracellullar ligand signals are interpreted by target cells by cell surface receptors and cytoplasmic signal transducers causing transcriptional change as a read-out for cell differentiation. Polypeptide growth factors including the TGF-β superfamily of ligands have been implicated as morphogens in the patterning of tissues during early embryogenesis. This is well explained in *Xenopus* embryo; activin that belongs to the TGF-β superfamily, induces *goosecoid*, an organizer-specific homeobox gene that marks dorsal most mesoderm at high concentrations and *Xenopus brachyury* (*Xbra*), a pan mesodermal marker gene encoding T-box protein at lower doses. Thus, it is suggested that a concentration gradient of activin patterns mesoderm. More recently, Cyclops and Squint, two Nodal-related TGF-β signals have been found to be required for mesoderm formation and patterning in zebrafish [24]. It has also been shown that only Squint can function as a direct long-range signal [8], whereas different levels of both Squint and Cyclops can induce different downstream genes. In addition, overexpression of increasing doses of antivin progressively deleted posterior fates within the ectoderm. Together, these findings suggest that both mesoderm and ectoderm are patterned by the concentration gradient of TGF-β family members. In this model, activin/nodal and antivin are assumed to be diffusible and long-range signals forming a shallow gradient along the mesoderm and ectoderm, respectively. On the other hand, it is interesting to note that not all TGF-β family ligands appear to have long range actions. Bone morphogenetic

activin, "long range"   **BMP-4, "short range"**

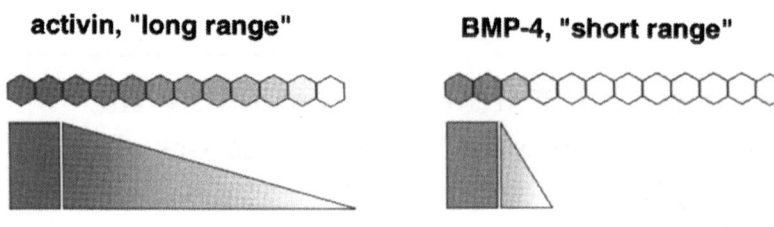

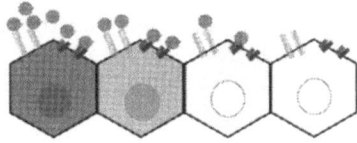

Fig. 6.1. Action range of activin and BMP. Activin is thought to act over many cell diameters in *Xenopus* mesoderm, while BMP is believed to be a short range signal that acts locally. This is possibly because BMP is trapped on the cell surface by ECMs (indicated by green).

proteins (BMPs) which constitute the largest subfamily in the TGF-β family are also implicated in the determination of ventral fate of both mesoderm and ectoderm. In contrast to the predicted long range action of activin and a nodal-related ligand Squint, BMP-2 and BMP-4 appear to diffuse inefficiently, and thus they appear to be short range signals. In fact, despite their secretory nature, both proteins are barely secreted into the medium when they are produced in mammalian cultured cells. In early embryogenesis, the presumptive neural region marked by neural-specific genes is just complementary to the BMP expression domain and no gap is observed, suggesting that BMP action is restricted to nearby BMP-producing cells. This supports the idea that, unlike activin and Squint, BMPs act over a short range. Therefore, it is predicted that some mechanisms regulate BMP diffusion creating a steep concentration gradient of BMPs, and pattern ectoderm in a nearly cell-autonomous manner (Fig. 6.1).

The above-mentioned difference in diffusion properties of polypeptide growth factors is directly linked to how the morphogen gradient is established and how tissues are patterned according to the gradient of morphogen activity. Thus, regulation of ligand diffusion is a key step for proper tissue patterning and morphogenesis. In this chapter, we address this problem by taking BMPs and Wingless-related ligands as examples of morphogens essential during early embryogenesis.

6.2 Basic amino acid core restricts diffusion of BMPs in *Xenopus* embryo

In comparing the primary structures of the TGF-β superfamily members, we noticed that some of the BMP subfamily ligands, particularly BMP-2 and BMP-4, contain a unique core of basic amino acids in their N-terminus that is highly conserved among species (Fig. 6.2A). Interestingly, a similar sequence was not found in the N-terminus of activin-βB or TGF-β2, both of which are assumed to be highly diffusible [16, 19]. Xnr-2, which is a short-range signal, contains some basic amino acids, although they are not consecutive (Fig. 6.2A). Therefore, we hypothesized that the distinct basic amino acid sequence in the BMPs might play a role in regulating the diffusion rate of these ligands in the embryo. To test this possibility, we performed a structure-function analysis of BMP-4 [9, 12], using the animal cap of *Xenopus laevis* as a model system. We constructed two cDNAs encoding *Xenopus* BMP-4 with deletions in the N-terminal region: Δ1BMP-4, lacking eight amino acids (KQQRPRKK), and Δ2BMP-4, lacking only three amino acids (RKK; Fig. 6.2B). To compare the action range of WTBMP-4 and the BMP-4 variants, we used an animal cap conjugation assay in which the animal cap from embryos injected with the mRNA for WTBMP-4 or one of the BMP-4 variants was conjugated with an untreated animal cap (Fig. 6.3A). The effective range of BMP was then examined by visualizing the nuclear transition of the phosphorylated BMP signal transducers Smad1 and Smad5, which are known to faithfully represent the activation of the BMP signal [11, 17] with an antibody (PS1) that preferentially recognizes the BMP-driven phosphorylated forms of Smad (pSmad) 1, 5, and 8 [22]. WTBMP-4-expressing donor animal caps not only induced the nuclear translocation of BMP-driven pSmads within the cap, but also influenced the juxtaposed, conjugated animal caps in the few rows of cells closest to the donor cap (Fig. 6.3B, upper left). This short-range effect was also confirmed by staining for *Xbra* mRNA (data not shown). Because essentially the same results were obtained with mouse wild-type BMP-4 and *Xenopus* BMP-2 mRNA (data not shown), the fairly short-range action of BMP-4 is likely to be conserved among animal species and the BMP subfamily. When the donor animal cap was injected with the same dose of Δ1 or Δ2BMP-4 mRNA, each producing a lower or similar amount of processed mature BMP ligand (data not shown), the effects on the recipient animal caps were pronounced: nuclear staining of pSmads (Fig. 6.3, upper right and bottom left) and induction of *Xbra* transcripts (data not shown) were observed in almost the entire recipient animal cap, suggesting that Δ1 and Δ2BMP-4 had acquired long-range effects. This long-range effect of the BMP-4 variants is reminiscent of the effect of activin (data not shown), as assessed with PS2 antibodies, which detect activin activity [22] by preferentially recognizing activin/nodal/Vg1-driven phosphorylated Smad2. These results led us to propose that the basic amino acids in the N-terminal region of BMP-4 and,

A

```
          QARHKQRKRLKSSC BMP-2
          SPKQQRPRKKNKHC BMP-4
VSGGEGGGKGGRNKRHARRPTRRKNHDDTC Dpp

          NQKTKNTIVMNTIPSRSVGKTLC Xnr-2
ECDGRTSLCCRQQFYIDFRLIGWNDWIIAPAGYYGNYC ActivinβB
```

B

SPKQQRPRKKNKHC   **BMP-4**

SPKQQRPRKKNKHC   **Δ1 BMP-4**

SPKQQRPRKKNKHC   **Δ2 BMP-4**

Fig. 6.2. Conserved N-terminal basic amino acid core in BMP family proteins. (A) BMP-2, BMP-4 and *Drosophila* Dpp have consecutive basic amino acids in their N-termini of mature proteins. (B) To test the possible role of the basic amino acid core in extracellular behaviour of BMPs, 8 amino acids including the basic core (Δ1BMP-4), or only three basic amino acids RKK (Δ2BMP-4), were deleted from BMP-4.

particularly, the RKK residues play an essential role in conferring on BMP-4 a short-range action in the animal cap.

6.3 Interaction of heparan sulfate proteoglycans with BMP through the basic amino acid core

The positively charged basic core of BMP-4 [23] may interact with the negatively charged glycosaminoglycans (GAG) of proteoglycans, which are produced in the early *Xenopus* embryo. Therefore, we next examined the possibility that the restriction of BMP-4 diffusion may depend on its interaction with the extracellular matrix (ECM). When BMP-4-expressing animal cap cells were cultured in a dissociated condition, mature WTBMP-4 and Δ2BMP-4 immunoreactivities were efficiently recovered in the medium, as determined by immunoreactivity (Fig. 6.4A). Interestingly, however, when the animal cap was kept intact and cultured in an undissociated condition, mature WTBMP-4 was not recovered in the medium, although Δ2BMP-4 was. This result demonstrates that despite having similar synthesis and secretion rates to Δ2BMP-4,

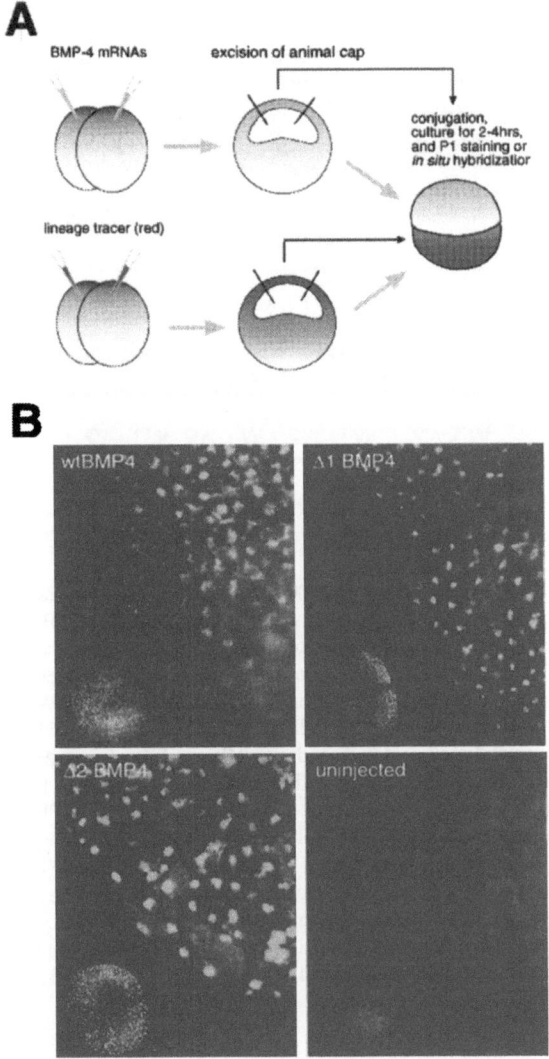

Fig. 6.3. BMP-4 variants acquired a more long-range effect compared with wild-type. (A) Schematic representation of the animal cap conjugate assay. (B) Staining with PS1 antibody showing the action range of wild type BMP-4 and its variants. Donor animal caps were either uninjected (bottom right) or injected with 300 pg mRNA of wild-type BMP-4 (upper left, wtBMP4), Δ1BMP-4 (upper right), or Δ2BMP-4 (bottom left). Recipient animal caps were injected with the lineage tracer RLDx (red) only. Nuclear staining in entire recipient animal cap conjugated Δ1BMP4, or Δ2BMP4-expressing animal cap, is evident.

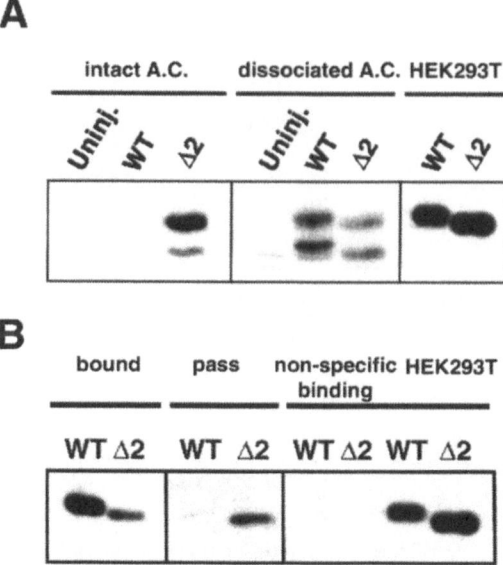

Fig. 6.4. WTBMP-4 is trapped by the ECM but the BMP-4 variants are not. (A) Mature (20 kDa) and degraded (19 kDa) forms of WT or Δ2BMP-4 in the culture medium from animal caps (A.C.) or dissociated animal cap cells were detected by Western blot analysis. For the negative controls, uninjected dissociated or undissociated animal caps were used (Uninj.). For the positive controls, WT or Δ2BMP-4 in the culture medium from HEK 293T cells was used. Although WT and Δ2BMP-4 were almost equally produced and secreted into the medium from dissociated animal cap cells (lanes 2 and 3), more Δ2BMP-4 diffused into the medium from the ECM of the cell surface of the intact (undissociated) animal cap than WTBMP-4. (B) WT or Δ2BMP-4 obtained from HEK293T cells was incubated with heparin-sepharose. The level of the mature form of the BMPs in the bound/or passed (3 and 4) fraction was detected by Western blot analysis. For the negative control lanes (non-specific binding), sepharose beads with no immobilized protein were used. WTBMP-4 was detected in the heparin bead binding fraction (bound) but not in the passed fraction (pass), while a large portion of Δ2BMP-4 was detected in the passed fraction and less in the heparin bead binding fraction.

WTBMP-4 tended to be trapped by the ECM of the cell surface. On balance, these results suggest that Δ2BMP-4 may have escaped the ECM because it lacked the three basic amino acids, RKK. Next, we analyzed the direct binding of BMP-4 to heparin-bound beads to test whether BMP-4 binds heparan sulfate proteoglycans (HSPGs), which are synthesized at high levels in early *Xenopus* embryogenesis [7] and known to affect the signaling of several growth factors [15]. We detected WTBMP-4 in the heparin-bead binding fraction but not in the passed fraction, while the Δ2BMP-4 (Fig. 6.4B) or alaBMP-4 immunoreactive protein (data not shown) was recovered in the passed fraction to an almost equal extent as in the heparin bead binding fraction. These results demonstrated that WTBMP-4 has a much higher capacity to bind heparin than does Δ2BMP-4. To examine whether WTBMP-4 binding to the GAG of HSPGs is responsible for BMP-4 being a short-range signal, we performed a conjugate assay using an animal cap from which heparan sulfate GAG was depleted. To cleave heparan sulfate GAG from the core proteins, heparitinase I, which specifically catalyses the cleavage of the glycosaminidic linkage in heparan sulfate GAG but not in heparin GAG [14], was injected into the blastocoele of the donor, the recipient, or both animal caps at stage 6.5. When either the donor or recipient animal cap was dissected from an untreated (buffer-injected) embryo, WTBMP-4 was unable to induce *Xbra* expression in the entire recipient animal cap. However, when both caps were dissected from heparitinase-treated embryos, even WTBMP-4 could induce *Xbra* expression in the opposite end of the donor cap at a high frequency (data not shown). These results demonstrated that there is at least one type of molecule trapping WTBMP-4 on the surface of animal cap cell HSPG.

6.4 Proteoglycan regulates a non-canonical Wnt signaling

Vertebrate Wnts are a group of growth factors related to *Drosophila* Wingless. Wnts regulate a number of processes during early embryogenesis including cell proliferation, differentiation, cell movement and morphogenesis. In *Xenopus* embryos, gastrulation cell movements, which include involution and convergent extension, are driven predominantly by mesodermal cells. This is particularly obvious during convergent extension, when polarized axial mesodermal cells intercalate in radial and mediolateral directions to cause dramatic elongation of the dorsal marginal zone along the antero-posterior axis. Recently, it has been reported that a non-canonical Wnt signaling cascade, which is known to regulate planar cell polarity (PCP) in *Drosophila* [1, 20] also participates in the regulation of convergent extension movements in *Xenopus* as well as in the zebrafish embryo [10, 13, 26, 30]. The zebrafish *silberblick* (*slb*) locus encodes Wnt11 and Slb/Wnt11 activity is required for cells to undergo correct convergent extension movements during gastrulation. The signal transducer Dishevelled (Dsh) acts downstream of Slb/Wnt11 through the domains specific to the non-canonical Wnt/PCP signaling cascade and directly regulates cell polarity within cells undergoing convergent extension. In addition, the relocalization of Dsh to the cell membrane is required for convergent extension movements in *Xenopus* gastrulae, as recruitment of Dsh to the membrane through the Frizzled receptor is required for the PCP pathway in *Drosophila* [30, 2]. In addition to the intracellular signaling mechanism, intercellular modulators are involved in regulating coordinated movements of large cell populations. HSPGs have been implicated in the modulation of intercellular signaling in vertebrates and in *Drosophila* and have been shown to be required for gastrulation movements in *Xenopus* embryo [7, 15]. Glypican, a member of the membrane-associated HSPG family, is known to regulate Wnt signaling in *Drosophila* [28, 4] and zebrafish embryos [27]. The molecular function of the HSPGs is, however, largely unknown. Here we investigate a proteoglycan *Xenopus* glypican-4 (Xgly4), a member of the HSPG family, and discuss its role in the non-canonical Wnt/PCP signaling pathway during gastrulation.

6.5 Loss-of-function of Xgly4 perturbs gastrulation cell movements and causes phenotypes similar to a zebrafish mutant *knypek*

Xgly4 is expressed in the dorsal mesoderm and ectoderm during gastrulation. To clarify the in vivo function of Xgly4, we performed loss-of-function and gain-of-function analyses. For the loss-of-function analysis, the translation of Xgly4 mRNA was blocked by a specific antisense morpholino-oligonucleotide (Xgly4Mo) directed against the 5' untranslated (5'UTR) and the first methionine region. The specificity and efficacy of Xgly4Mo was confirmed by noting that Xgly4Mo specifically reduced the protein level of the Xgly4-flag with the native 5'UTR (data not shown). Interestingly, a dorsal injection of Xgly4Mo at the 4-cell stage caused severe defects. Embryos at the tail-bud stage showed *spina bifida*, shortening of the antero-posterior(AP) axis, and a reduction of the head that especially affected eye development (Fig. 6.5A, bottom left), while ventral injection of Xgly4Mo or wild-type Xgly4 mRNA did not perturb gastrulation movements (data not shown). Importantly, these defects caused by Xgly4Mo were similar to those of the zebrafish mutant *knypek*. It is noteworthy that overexpression of Xgly4 also caused severe defects in gastrulation cell movement and caused *spina bifida* (Fig. 6.5A, bottom right).

6.6 Xgly4 acts in a non-canonical Wnt pathway as a coreceptor for Frizzled

To analyze the functional homology of Xgly4 to *knypek*, we injected Xgly4 mRNA into zebrafish *knypek* mutant embryos to examine whether Xgly4 could rescue the *knypek* phenotype. The embryonic axis and organ primodia of *knypek* mutants are shorter along the AP axis and broader medio-laterally compared with their wild-type (WT) siblings. These phenotypes were significantly rescued by the injection of Xgly4 mRNA (data not shown), suggesting that Xgly4 is a functional homologue of the zebrafish *knypek* gene. We also confirmed that these defects were due to impaired gastrulation cell movement but not the inhibition of mesodermal differentiation. Convergent extension movements have been analyzed in a simple assay in which naive animal cap

A

B

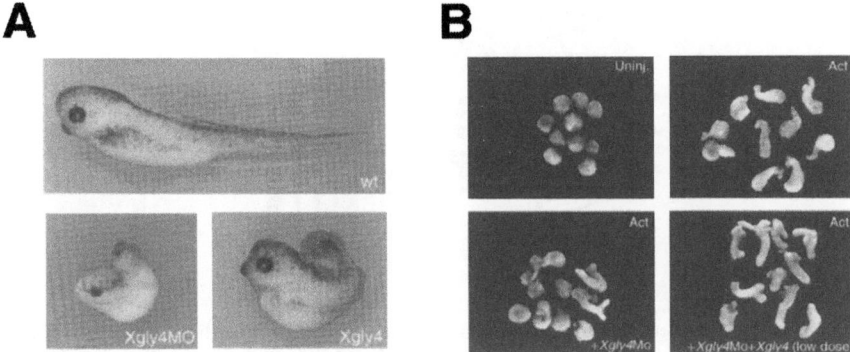

Fig. 6.5. Inhibition of Xgly4 translation blocks gastrulation movements. (A) Embryos injected with Xgly4Mo (41 ng) and Xgly4 mRNA (1ng) at stage 35/36. Xgly4 Morpholino oligo, which was injected into the dorsal region, blocked gastrulation movements (bottom left). Overexpression of wild type Xgly4 also perturbed gastrulation and caused *spinabifida* (bottom right). (B) Activin (0.25 pg)-induced elongation of animal cap explants (upper right) mimics the convergent extension movements seen during gastrulation. This elongation was inhibited by Xgly4Mo (41 pg, bottom left). The inhibition was rescued by Xgly4 injection at a low dose (25 pg, bottom right).

cells can elongate in response to activin closely in association with the induction of cell populations that give rise to notochord and muscle. To analyze the effects of Xgly4 on activin-induced animal cap elongation, Xgly4Mo was injected with activin mRNA into the animal pole and the animal cap assay was performed. Although tissue elongation occurred in animal caps that received activin mRNA injection (Fig. 6.5B, upper right), the elongation in response to activin was strongly inhibited by co-injection of Xgly4Mo (Fig. 6.5B, bottom left). In addition, Xgly4Mo-induced inhibition was rescued with a low dose of wild-type Xgly4 mRNA without the native 5'UTR region (Fig. 6.5B, bottom right). To analyze the pathway specific contribution of Xgly4 to Wnt signaling, we tried to rescue the inhibition of activin-induced elongation of animal caps that is caused by injection of Xgly4Mo with pathway-specific Xdsh mutants. The inhibited elongation caused by Xgly4Mo (Fig. 6.5B) was rescued by wild-type Xdsh (data not shown) or by a Xdsh mutant lacking the DIX but not the DEP domain. Because the DIX domain of Xdsh protein is required for the canonical Wnt/-catenin pathway and the DEP domain for the non-canonical Wnt/PCP pathway [2, 6, 26, 30], the results indicate that Xgly4 would participate in activin-induced elongation movements through the non-canonical Wnt/PCP pathway. Our results support the idea that Xgly4 affects gastrulation movement through the non-canonical Wnt/PCP pathway upstream of Dsh. Based on the above functional analysis, Xgly4 seems very likely to play a role as a positive modulator of Xwnt11.

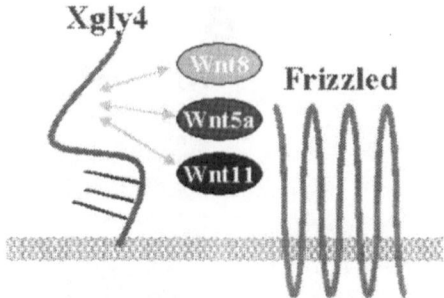

Fig. 6.6. Proposed role of Xgly4 in non-canonical Wnt signalling. Although Xgly4 is capable of binding Wnt5A, Wnt8 and Wnt11, Wnt 8 is not likely to be the endogenous ligand. Xgly4 is thought to physically interact with and present Wnt5A and Wnt11 to Frizzled receptors to activate a non-canonical Wnt signalling that regulates gastrulation cell movement. In this model, Xgly4 acts as a coreceptor for non-canonical Wnts.

6.7 Xgly4 binds Wnt ligands

Consistent with this functional interaction, we found that the HA-tagged Wnt11 ligand coimmunoprecipitated with the extracellular domain of Xgly4 protein (Fig. 6.5A). These results suggest that Xgly4 protein in the extracellular matrices binds the Wnt11 ligand to modulate its action on the Frizzled7 receptor [10] (Fig. 6.6). We also found that HA-tagged Wnt5A and Wnt8 ligands coimmunoprecipitated with Xgly4 protein. Therefore, Xgly4 may also regulate Xwnt5A's activity in vivo, because these two molecules are colocalized during early *Xenopus* development. However, Xgly4 is not likely to affect Xwnt8 signaling in a dominant manner, given that their expression patterns are complementary to each other (data not shown). We next examined which domain is required for binding to Wnt11. To address this question, we constructed Xgly4 mutant cDNAs designed to encode Gly4 proteins lacking the C-terminal region (Gly4ΔC) of cystein-rich domain (Gly4ΔCRD), or replaced with Ala residues the three putative glycosylated Ser residues (Gly4ΔGAG). We found that activin-induced elongation was strongly inhibited by injection of Xgly4ΔC and Gly4ΔGAG as wild-type Gly4, but not XglyΔ4CRD mRNA. Because we also found that HA-tagged Wnt11 ligand coimmunoprecipitated with the Gly4 lacking C-terminal region (GlyΔ4C) (data not shown), Gly4 may interact with Wnt11 through the CRD domain. Together, the CRD domain is thought to be essential for Xgly4 function in convergent extension movements. It is interesting to note that Wnt11 receptor Fz and a coreceptor Gly4 share a common structural feature (CRD) for binding the ligand.

6.8 Discussion

In this chapter, we would like to emphasize that extracellualar behavior and function of polypeptide growth factors that regulate pattern formation and morphogenesis are ingeniously controlled by cell-surface proteoglycans. First, we have shown that diffusion but not signaling of BMP is controlled by HSPGs. The N-terminal basic amino acid core conserved among BMP-2, BMP-4 and *Drosophila* Dpp is required for the high affinity binding between BMP and HSPGs. It is interesting to note that each of the TGF-β family has a different characteristic to its N-terminal region, which suggests that, during evolution, TGF-β family ligands acquired distinct diffusion properties through the editing of their N-terminal regions, which might be a motive force for morphological diversity. A growing body of evidence also indicates that HSPGs have critical roles in the regulation of other growth factors, including FGF and TGF-β [5, 29]. Thick veins (Tkv), a receptor for *Drosophila* Dpp, a fly orthologue of vertebrate BMP, was reported to restrict Dpp diffusion by a direct physical interaction. It is also known that interaction of HSPGs with growth factors not only regulates growth factor diffusion but also their activity depending on the context as discussed below.

In the second part of this chapter, we have shown that Xgly4 is required for proper signaling of Wnts. It has been shown genetically that mutations of the *Drosophila dally, dally-like*, or zebrafish *knypek* proteins perturb normal patterning or cell movements controlled by Wingless or Wnt, respectively. It has been proposed that the proteoglycans Dally, Dally-like, and Knypek act as a component of a receptor complex serving as a co-receptor for Frizzled. In addition to genetic evidence, we have been able to show successfully that Xgly4 physically interacts with the Wnt11 ligand (Fig. 6.5A) and Fz7 receptor (data not shown), which activate the non-canonical Wnt/PCP pathway. These results suggest that Xgly4 may present Wnt ligands to Fz receptors as the FGF low-affinity HSPG receptor does with FGF high-affinity signalling receptors [21, 25].

Taken together, it seems that cell-surface proteoglycans can serve as negative regulators for ligand diffusion and positive regulators for receptor activation. This was well documented for the Dpp receptor Tkv. Tkv is not uniformly expressed along the anterior-posterior axis of the wing imaginal disc; receptor levels are low where Dpp induces its target genes in the wing pouch. Receptor levels become higher in cells farther from the source of Dpp in the lateral regions of the disc as Dpp signaling negatively regulates tkv expression. Interestingly, the level of receptor influences the effective range of the Dpp gradient[18]. Importantly, high levels of tkv sensitize cells to low levels of Dpp, while also limiting the movement of Dpp outside the wing pouch shaping the Dpp gradient. In the *Drosophila* wing disc, it was also reported that a glypican Dally can regulate the wingless protein distribution in the extracellular spaces and, in some contexts at least, block wingless signaling [4]. These findings suggest that the level of proteoglycans in relation to ligand, and the environment of the interaction, are critical for how proteoglycan affects ligand activity and diffusion. Thus, this process is likely to be highly tissue- and context-dependent as has been shown from genetic interactions between Dally and Dpp that Dally does not serve the same role in all tissues.

Acknowledgements

We thank Drs. Yamamoto, Iemura, and Tada for helpful discussions and technical support, and Dr.ten Dijke for supplying us with the PS antibodies. This work is supported by 'Research For the Future' grant from Japanese Society for the Promotion of Science.

Appendix

The data presented in this chapter were originally reported in the following articles.

Ohkawara, B., Iemura, S.-I., ten Dijke, P. and Ueno, N. (2002). Action range of BMP is defined by its N-terminal basic amino acid core. *Curr. Biol.* **12**, 205-209.

Ohkawara, B., Yamamoto, T., Tada, M. and Ueno, N. (2003). Role of glypican 4 in the regulation of convergent extension movements during gastrulation in *Xenopus laevis. Development* **130**, 2129-2138.

References

1. Adler, P. N. (1992). The genetic control of tissue polarity in *Drosophila. Bioessays*, **14**, 735-741.
2. Axelrod, J. D. (2001). Unipolar membrane association of Dishevelled mediates Frizzled planar cell polarity signaling. *Genes Dev*, **15**, 1182-1187.
3. Axelrod, J. D., Miller, J. R., Shulman, J. M., Moon, R. T., and Perrimon, N. (1998). Differential recruitment of Dishevelled provides signaling specificity in the planar cell polarity and Wingless signaling pathways. *Genes Dev*, **12**, 2610-2622.
4. Baeg, G. H., Lin, X., Khare, N., Baumgartner, S., and Perrimon, N. (2001). Heparan sulfate proteoglycans are critical for the organization of the extracellular distribution of Wingless. *Development*, **128**, 87-94.
5. Bernfield, M., Gotte, M., Park, P. W., Reizes, O., Fitzgerald, M. L., Lincecum, J., Zako, M. (1999). Functions of cell surface heparan sulfate proteoglycans. *Annu Rev Biochem*, **68**, 729-777.
6. Boutros, M., and Mlodzik, M. (1999). Dishevelled: at the crossroads of divergent intracellular signaling pathways. *Mech Dev*, **83**, 27-37.
7. Brickman, M.C., and Gerhart, J.C. (1994). Heparitinase inhibition of mesoderm induction and gastrulation in *Xenopus* laevis embryos. *Dev. Biol.*, **164**, 484-501.

8. Chen, Y., and Schier, A.F. (2001). The zebrafish Nodal signal Squint functions as a morphogen. *Nature*, **411**, 607-610.
9. Dale, L., Howes, G., Price, B.M., and Smith, J.C. (1992). Bone morphogenetic protein 4: a ventralizing factor in early *Xenopus* development. *Development*, **115**, 573-585.
10. Djiane, A., Riou, J., Umbhauer, M., Boucaut, J., and Shi, D. (2000). Role of frizzled 7 in the regulation of convergent extension movements during gastrulation in *Xenopus* laevis. *Development*, **127**, 3091-3100.
11. Faure, S., Lee, M.A., Keller, T., ten Dijke, P., and Whitman, M., (2000). Endogenous patterns of TGFbeta superfamily signalling during early *Xenopus* development. *Development*, **127**, 2917-2931.
12. Graff, J.M. (1997). Embryonic patterning: to BMP or not to BMP, that is the question. *Cell*, **89**, 171-174.
13. Heisenberg, C. P., Tada, M., Rauch, G. J., Saude, L., Concha, M. L., Geisler, R., Stemple, D. L., Smith, J. C., and Wilson, S. W. (2000). Silberblick/Wnt11 mediates convergent extension movements during zebrafish gastrulation. *Nature*, **405**, 76-81.
14. Hovingh, P., and Linker, A. (1974). The disaccharide repeating-units of heparan sulfate. *Carbohydr.Res.*, **37**, 181-192.
15. Itoh, K., and Sokol, S.Y. (1994). Heparan sulfate proteoglycans are required for mesoderm formation in *Xenopus* embryos. *Development*, **120**, 2703-2711.
16. Jones, C.M., Armes, N., and Smith, J.C. (1996). Signalling by TGF-beta family members: short-range effects of Xnr-2 and BMP-4 contrast with the long-range effects of activin. *Curr. Biol.*, **6**, 1468-1475.
17. Kurata, T., Nakabayashi, J., Yamamoto, T.S., Mochii, M., and Ueno, N. (2001). Visualization of endogenous BMP signalling during *Xenopus* development. *Differentiation*, **67**, 33-40.
18. Lecuit, T. and Cohen, S. M. (1998). Dpp receptor levels contribute to shaping the Dpp morphogen gradient in the Drosophila wing imaginal disc. *Development*, **125**, 4901-4907.
19. McDowell, N., Zorn, A.M., Crease, D.J., and Gurdon, J.B. (1997). Activin has direct long-range signalling activity and can form a concentration gradient by diffusion. *Curr. Biol.*, **7**, 671-681.
20. Mlodzik, M. (1999). Planar polarity in the *Drosophila* eye: a multifaceted view of signaling specificity and cross-talk. *Embo J*, **18**, 6873-6879.
21. Pellegrini, L., Burke, D. F., von Delft, F., Mulloy, B., and Blundell, T. L. (2000). Crystal structure of fibroblast growth factor receptor ectodomain bound to ligand and heparin. *Nature*, **407**, 1029-1033.
22. Persson, U., Izumi, H., Souchelnytskyi, S., Itoh, S., Grimsby, S., Engstrom, U., Heldin, C.H., Funa, K., and ten Dijke, P. (1998). The L45 loop in type I receptors for TGF-beta family members is a critical determinant in specifying Smad isoform activation. *FEBS Lett.*, **434**, 83-87.
23. Ruppert, R., Hoffmann, E., and Sebald, W. (1996). Human bone morphogenetic protein 2 contains a heparin-binding site which modifies its biological activity. *Eur. J. Biochem.*, **237**, 295-302.
24. Schier, A.F., and Shen, M.M. (2000). Nodal signalling in vertebrate development. *Nature*, **403**, 385-389.
25. Schlessinger, J., Lax, I., and Lemmon, M. (1995). Regulation of growth factor activation by proteoglycans: what is the role of the low affinity receptors? *Cell*, **83**, 357-360.

26. Tada, M. and Smith, J. C. (2000). Xwnt11 is a target of *Xenopus* Brachyury: regulation of gastrulation movements via Dishevelled, but not through the canonical Wnt pathway. *Development*, **127**, 2227-2238.

27. Topczewski, J., Sepich, S. D., Myers, C. D., Walker, C., Amores, A., Lele, Z., Hammerschmidt, M., Postlethwait, J., and Solnica-Krezel, L. (2001). The Zebrafish Glypican Knypek controls cell polarity during gastrulation movements of convergent extension. *Developmental Cell*, **1**, 251-264.

28. Tsuda, M., Kamimura, K., Nakato, H., Archer, M., Staatz, W., Fox, B., Humphrey, M., Olson, S., Futch, T., Kaluza, V., Siegfried, E., Stam, L. and Selleck, S. B. (1999). The cell-surface proteoglycan Dally regulates Wingless signalling in *Drosophila*. *Nature*, **400**, 276-280.

29. Tumova, S., Woods, A., and Couchman, J. R. (2000). Heparan sulfate proteoglycans on the cell surface: versatile coordinators of cellular functions. *Int J Biochem Cell Biol*, **32**, 269-288.

30. Wallingford, J. B., Rowning, B. A., Vogeli, K. M., Rothbacher, U., Fraser, S. E., and Harland, R. M. (2000). Dishevelled controls cell polarity during *Xenopus* gastrulation. *Nature*, **405**, 81-85.

Changing Roles of Homeotic Gene Functions in Arthropod Limb Development

Shigeo Hayashi[1], Hideo Yamagata[2] and Yasuhiro Shiga[2]

[1] *Morphogenetic Signaling Group, Riken Center for Developmental Biology, 2-2-3, Minatojima-Nakamachi, Chuo-ku, Kobe, Hyogo, 650-0047 Japan*
[2] *School of Life Science, Tokyo University of Pharmacy and Life Science 1432-1, Horinouchi, Hachioji, Tokyo 192-0392, Japan*
e-mail: shayashi@cdb.riken.go.jp
telephone: 81-78-301-306-3184
fax: 81-78-306-3183

Summary. The extensive diversification of limb morphologies during arthropod evolution has aided the spread of new species to distant locations and adaptation for distinct niches. Functional evolution of homeotic genes is thought to have played a key role in this diversification process, but the molecular mechanisms underlying this process remain to be elucidated. Insects and crustaceans are close relatives within arthropods and possess a variety of diversities in the shape of their limbs. Comparison of genetic circuitries underlying the formation of insect and crustacean limbs should reveal the genetic history of their morphological evolution. We have recently analyzed the functional role of ANTENNAPEDIA (ANTP) homeotic protein of the crustacean *Daphnia magna* and compared its properties to its well studied counterpart, the insect *Drosophila melanogaster*. Our results suggest that highly restricted expression of ANTP in the first leg of *Daphnia* specifies its morphology. Furthermore, while the core ANTP function of specifying thoracic identity is conserved, changes in the protein region outside of the homeodomain altered the target gene specificity. Based on these findings, we discuss the role of homeotic gene evolution in diversification of arthropod limb morphology.

7.1 Introduction

Homeotic or Hox genes are a group of genes encoding a family of transcription factors containing the homeodomain DNA binding domain [14, 17]. Differential expression of Hox genes is central to the specification of body patterns along the anterior-posterior axis in the metazoan bodyplan. Hox genes are clustered in one or several chromosomal loci, forming a HOM complex, and their anterior expression boundaries and sites of their actions are colinear with the order of their chromosomal locations. Genetic analyses of homeotic genes in *Drosophila* have demonstrated that homeotic genes function as a key determinant of segment-specific morphological characteristics. For example, genetic removal of *Antennapedia (Antp)* homeotic gene activity from the second thoracic leg transforms it to antennal tissue [23], and ectopic activation of *Antp* in the head causes a striking transformation of antenna to leg [21]. These functional demonstrations of the HOM genes and the evolutionary conservation of the colinearity of HOM complex organization and expression suggest that HOM complex serves as a groundplan for metazoan body patterning. In contrast to the high conservation of HOM complex organization, the number of morphological pattern elements that are under the control of HOM genes far exceeds the complexity of HOM gene organization. One example is the shape of arthropodan appendages. Insect legs and wings exhibit a great deal of diversity to adapt to a variety of locomotive functions (walking, flying, swimming), and to escape from predators. How the function of HOM genes has diverged and contributed to diversification of animal morphologies is a question that has been a major focus of investigation [7]. Alteration of HOM gene expression patterns has been suggested as one mechanism that has diversified HOM gene function [25]. Ultrabithorax and abdominal A genes of *Drosophila* are expressed in posterior T3 and abdominal segments and repress leg development, thereby limiting the number of larval and adult legs to three pairs. Lepidopteran larvae bear rudimental legs in the abdomen which express Ubx/abd A. The abdominal legs express Distal-less gene, a universal marker for appendage development. It has been shown that at the onset of abdominal leg development, Ubx/abd A expression is down-regulated locally at the region where DLL becomes activated [25]. Such a change in Ubx/abd A expression has never been observed in *Drosophila*. This observation suggests that down-regulation of Ubx/abdA plays a permissive role in the formation of abdominal leg development in Lepidopterans. Another hypothesis is that structural changes in homeotic proteins have altered their regulatory capabilities and contributed to evolutional changes in the number of arthropods limbs [15]. Recent comparative work on homeotic genes of arthropod and lower animals has revealed new aspects of homeotic gene evolution. In this review, we summarize our recent work on the function of the *Antp* gene of *Daphnia magna* and its implication in diversification of crustacean limb patterns.

7.2 Homeotic gene expression boundaries and crustacean limb patterns

Crustaceans possess several pairs of limbs in a variety of sizes and branching patterns [6]. Their thoracic limbs are grouped into two major types. Larger ones serve as legs for locomotive functions. Another type of thoracic limb, termed the maxilliped, is smaller and its branching is suppressed. It resembles the head appendages used for feeding functions. The number of maxillipeds varies from three pairs in Homarus to one in Mysidium and none in *Artemia*, and has been used as a major landmark to classify crustacean species [6].

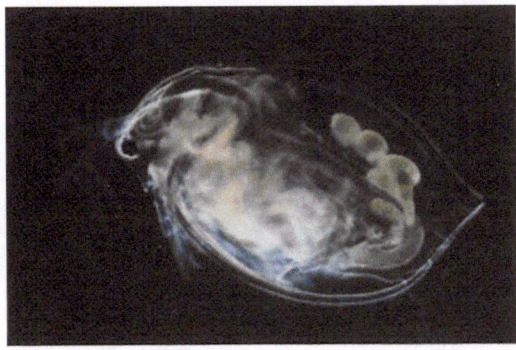

Fig. 7.1. Adult of *Daphnia magna*. It swims with antennae. Thoracic legs are enclosed in the carapace. Fertilized eggs are seen in the back side of the animal.

In [4], ten crustacean species were examined for expression of Ubx/abdA, and it was found that the anterior borders of Ubx/abdA expression varied from one species to another. The changes correlated well with the borders of transition from a small type of thoracic limb (maxilliped) to the larger one (leg). The authors proposed that the anterior thoracic segments which do not express Ubx/abdA develop maxillipeds, and anterior expression boundaries of Ubx/abdA vary from one species to another. The results have led to a model showing that shifts of anterior Ubx/abdA expression boundaries took place in relatively short periods of crustacean evolution, and these shifts have driven the change in the number of segments bearing maxillipeds. The proposal in [4] had a profound impact on our thinking on the evolution of the function of the homeotic gene. However, several predictions have to be examined before the model can be fully accepted. What is the mechanism that specifies the morphological difference of the leg and maxilliped? Does another homeotic gene which is different from Ubx/abdA participate in this process? What kind of molecular change in the homeotic gene is responsible for the change in its activity? We have examined these questions through the analyses of *Daphnia Antp* and *Dll* gene functions in *Daphnia* limb development [22], which are described below.

7.3 Segmental differences in *Daphnia* leg patterns correlate with a change in Distal-less expression

We have chosen the water flea, *Daphnia magna* (Cladcera, Brancheopoda, Fig. 7.1), to study the molecular basis for structural diversification of branched crustacean limbs. *Daphnia* has five pairs of multiple branched thoracic limbs with a variety of forms (Fig. 7.2, upper panels) which are not evident in its close relative *Artemia* (Brancheopoda) which possesses homonomous trunk

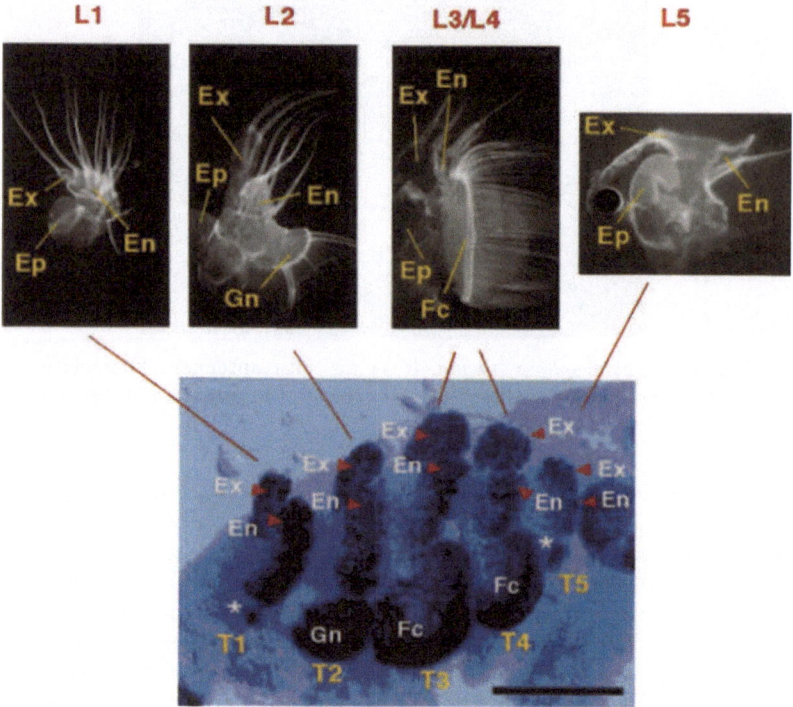

Fig. 7.2. First to fifth trunk limbs of *Daphnia* and their relationship to DLL expression. Ep, Ex, En and Gn or Fc are arranged in a dorsal to ventral orientation. Endites of second to fourth leg bear prominent comb-like structures (Gn and Fc) that correspond to strong expression of DLL. No such branches or DLL expression are present in T1 or T5 (asterisks). T1-T5, first to fifth thoracic segment; L1-L5, first to fifth thoracic limb; Ex, exopod; En, endopod; Ep, epipod; Gn, gnathobase; Fc, filter comb. Scale bar: 100 micrometers.

limbs. All the thoracic limbs of *Daphnia* are covered with carapace and are not used directly for locomotive functions. Instead, *Daphnia* swims with its prominent second antenna located on the head. Classical literature has considered all these thoracic limbs to be the leg, and no maxilliped is thought to be present in *Daphnia*. We have carried out detailed morphological analyses of *Daphnia* limbs with molecular markers to clarify their identity.

The second to fourth legs (L2-L4) of *Daphnia* bear prominent comb-like structures (gnathobase: Gn, filter comb: Fc) associated with endites. These structures are used for feeding functions. In addition, L2-L4 are significantly larger than L1 and L5 (Fig. 7.2, upper panels). To further investigate the basis for the segmental difference in the leg structures, we followed the expression of the homeodomain protein DLL (Fig. 7.2, lower panels). DLL is implicated

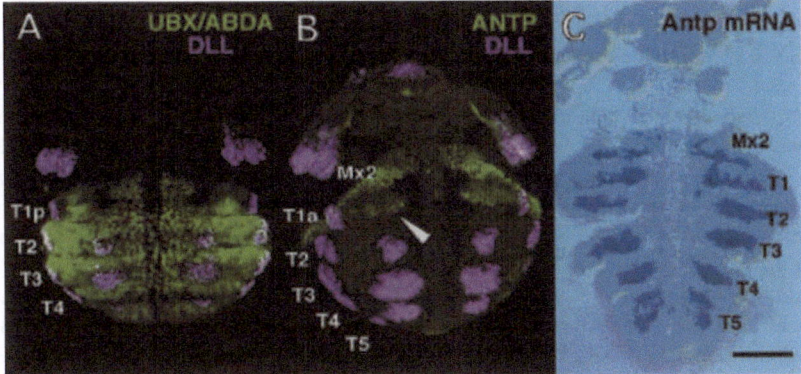

Fig. 7.3. Expression patterns of UBX/ABDA, ANTP and DLL proteins (A, B), and that of *Antp* mRNA (C). Confocal images of embryos double labeled with DLL (purple) and UBX/ABDA (A, green) or ANTP (B, green). ANTP expression was limited to Mx2 and the anterior of the first thoracic segment (T1a). Note that prominent expression of ANTP in ventral T1 corresponds to reduced expression of DLL (arrowheads in B). C. Expression pattern of *Antp* mRNA. It was detected broadly in Mx2 and more posterior segments. Scale bar: 100 micrometers.

in the development of distal parts of appendages of several taxa [18, 19]. DLL expression detected by an antibody starts early in limb development. Since DLL expression persists until a late stage when each limb primordium acquires the branched morphology characteristic of each segment, we could correlate each domain of DLL expression to specific limb branches (Fig. 7.2, lower panel). In the late stage embryo, DLL is highly accumulated in endites in T2-T4, which form the comb-like feeding structures. In T1 and T5, where the comb-like structures from endites are not formed, ventral DLL expression is reduced to a few cells. Since DLL expression is synonymous with distal leg development in several model organisms, the lack of DLL in ventral L1 and L5 indicates they are a reduced form of the leg. Next we examined the expression of posterior homeotic genes Ultrabithorax (Ubx) and abdominal A (abdA) with monoclonal antibody FP6.87 which recognizes an epitope (UBX/ABDA) common to UBX and ABDA [13]. Expression of UBX/ABDA was strong in posterior T1 through T4, and weak in T5 and postthoracic segments (Fig. 7.3A). The anterior borders of UBX/ABDA expression correlate with the change in leg patterns.

This molecular evidence and the morphological characteristics suggest that the L1 of *Daphnia* should be considered as the maxilliped as opposed to the legs in L2-L4. We focused our analyses on the difference between L1 and L2-4.

7.4 Post-transcriptional mechanism limits *Daphnia* ANTP expression to the segment bearing maxillipede

Expression patterns of homeotic genes have been studied in several crustacean species (*Porcellio*: [1]; *Artemia*: [3]; *Crayfish* [2]), in which expression domains of *Antp*, *Ubx* and *abdA* overlap in a broad region covering the trunk. It was argued that coexpression of the same set of homeotic genes in the trunk segments accounts for the homonomous morphologies of the segments. We examined the expression patterns of *Antp* and *Ubx/abdA* in *Daphnia* embryos. The amino acid sequence of *Daphnia* ANTP (DapANTP) was highly homologous to *Drosophila* ANTP (DmANTP) in the region spanning the YPWM motif and the homeodomain, but the remaining protein coding region was highly divergent. In situ hybridization revealed that DapANTP mRNA was distributed broadly from the second maxilla (Mx2) to the postthoracic segments (Fig. 7.3C). This pattern was similar to those found in *Artemia* and *Porcelio* [1, 3]. In contrast, the expression of DapANTP protein was detectable only in Mx2 and the anterior portion of the first thoracic segment (T1a) including L1 (Fig. 7.3B) suggesting that DapANTP protein expression is restricted to Mx2 and T1 by a post transcriptional mechanism.

7.5 Down-regulation of DLL in the region of ANTP expression in *Daphnia* L1

To understand the molecular basis for the L1-specific morphological characteristics, the expression patterns of DLL and DapANTP in L1 were compared. DapANTP was highly expressed in prospective ventral L1 where DLL expression was very low (Fig. 7.3B). Such an inverse correlation between DapANTP and DLL was also observed in Mx2 which does not develop distal outgrowth (data not shown). On the other hand, expression of UBX/ABDA overlapped extensively with DLL (Fig. 7.3A), as has been reported for other crustaceans, millipedes and insects [11, 18]. Therefore, the expression pattern complementary to that of DLL is a unique feature of ANTP. Given the general role of homeotic genes on segmental specification, these observations suggest a hypothesis that DapANTP represses DLL in L1 and modifies its morphology to that of maxilliped. On the other hand, the overlap of UBX/ABDA and DLL suggests that UBX/ABDA do not repress limb development in *Daphnia*. These findings are a striking contrast with the observation made in *Drosophila* where functional roles of homeotic genes on limb development have been extensively studied. In *Drosophila*, UBX/ABDA represses DLL expression and limb development [24] in the embryo, while ANTP has no repressive function on DLL. Therefore, the way in which DLL is regulated by ANTP and Ubx/ABDA appears to have changed after diversification of crustaceans and insects.

7.6 Novel activity of *Daphnia* ANTP revealed by assays in *Drosophila*

To investigate the molecular basis for the functional difference of *Daphnia* and *Drosophila Antp* genes, the activities of ANTP proteins from the two species were compared in the context of *Drosophila* development. DapANTP caused transformation of the antennae to the leg and of the dorsal head to the thorax when ectopically expressed in eye-antennal imaginal discs (Fig. 7.4, upper panels). A similar result was observed for DmANTP, and for mouse Hox-B6 [16]. DapANTP also induced ectopic expression of the trunk-specific marker gene *teashirt* (*tsh*) [9] when misexpressed in the embryonic head (Fig. 7.4, lower panels). These results demonstrated that *Daphnia* ANTP is a functional homolog of *Drosophila* ANTP. Unexpectedly, assays in thoracic segments revealed an activity unique to *Daphnia* ANTP. When *Antp* P1 promoter was used to drive expression, DapANTP inhibited development of the ventral thorax (Fig. 7.5, Antp > DapAntp). Larvae showed various defects in ventral epidermis, including reduction of ventral denticle belts, loss of Keilinfs organ and loss of ventral cuticle. Dorsal landmarks such as dorsal hairs and dorsal black dots formed normally. Similar defects were observed when another driver *(ptc-Gal4)* promoting equal levels of expression in the dorsal and ventral sides of each segment was used (data not shown), suggesting that the effect of DapANTP was biased toward the ventral side.

7.7 *Daphnia* Antp represses *Dll* transcription

The *in vivo* assay in *Drosophila* revealed a novel activity of DapANTP not present in DmANTP. Whether the activity unique to DapANTP has a role in modulating limb development was investigated. DLL expression was used as a marker for limb development in the embryo and was found to be eliminated upon expression of DapANTP, suggesting that DapANTP inhibits limb development (Fig. 7.6). To test whether DapANTP directly regulates *Dll* transcription, we examined the transcriptional enhancer of *Dll* [24] (Dll304: Fig. 7.7). Dll304 enhancer promotes transcription in the limb primordia in the *Drosophila* embryo and mediates repression by homeotic genes in the bithorax complex that includes Ubx, abdA and AbdB. UBX and ABDA bind to multiple sites in Dll304. While ectopic DmANTP has no effect on Dll304, DapANTP strongly repressed Dll304 expression, leaving only a small number of Dll304 expressing cells (Fig. 7.7, upper panels). Dll305 is a derivative of Dll304 in which all but one of the homeodomain binding sites from Dll304 were deleted. Dll305 was no longer repressed by the bithorax complex genes, suggesting that those sites are functional targets for UBX and ABDA [24]. DapANTP failed to repress Dll305 (Fig. 7.7, lower panel), suggesting that the homeodomain binding sites present in Dll304, but absent in Dll305, are

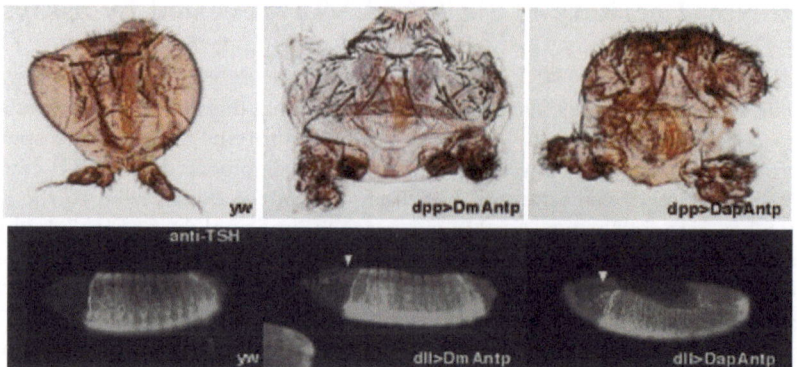

Fig. 7.4. ANTP proteins from *Daphnia* and *Drosophila* specify thoracic identities in the *Drosophila* head. From left to right, animals of wild-type, DmAntp over expression, and DapAntp over expression are shown. Upper panels. Dorsal views of adult head. Transformation of the adult head into the thorax by ectopic expression of ANTP. dppGal4 was used to drive expression. Transformation of antenna to leg and dorsal head into notum was observed. Compound eyes were also eliminated. Lower panel. Induction of an ANTP target gene Teashirt (TSH). *Dll*Gal4 was used as a driver. Ectopic head expression of TSH was induced by ANTP proteins (arrowheads).

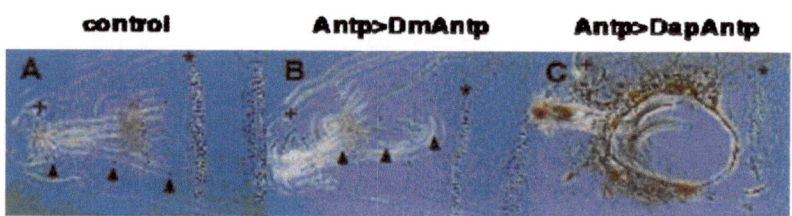

Fig. 7.5. Effects of ectopic ANTP expression on cuticle development. DapANTP disrupts ventral epidermal development in *Drosophila* embryo. Here and in Figs. 7.6 and 7.7 *Antp*-Gal4 was used to drive ectopic expression. Positions of Keilin's organ, T1, and A1 denticle belts are indicated by arrowhead, cross, and asterisk, respectively.

the target of DapANTP in the *Drosophila* embryo. Taking these results together, we have demonstrated a striking difference in the way homeotic genes of *Drosophila* and *Daphnia* regulate the *Dll* gene. In *Drosophila* embryogenesis, products of bithorax complex genes bind Dll304 enhancer and repress *Dll* expression in the abdomen while ANTP has no repressive function on *Dll*. When expressed in *Drosophila*, DapANTP can take over the role of *Drosophila*

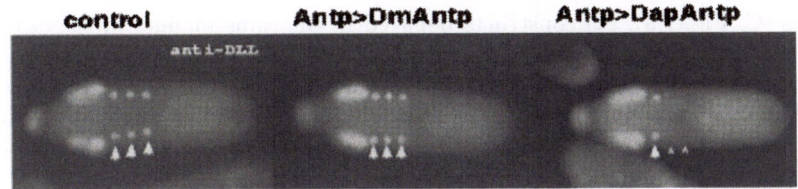

Fig. 7.6. Effect of ectopic ANTP expression on DLL expression. DapANTP eliminated DLL expression in T2 and T3 (asterisk).

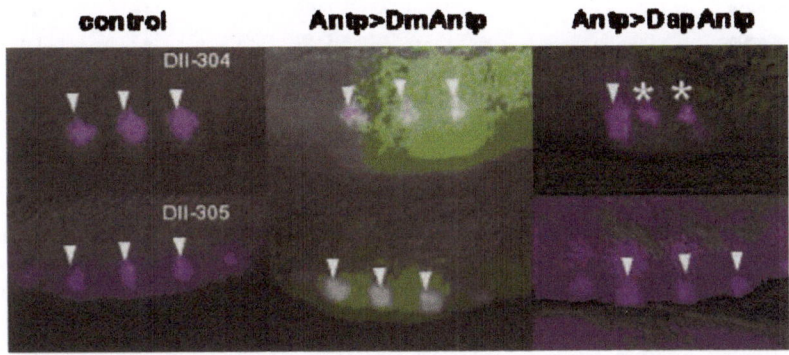

Fig. 7.7. DapANTP represses *Dll* through a DNA element that is a target for *Drosophila* UBX and ABDA. Upper panels. Expression pattern of Dll-304 (purple). It reproduces DLL expression in the thorax and is repressed by DapANTP (asterisks). Lower panels. Expression of Dll-305 was not affected by either DmANTP or DapANTP. DmANTP was stained green to indicate the domain of ANTP expression which overlaps with T2 and T3. Leg primodia are indicated by arrowheads.

bithorax complex genes and repress *Dll* enhancer through the region normally mediating repression by UBX and ABDA.

7.8 Domains of *Daphnia* ANTP responsible for the limb repressing activity

The activity of *Daphnia* ANTP to repress *Dll* in *Drosophila* most likely resides in its coding region. To map the region responsible for the functional differences in ANTP from *Daphnia* and *Drosophila*, we constructed a series of chimeric genes and tested their activities in the assays described above. We divided ANTP proteins into three regions: the diverged N-terminal region (N), highly conserved YPWM motif and homeodomain (HD), and C-terminal tails (C, Fig. 7.8). All constructs behaved in the same manner in the assays of the head to the thorax transformation, and induction of TSH expression in the embryonic head. However, in the assay to test their ability to repress *Dll* expression, they showed distinct activity and allowed us to assign specific activity to each region. The results are summarized as follows:

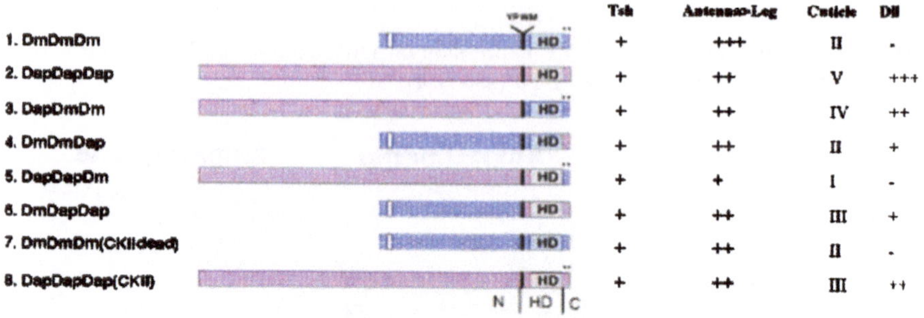

		Tsh	Antenna>Leg	Cuticle	Dll
1. DmDmDm		+	+++	II	-
2. DapDapDap		+	++	V	+++
3. DapDmDm		+	++	IV	++
4. DmDmDap		+	++	II	+
5. DapDapDm		+	+	I	-
6. DmDapDap		+	++	III	+
7. DmDmDm(CKIIdead)		+	++	II	-
8. DapDapDap(CKII)		+	++	III	++

Fig. 7.8. Structures of *Drosophila* (lane 1) and *Daphnia* ANTP (lane 2), and their chimera (lane 2-6). Mutants with altered CKII sites are shown below (lane 7, 8). Borders of N, HD and C regions are indicated below lane 8. Amino acid sequences derived from DapANTP are shown in shaded pink and those from DmANTP in dotted blue. Black bars, blocks and open boxes indicate the YPWM motifs, homeodomains and the homology found in several insects, respectively. Asterisks indicate putative CKII sites in the DmANTP. To the right, activities of the proteins in four different assays are shown. Tsh, activity to induce teashirts expression in the embryonic head. Antenna > Leg, transformation of adult antenna into leg. Cuticle, severity of cuticle defect was classified into five classes which range from normal (I, Fig. 7.5 left panel) to severe (V, Fig. 7.5 right panel). Dll, repression of DLL expression. - indicates no repression and +++ indicates strong repression.

1. The N terminal region of *Daphnia* ANTP (NDap) is responsible for the activity to repress *Dll*. Replacement of NDm with NDap conferred the *Daphnia*-specific activity (compare constructs 1 and 3), and a reciprocal replacement of NDap with NDm greatly compromised DapANTP activity (constructs 2, 6).
2. The C terminal region of *Drosophila* ANTP (CDm) antagonizes NDap. When combined with the rest of DapANTP (construct 5), CDm greatly compromised the *Daphnia*-specific activity. However, the activity of CDm appears to be limited to the modulation of *Daphnia* ANTP specific activity, since replacement of CDm with CDap did not alter the activity of *Drosophila* ANTP (construct 4).
3. Phosphorylation by casein kinase II (CKII)-like activity may regulate the activity of CDm. In CDm, there are two CKII phosphorylation consensus

sites. These sites are conserved in several insect ANTPs, but are absent in DapANTP. If phosphorylation at the CKII sites of CDm is important in modulating DapANTP activity, introduction of CKII sites into CDap is expected to modulate *Daphnia* ANTP activity. As expected, the mutation of CDap that creates CKII sites compromised DapANTP activity (construct 8). On the other hand, the mutation of CDm that disrupts CKII sites failed to provide *Daphnia*-specific activity to DmANTP (constructs 7).

4. The assays did not distinguish the activity of the central region of the proteins (HDDap and HDDm). The regions differ in two places: one is conservative F to Y substitution in the homeodomain and the other in the region connecting YPWM motif and homeodomain, the latter being affected by alternative use of splicing accepter sites [5]. The significance of these differences remains to be determined.

Collectively the results suggest that two separate alterations of ANTP protein occurred after separation of insects and crustaceans. The first one is extensive change in the N terminal region. NDap is twice as long as NDm and no longer retains obvious sequence homology, except for a short region at the N-terminus. Through this process NDap must have acquired a domain that is able to repress *Dll* transcription. Second, the C terminus region of *Drosophila* ANTP (CDm) becomes different from CDap in its activity to modulate NDap, and phosphorylation by CKII-like activity is most likely responsible for this activity. There is evidence that a CKII-like phosphorylation plays a role in down-regulating *Drosophila* ANTP. It was reported that a mutant of DmANTP with amino acid alterations that change the two CKII sites to non-phosphorylatable gained a novel activity to disrupt ventral epidermal development [12]. The results suggest that NDap is a major determinant of *Daphnia* specific activity of ANTP to repress *Dll*, and CDm can interfere with this activity, possibly through phosphorylation of CKII sites.

7.9 Evolution of *Antp* proteins

The results of the work discussed here prompt several interesting speculations on the evolution of homeotic gene functions. The three pairs of *Drosophila* legs are very similar in shape and size, and only a few morphological landmarks such as sex combs in male L1 mark the differences between each leg. Consequently, the role of ANTP in differentiating thoracic leg identity has not been noted so far. The study of *Daphnia* ANTP provides an interesting hypothesis that DapANTP expressed in a highly restricted manner in L1 modifies the morphology of the T1 leg to a smaller one. This modification may involve a transcriptional repression of *Dll*, as suggested by its activity to repress *Dll* enhancer in the *Drosophila* assay. This hypothesis may be applied to a wider range of crustacean species, since the observation in ten crustacean species demonstrated that crustacean legs anterior to the domain of UBX/ABDA, where ANTP should be expressed, are in general small and resemble feeding appendages when compared to more posterior limbs specialized for locomotive functions [4]. A test of a causal relationship between the expression of ANTP and DLL in *Daphnia* would require genetic approaches to modify *Antp* activity. RNA interference to knock down gene expression, and virus mediated gene transfer to misexpress exogenous genes, are promising techniques that can be tested in *Daphnia*. We have shown here that diversification of the ANTP protein outside the homeodomain contributed to its functional variation in modifying limb patterns. The region responsible for *Daphnia*-specific activity was mapped to the N terminal region of ANTP that is highly diverged. Two recent works on Ubx proteins [10, 20] reported that functional alteration of homeotic proteins plays a significant role in restricting the number of insect limbs. Our work demonstrated that an evolutional change in ANTP protein contributed to the modification of the shape of T1 leg as opposed to T2-4 legs of *Daphnia*, which is a microevolutionary event. Taken together, homeotic proteins have undergone a number of alterations in regions outside the homeodomain to change their target specificity and the way they control limb development. Another aspect of ANTP function is shared by the *Drosophila* and *Daphnia* proteins. The major function of *Drosophila* ANTP is thought to specify leg identity. This leg specifying activity of ANTP involves transcriptional repression of homothorax which plays a key role in specifying antennal identity [8]. Since *Daphnia* ANTP is also capable of transforming the head into the thorax, the thorax specifying activity of ANTP is likely to be conserved between insects and crustaceans. Furthermore, the ability of human Hox-B6 gene to transform *Drosophila* antenna into leg suggests that determination of thoracic identity is a core function of ANTP that is retained within the homeodomain: the only region with significant amino acid sequence homology among ANTP from the three species. It follows that in the course of animal evolution, there is strong selective pressure to preserve ANTP type homeodomain while other parts of the proteins are relatively free to diverge, and allow for the creation of novel interaction with different sets of transcription factors to acquire species-specific subfunctions. Such diversifications are very likely to have catalyzed evolution of animal morphology.

Acknowlegements

This work was supported by Grants-in-Aid for Scientific Research from the MEXT Japan to YS, HY, and SH and from JSPS (SH, Research for the Future), and NIG Cooperative Research Program to YS.

References

1. Abzhanov, A. and Kaufman, T.C. (2000). Crustacean (malacostracan) Hox genes and the evolution of the arthropod trunk. *Development*, **127**:2239-2249.
2. Abzhanov, A. and Kaufman, T.C. (2000). Embryonic expression patterns of the Hox genes of the crayfish Procambarus clarkii (Crustacea, Decapoda). *Evol Dev*, **2**:271-283.
3. Averof, M. and Akam, M. (1995). Hox genes and the diversification of insect and crustacean body plans. *Nature*, **376**:420-423.
4. Averof, M. and Patel, N.H. (1997). Crustacean appendage evolution associated with changes in Hox gene expression. *Nature*, **388**:682-686.
5. Bermingham, J.R.J. and Scott, M.P. (1988). Developmentally regulated alternative splicing of transcripts from the Drosophila homeotic gene Antennapedia can produce four different proteins. *EMBO J.*, **7**:3211-3222.
6. Brusca, R.C. and Brusca, G.J. (1990). *Invertebrates*. Sinauer Associates, Inc., Sunderland, Massachusetts.
7. Carroll, S., Grenier, J. and Weatherbee, S. (2001). *From DNA to Diversity-Molecular Genetics and the Evolution of Animal Design*. Blackwell Science Inc.
8. Casares, F. and Mann, R.S. (1998). Control of antennal versus leg development in Drosophila. *Nature*, **392**:723-726.
9. Fasano, L., Rder, L., CorE N., Alexandre, E., Vola, C., Jacq, B. and Kerridge, S. (1991). The teashirt gene is required for the development of Drosophila embryonic trunk segments and encodes a protein with widely spaced zinc finger motifs. *Cell*, **64**:63-79.
10. Galant, R. and Carroll, S.B. (2002). Evolution of a transcriptional repression domain in an insect Hox protein. *Nature*, **415**:910-3.
11. Grenier, J.K., Garber, T.L., Warren, R., Whitington, P.M. and Carroll, S. (1997). Evolution of the entire arthropod Hox gene set predated the origin and radiation of the onychophoran/arthropod clade. *Curr Biol*, **7**:547-553.
12. Jaffe, L., Ryoo, H.D. and Mann, R.S. (1997). A role for phosphorylation by casein kinase II in modulating Antennapedia activity in Drosophila. *Genes Dev*, **11**:1327-1340.
13. Kelsh, R., Weinzierl, R.O., White, R.A. and Akam, M. (1994). Homeotic gene expression in the locust Schistocerca: an antibody that detects conserved epitopes in Ultrabithorax and abdominal-A proteins. *Dev Genet*, **15**:19-31.
14. Krumlauf, R. (1992). Evolution of the vertebrate Hox homeobox genes. *Bioessays*, **14**:245-52.
15. Li, X. and McGinnis, W. (1999). Activity regulation of Hox proteins, a mechanism for altering functional specificity in development and evolution. *Proc Natl Acad Sci U S A*, **96**:6802-7.
16. Malicki, J., Schughart, K., McGinnis, W. (1990). mouse Hox-2.2 specifies thoracic segmental identity in Drosophila embryo and larvae. *Cell*, **63**:961-967.
17. Morata, G. (1993). Homeotic genes of Drosophila. *Curr Opin Genet Dev*, **3**:606-614.
18. Panganiban, G., Irvine, S.M., Lowe, C., Roehl, H., Corley, L.S., Sherbon, B., Grenier, J.K., Fallon, J.F., Kimble, J. and Walker, M. et al. (1997). The origin and evolution of animal appendages. *Proc Natl Acad Sci U S A*, **94**:5162-5166.
19. Panganiban, G., Sebring, A., Nagy, L. and Carroll, S. (1995). The development of crustacean limbs and the evolution of arthropods. *Science*, **270**:1363-1366.

20. Ronshaugen, M., McGinnis, N. and McGinnis, W. (2002). Hox protein mutation and macroevolution of the insect body plan. *Nature*, **415**:914-7.
21. Schneuwly, S., Klemenz, R. and Gehring, W.J. (1987). Redesigning the body plan of Drosophila by ectopic expression of the homoeotic gene Antennapedia. *Nature*, **325**:816-818.
22. Shiga, Y., Yasumoto, R., Yamagata, H. and Hayashi, S. (2002). Evolving role of Antennapedia protein in arthropod limb patterning. *Development*, **129**:3555-61.
23. Struhl, G. (1981). A homoeotic mutation transforming leg to antenna in Drosophila. *Nature*, **292**:635-8.
24. Vachon, G., Cohen, B., Pfeifle, C., McGuffin, M.E., Botas, J. and Cohen, S.M. (1992). Homeotic genes in of the bithorax complex repress limb development in the abdomen of the Drosophila embryo through the target gene Distal-less. *Cell*, **71**:437-450.
25. Warren, R.W., Nagy, L., Selegue, J., Gates, J. and Carroll, S. (1994). Evolution of homeotic gene regulation and function in flies and butterflies. *Nature*, **372**:458-461.

8

Variation, Adaptation and Developmental Constraints in the Mimetic Butterfly *Papilio dardanus*

Alexandra Cieslak[1,2,3], Richard I. Vane-Wright[1] and Alfried P. Vogler[1,2]

[1] Department of Entomology, The Natural History Museum, London SW7 5BD, U. K.
[2] Department of Biology, Imperial College at Silwood Park, Ascot, Berkshire SL5 7PY, U.K.
[3] Present address: Medizinische Hochschule Hannover, Department of Gastroenterology and Hepatology, Forschungzentrum Oststadt, 30655 Hannover, Germany

Summary. Wing patterns in mimetic butterflies can diversify rapidly to match a chemically defended model, and polymorphic species as the African Mocker Swallowtail, *Papilio dardanus*, even may mimic several different models. Evolutionary geneticists have ascribed the accurate control of complex differences in wing patterns to the action of 'supergenes', i.e. tightly linked multiple genes each specifying particular elements of the wing pattern. However, this concept appears less plausible in the light of modern developmental biology. Instead, we propose that Turing type mechanisms of morphogen gradients may account for a co-ordinate system that, while largely buffered from variation, can be modified to produce new or alternate phenotypes by changing a small set of parameters during wing development. The sequential specification of cells in the developing wing allows for the repeated intervention of regulatory components to affect the phenotype, producing complex variation even if genetic differences are small.

8.1 Introduction

Butterfly wing patterns are immensely diverse and therefore present an ideal system for studying the evolution of phenotypic diversification [2, 21]. The wing patterns of Lepidoptera, other than a few exceptions involving the additional use of membrane effects, are produced by mosaics of numerous scales on the wing surface, each of which is regulated individually to produce a particular colour. Scales are modified setae that arise from individual epidermal cells. The different wing patterns that are produced are subject to a range of different selection pressures, and are generally considered to be highly functional.

Butterfly wings have become an increasingly popular system for the study of pattern formation and morphogenesis at a molecular level [2, 3, 4, 13, 15, 16, 18, 21]. It is now well established that wing pattern expression is achieved through a series of sequential steps whereby gene expression cascades, initiated early in larval development, determine the colour pattern by controlling colour expression of the individual scales. These processes are capable of generating the phenotype with great reliability [14, 23, 24].

The species-specific morphology is evidently determined by stringent onto-genetic programmes of patterning and organogenesis. As initially proposed by Waddington, [36, 37], these developmental processes are thought, in general, to be highly co-selected, finely tuned and strongly buffered against change. Any specified variation to the phenotype requires the modulation of these interactive processes ("developmental reprogramming"), with the risk that tight co-ordination cannot be maintained and a non-functional structure is produced. Changes in phenotype would appear to be even more unlikely if we consider that developmental components function in multiple contexts [5]. Yet, while many parts of butterfly wing patterns are stringently controlled, they also exhibit highly variable elements. For example, the Indian leaf butter-fly (*Kallima inachus*) offers a striking contrast between the highly constant, brightly blue-and-orange patterned upperside, and the cryptic, dead-leaf-like underside, which appears to be expressed as individually as a human finger-print. Different compartments of the same individual can thus be very tightly regulated or not, in this case depending on biological function.

Evolutionary ecologists similarly have struggled with the problem of mor-phological diversification, as the multiple functions of wing patterns for de-fence, thermoregulation and intraspecific signalling would also seem to pre-clude ready alteration. This is particularly problematic for understanding the evolution of mimicry, where unrelated species converge on a common pheno-type [11, 12, 29]. Effective protection from predators can only be achieved when there is very close resemblance to the model. However, starting from a dissimilar phenotype, it seems very unlikely that a single mutation could produce the required close resemblance to the pattern of another species. The problem is compounded in polymorphic species, where several divergent morphs can coexist and interbreed. In such cases the genome of a single species has to provide the possibilities for multiple developmental programs to express the phenotype. How have these precisely matching colour patterns arisen, if not in a single evolutionary step? And, how can such complexity be controlled genetically?

Here we discuss these questions focussing on the African Mocker Swallow-tail, *Papilio dardanus* Brown, which exhibits well over a dozen distinct mimetic forms in the females, plus a greatly divergent male phenotype. This species has received much attention from geneticists attempting to understand how colour pattern control is switched between morphs [6, 7, 9, 12, 19, 25, 29, 31, 32, 34]. *Papilio dardanus* has been a focal system to address the evolution of Batesian mimicry and has provoked evolutionary theories to explain the evolution of complex traits more widely. It is also of great interest to developmental biology due to the polymorphic expression of several phenotypes. The aim of this article is to provide an historical perspective on previous studies on the evolution of mimicry patterns, with a focus on *P. dardanus*. The traditional view, conceived in the framework of evolutionary genetics, has significant shortcomings to explain the origin of phenotypic diversity in this system. We therefore discuss alternatives in the light of modern developmental biology, including the possible role of morphogens [30] in controlling morphogenesis and providing rapid evolutionary change.

8.2 The traditional view of evolutionary genetics

The striking resemblance of wing patterns between unrelated species in mimetic butterflies is due to convergences that apparently result from evolutionarily fast changes, as great differences in wing pattern exist between closely related species and populations. This indicates that the evolution of at least some of these wing patterns is more recent than the divergence of species. Hence, novel patterns can arise over evolutionarily short time scales, even if phenotypes are greatly divergent. Yet the divergent phenotypes are very precise in their expression.

The complexity of the pattern requires precise determination of positioning of colour elements, specification of colour, and in many cases differences in wing shape (including the presence of long hindwing tails in *P. dardanus* males, but not in females). Producing precisely tuned differences in the wing patterns requires concerted variation in multiple parameters, making evolutionary change seem unlikely. In addition, the complexity of changes led evolutionary geneticists to believe that there is little chance that the mimetic patterns are achieved in a single step by mutation [12, 29, 33]. Instead, they postulated the involvement of several genes to accommodate models of gradual evolutionary change. Their theory proposed that the evolution of mimicry involves, first, a major genetic event specifying a large shift in phenotype to confer a vague resemblance to the model sufficient to give a selective advantage, followed by the selection of minor-effect or fine-tuning genes ('modifiers'), to bring about closer and closer resemblance to the model [29].

The conventional evolutionary genetics model further postulates that the wing pattern and its many variable components are controlled by several genes (besides the modifiers). As a polygenic trait cannot spread in a population unless all genes involved segregate in concert, it was expected that these genes would become closely linked, i.e. they would be selected to operate as a 'supergene' consisting of several genes which act together to specify the phenotype (see review by [26]). The supergene specified phenotype is then further elaborated by modifiers [29]. Support for this model was obtained experimentally by the classical genetic studies of Clarke and Sheppard [8, 9]. They found that the control of colour pattern in *P. dardanus* maps to a single locus, termed

H, consistent with the theory of major-effect genes determining the mimicry morphs. Further, they found that crosses between geographically distant races (subspecies) gave rise to hybrids exhibiting imperfect mimicry, interpreted as a malfunction or an unselected and inappropriate constellation of the locally adapted modifiers to a non-indigenous *H* allele. Thus the *H* allele responsible for 'hippocoonides' in East Africa produces a phenotype with a narrow hindwing dark marginal band, whereas the same allele in West Africa produces the broader-margined 'hippocoon'. The difference between these two morphs, corresponding to the major exophenotypic difference between western *Amauris niavius niavius* and the eastern *A. niavius dominicanus*, is thus produced by the modifiers, not differences in the *H* alleles.

This explanation of Clarke and Sheppard's data has been widely accepted, but it rests more on unproven assumptions about the evolutionary process leading to Batesian mimicry than on the data themselves. For example, the apparent complexity of the *H*-dependent phenotype was the basis for postulating that *H* consists of multiple, closely linked genes (the 'supergene') acting together. However, this conclusion is not based on any direct evidence, as recombination or segregation has never been shown convincingly [21, 22] [1]. Thus there is also no empirical supporting evidence for the polygenic nature of *H* and the underlying evolutionary process of its origination. Similarly, the modifiers have never been characterised genetically and, as the breakdown of phenotypes has not been attributed directly to a particular gene or function, it is not justified to use them as corroborating evidence for the evolutionary process that presumably led to their existence. Furthermore, Clarke and Sheppard argued that each polymorphic form is determined by a different cistron within the supergene. If this hypothesis is correct, however, this single cistron would control the variation in several parameters (pattern, colour, wing shape), which contradicts precisely the postulate of a polygenic basis for the expression of the colour pattern, and hence the reason for assuming a supergene in the first place.

8.3 Alternatives to the conventional evolutionary genetics model

The ecological genetics school of Ford and his followers was mostly concerned with the evolution of divergence in wing patterns [12]. They did not analyse the specific differences in wing patterns between mimetic morphs beyond the question of similarity, or lack thereof, with the models. Despite their great diversity, butterfly wing patterns do not vary without limit. Instead the distinct pattern elements, such as eye spots (border ocelli) and marginal

---

[1] In another polymorphic swallowtail, *Papilio memnon*, there is some evidence of rare recombinants that could have arisen as a result of crossing-over within a putative supergene [10, 32].

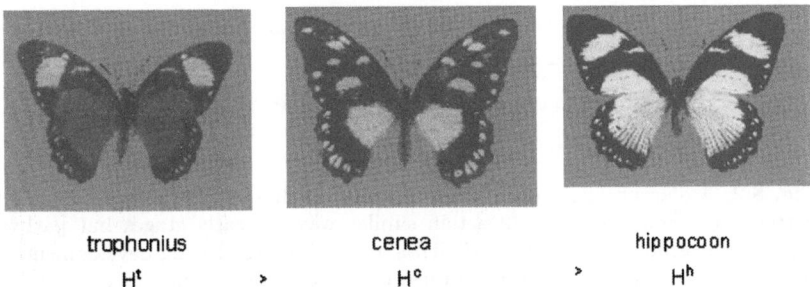

trophonius   cenea   hippocoon

H^t > H^c > H^h

Fig. 8.1. Dominance hierarchy in *P. dardanus*. In genetic segregation studies the wing pattern of female form 'trophonius' is dominant over the 'cenea' phenotype which itself is dominant over the 'hippocoon' form. Note that the dominance hierarchy is different for the individual colour elements. For example, the wide band and large posterior blotch on the forewing and the narrow black band on the hindwing are dominant in crosses of 'trophonius' and 'cenea', but are recessive in crosses of 'cenea' and 'hippocoon'.

bands, are confined to particular areas of the wing — but they can occur in many different combinations. Detailed analysis of this variation led to insights into the building blocks of the wing colour patterns and revealed their modular design. Butterfly wing patterns are now understood to be composed of autonomously varying elements differing in size, position and coloration [21, 22]. These pattern elements are apparently controlled independently, and they display idiosyncratic dominance patterns (Table 6.2 in [21]) (Fig. 8.1). The independence of pattern elements suggests that they are controlled through specification of spatially defined fields in development, where their extent and colour can be regulated independently from other portions of the wing.

From this perspective, H constitutes a decision point for the variation of the phenotype that regulates autonomous elements differently. Currently we do not know the nature of the switch in which H is involved. But if indeed the global colour pattern is a composite of largely independently expressed elements, then each will have alternative states that need to be specified by the H-dependent control. Hence, the great diversity of *P. dardanus* phenotypes, and its great capacity for evolutionary change, may result from the capabilities of H to control the expression of multiple portions of the wing independently.

The mechanism by which H can exert such control is entirely open, as long as it has not been characterised on the molecular level. Waddington [37] long ago suggested that one cannot necessarily point to any particular gene as being responsible for switching development into a particular pathway. Butterfly wing development is a complex process of spatial and temporal differentiation involving several steps during which subdividing populations of cells are increasingly determined [14]. The effects of these processes can be

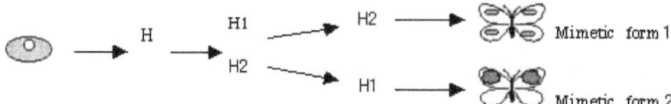

Fig. 8.2. A diagram illustrating the action of *H* at different decision points in wing patterning. Alleles H1 and H2 act in similar ways at early stages but pathways separate in response to different *H* alleles at later stages of wing development, and the separation of pathways is maintained by these alleles through metamorphosis.

subdivided notionally into those affecting wing shape, pattern elements and colour. Although specification of these parameters is not developmentally separable because specification during earlier stages irrevocably determines the repertoire of future possibilities, it is useful to think of them as a temporal succession of decision points (Cieslak et al., in preparation) (Fig. 1.2). Generally, specification of wing shape and major pattern elements precedes the specification of colour [2, 15, 22]. As the various *P. dardanus* morphs differ in many parameters, *H*-dependent control is expected to operate at several stages throughout morphogenesis and colour pattern formation, 'micromanaging' the development of the phenotype at subsequent stages of wing development.

8.4 Specifying and varying the phenotype

In a scenario where *H* controls the sequence of successive steps in morphogenesis, two requirements remain: the faithful replication of the phenotype, and the possibility to specify radically different colour patterns. Reaction diffusion models have been proposed to model butterfly wing patterns. These assume that the anterior and posterior margins of the wing are sources from which a wave of morphogen concentration emanates. Early applications of such models [20] were able to account for the cross-bands of wing pigmentation commonly observed in Lepidoptera. Recently, modelling studies have been conducted specifically to apply this approach to the wing patterns of *P. dardanus*. These studies recovered two-dimensional patterns showing great similarity to the various forms produced by this species. The modelled patterns are effectively buffered against a range of parameter variation, but by changing critical threshold values and the mode of the wave function, several discrete states of the wing pattern can be generated which are reminiscent of the *H*-dependent switch between morphs [17, 28]. While the biological reality of these models remains to be established, the exercise suggests the possibility of a mechanism where a single parameter can control wing pattern elements

Fig. 8.3. Inter-species hybrids of *P. phorcas* x *P. dardanus* segregating two forms of female offspring. Note the resemblance with the form 'hippocoon' (top) and 'cenea' (bottom). From the Clarke, Sheppard and Gill collection (in The Natural History Museum, London).

across the entire wing, and certain changes in these parameter values permit switches between discrete states of the global wing pattern.

The precision with which the phenotype is replicated suggests that specific interactions between components of the signalling networks have to be maintained. Similarly, the mechanism that permits switching between different states requires discrete effects on the signalling systems. The breakdown of phenotypes in many broods between distantly related subspecies of *P. dardanus* is an indication that these signalling networks are subtle, and subject to perturbation when divergent genotypes recombine. A clear example is seen in crosses between *P. dardanus* and its putative sister species, *P. phorcas*. These inter-specific crosses produce viable offspring, although of reduced body size, which clearly differ from either of the parental species. In a brood from the Clarke , Sheppard and Gill collection the hybrids segregate as two forms, apparently corresponding to the 'hippocoon' and 'cenea' morphs of *P. dardanus* (Fig. 8.3). This demonstrates that the *H*-locus is active and can be read-out in related species, including the polymorphic *P. phorcas*, and another closely related but monomorphic swallowtail species, *P. constantinus* (not shown). Hence the basic signalling mechanism is present also in these species, but the effect of the *P. dardanus H* system (any allele) is semi-dominant, resulting in recognisable, if noisy, *H* phenotypes.

An interesting feature of the *P. phorcas* x *P. dardanus* hybrids is a banding pattern in both fore and hind wings, at some distance from the base, that is reminiscent of bands seen in the wing patterns of many moth species. This band is apparently unaffected by the *H*-dependent expression of elements specific to the 'hippocoon' and 'cenea' morphs, recognisable as areas of white or orange coloration, respectively, near the wing base. These bands are not apparent in the parental species, the patterns of both of which are more highly structured. Conceivably, these wave-like patterns are an expression of a basic feature of the wing pattern which, however, are not clearly recognisable in the *P. dardanus* morphs, as they are further elaborated by the expression of the various elements in subsequent stages of pattern formation. In the hybrids, this integration of primary gradient with the downstream co-ordination mechanism may be imprecise, resulting in an incorrect placement of the various elements, but permitting insight into the earlier processes of wing development.

8.5 The role of *H* in modifying signalling pathways

If we accept that *H* controls the pattern formation at successive stages of wing development, then these stages can perhaps be further dissected into those that produce the primary morphogen gradients and those responsible for the subsequent expression of the colour elements themselves. The integration of these various stages is critical for the fidelity of the pattern expression, and has to be maintained throughout wing development. Initial patterning during the development of the wing discs involves the dorsal-ventral, anterior-posterior and proximal-distal axes, and these processes probably involve highly conserved signalling pathways common to holometabolous insects. By analogy with *Drosophila*, the proximal-distal patterning would require Wingless (Wg) and Distal-less (Dll) gradients [18]. Gene expression of subordinate pathways specifying particular pattern elements would respond in a threshold-like manner to these primary gradients.

To produce the observed differences in phenotype, the various *H* alleles could act either by producing differences in the gradients, or by differences in the way those primary gradients are interpreted. To decide between these possibilities, modelling studies could be extremely helpful. For example, Turing-type equations can be altered to produce changes in the wave function, the mathematical analogue to changes in reaction-diffusion kinetics of long-range morphogens. Similarly, threshold values can be varied, equivalent to changing the interactions of downstream regulatory cascades with the primary patterning gradients [28]. In addition, careful studies of co-variation between portions of the wing will help to assess the extent and position of semi-autonomously varying fields (Nijhout et al., in prep.). This should include the analysis of hybrids which may be deficient in some aspect of the regulatory networks. As the example of the *P. phorcas* x *P. dardanus* hybrids demonstrates, analysis of 'mutant' features can reveal patterns not otherwise apparent.

While we do not understand the specific mechanisms without a molecular analysis of *H* and the downstream components, a mechanism as proposed here based on independently varying pattern elements specified in a cascade-like manner (Fig. 8.2) is attractive for several reasons. The great exophenotypic diversity of *P. dardanus* requires that the developmental processes responsible for generating the phenotype are potentially independent and versatile, as they participate in different associations which are easily shuffled. The hierarchical nature of the process would permit that the early steps are less variable in order to provide accurate signalling for the downstream processes which rely on these signals. With progressive development the signalling is ever more limited in options, and spatially more restricted areas are affected. Compartments also become increasingly independent from other parts of the developing organism. As changes in a pathway have no impact on other elements, their genetic segregation may be independent and they may be subject to independent evolutionary change.

Hence, butterfly wing development should be less constrained than other body parts, and perhaps this is the reason for the great diversity of wing patterns. However, the internal mechanisms of patterning and the finite set of developmental programmes limit the direction and pattern of evolution. Additional constraints result from the link to the next step of development, when the determined state is realised by morphogenesis and differentiation. (In the case of holometabolous insects this involves the development of patterned imaginal discs during metamorphosis.) This leads to the question of which changes are possible, and in which way the developmental programmes limit variation of wing patterns, within *P. dardanus* and more generally throughout the Papilionoidea (day-flying butterflies). It has long been known that there are basic features of the pattern (a 'groundplan') [27, 21] indicating a stringent developmental architecture. However, the great diversity of wing patterns is perhaps an indication that developmental constraints, if they exist, can be overcome and actually do not greatly restrict the morphological possibilities. Yet, the rather limited repertoire of underlying developmental pathways makes it possible that phenotypes are canalised to converge on similar patterns, and hence provides the fundamental basis for the evolution of wing pattern mimicry. The repeated convergence of distantly related butterfly species on a few aposematic patterns [35] may be an expression of a limited set of regulatory networks which, however, are shared widely within the Papilionoidea. The fact that the various *P. dardanus* morphs can link into several mimicry rings is perhaps indicative of a role of *H* in switching between these fundamental pathways.

8.6 Conclusions

To understand the role of *H* in producing phenotypic variation, we need to look for a mechanism where minor parameter differences can profoundly alter morphogenesis, affecting the entire wing or major portions of it. Turing [30] provided a highly contemporary view on this subject, which greatly enriched the models of evolutionary geneticists. As demonstrated in [21], butterfly morphogenesis involves certain fundamental shared processes affecting all members of the group (some 20,000 species currently recognised), together with species- or clade-specific components. Due to its switch function deciding between different morphs, H could perhaps inform about these deeply shared processes. A convincing theory for the evolution of mimicry is likely to emerge once we understand how mimetic wing patterns develop and how they are controlled at the molecular level.

Acknowledgements

This work greatly benefited from discussion with the members of the Human Frontier Science Program team, S. Collins, D. G. Heckel, H. F. Nijhout, and T. Sekimura. We thank H. F. Nijhout for the images used in Fig. 8.1. Funded by HFSP grant RG00323/1999-M.

References

1. Arthur, W. (2000). The concept of developmental reprogramming and the quest for an inclusive theory of evolutionary mechanisms. *Evol. Dev.*, **2**:49-57.
2. Beldade, P. and P.M. Brakefield (2002). The genetics and evo-devo of butterfly wing patterns. *Nat. Rev. Genet.*, **3**:442-452.
3. Brakefield, P.M. (1998). The evolution-development interface and advances with the eyespot patterns of *Bicyclus* butterflies. *Heredity*, **80**:265-272.
4. Carroll, S.B., Gates, J., Keys, D.N., Paddock, S.W., Panganiban, G.E.F., Selegue, J.E. and Williams, J.A. (1994). Pattern formation and eyespot determination in butterfly wings. *Science*, **265**:109-114.
5. Carroll, S.B., Grenier, J.K. and Weatherbee, S.D. (2002). *From DNA to Diversity: Molecular Genetics and the Evolution of Animal Design.* Blackwells, Oxford.
6. Charlesworth, D. and Charlesworth, B. (1976). Theoretical genetics of Batesian mimicry. III. Evolution of dominance. *J. Theor. Biol.*, **55**:325-327.
7. Clarke, C.A., Clarke, F.M.M., Collins, S.C., Gill, A.C.L. and Turner, J.R.G. (1985). Male-like females, mimicry and transvestism in swallowtail butterflies. *Syst. Entomol.*, **10**:257-283.
8. Clarke, C.A. and Sheppard, P.M. (1959). The genetics of *Papilio dardanus*, Brown. I. Race cenea from South Africa. *Genetics*, **44**:1347-1358.
9. Clarke, C.A. and Sheppard, P.M. (1963). Interactions between major genes and polygenes in the determination of the mimetic patterns of *Papilio dardanus*. *Evolution*, **17**:404-413.
10. Clarke, C.A. and Sheppard, P.M. (1971). Further studies on the genetics of the mimetic butterfly *Papilio memnon*. *Phil. Trans. Roy. Soc. B*, **263**:35-70.
11. Fisher, R.A. (1930). *The Genetical Theory of Natural Selection.* Clarendon, Oxford, UK.
12. Ford, E.B. (1936). The genetics of *Papilio dardanus* Brown (Lep.). *Trans. Roy. Entomol. Soc. Lond.*, **85**:435-466.
13. French, V. (1997). Pattern formation in colour on butterfly wings. *Curr. Opin. Gen. Develop.*, **7**:524-529.

14. Gerhart, J. and Kirschner, M. (1997). *Cells, Embryos, and Evolution*. Blackwell, Malden, MA.

15. Koch, P.B., Keys, D.N., Rocheleau, T., Aronstein, K., Blackburn, M., Carroll, S.B. and ffrench-Constant, R.H. (1998). Regulation of Dopa Decarboxylase expression during colour pattern formation in wild-typ and melanic Tiger Swallowtail Butterflies. *Development*, 125:2303-2313.

16. Koch, P.B., Lorenz, U., Brakefield, P.M. and ffrench-Constant, P.H. (2000). Butterfly wing pattern mutants: developmental heterochrony and co-ordinately regulated phenotypes. *Devel. Genes Evol.* 210:536-544.

17. Madzvamuse, A., Thomas, R.D.K., Sekimura, T., Wathen, A.J. and Maini, P.K. (2003). The moving grid finite element method applied to biological problems in T. Sekimura, S. Noji, N. Ueno, and P. K. Maini, eds. *Morphogenesis and Pattern Formation in Biological Systems*. Springer Verlag, Tokyo.

18. McMillan, O.W., Monteiro, A. and Kapan, D.D. (2002). Development and evolution on the wing. *Trends Ecol. Evol.* 17:125-133.

19. Murray, J. (1972). *Genetic Diversity and Natural Selection*. Oliver and Boyd, Edinburgh.

20. Murray, J.D. (1981). On pattern formation mechanisms for lepidopteran wing patterns and mammalian coat markings. *Phil. Trans. Roy. Soc. Lond. B*, 295:473-496.

21. Nijhout, H.F. (1991). *The Development and Evolution of Butterfly Wing Patterns*. Smithsonian Institution Press, Washington.

22. Nijhout, H.F. (1994). Developmental perspectives on evolution of mimicry in butterflies. *BioScience*, 44:148-157.

23. Nijhout, H.F. (1994). Symmetry systems and compartments in Lepidoptera wings: the evolution of a patterning mechanism. *Development Supp*:225-233.

24. Nijhout, H.F. (1999). When developmental pathways diverge. *Proc. Natl. Acad. Sci.*, 96:5348-5350.

25. O'Donald, P. and Barrett, J.A. (1973). Evolution of dominance in polymorphic Batesian mimicry. *Theoret. Pop. Biol.*, 4:173-192.

26. Robinson, R. (1971). *Lepidoptera Genetics*. Pergamon, Oxford.

27. Schwanwitsch, B.N. (1924). On the ground plan of wing-pattern in nymphalids and certain other families of rhopalocerous Lepidoptera. *Proc. Zool. Soc. Lond.*, 1924:509-528.

28. Sekimura, T., Madzvamuse, A., Wathen, A.J. and Maini, P.K. (2000). A model for colour pattern formation in the butterfly wing of *Papilio dardanus*. *Proc. Roy. Soc. Lond. B*, 267:851-859.

29. Sheppard, P.M. (1975). *Natural Selection and Heredity* (4th edn). Hutchinson, London.

30. Turing, A.M. (1952). The chemical basis of morphogenesis. *Phil. Trans. Roy. Soc. Lond. B*, 237:37-72.

31. Turner, J.R.G. (1963). Geographical variation and evolution in the males of the butterfly *Papilio dardanus* Brown (Lepidoptera: Papilionidae). *Trans R. ent. Soc. Lond.*, 115:239-259.

32. Turner, J.R.G. (1978). Why male butterflies are non-mimetic: natural selection, group selection, modification and sieving. *Biol. J. Linn. Soc.*, 10:385-432.

33. Turner, J.R.G. (1984). Mimicry: the palatability spectrum and its consequences. *Symp. R. ent. Soc. Lond.*, 11:141-161.

34. Vane-Wright, R.I. (1981). Mimicry and its unknown ecological consequences. pp. 157-168 in P. L. Forey, ed. *The Evolving Biosphere.* BMNH/ Cambridge UP, London.
35. Vane-Wright, R.I. and Boppré, M. (1993). Visual and chemical signalling in butterflies: functional and phylogenetic perspectives. *Phil. Trans. Roy. Soc. Lond. B,* **340**:197-205.
36. Waddington, C.H. (1942). The canalisation of development and the inheritance of acquired characters. *Nature,* **150**:563.
37. Waddington, C.H. (1957). *The Strategy of the Genes.* Allen and Unwin, London.

Formation of New Organizing Regions by Cooperation of *hedgehog*, *wingless*, and *dpp* in Regeneration of the Insect Leg; a Verification of the Boundary Model

Taro Mito[1], Yoshiko Inoue[2], Shinsuke Kimura[1], Katsuyuki Miyawaki[1], Nao Niwa[2], Yohei Shinmyo[1], Hideyo Ohuchi[1], and Sumihare Noji[1]

[1] Department of Biological Science and Technology, Faculty of Engineering, The University of Tokushima, 2-1 Minami-Jyosanjima-cho, Tokushima City 770-8506, Japan
telephone: +81-88-656-7528; fax: +81-88-656-9074;
email: noji@bio.tokushima-u.ac.jp
[2] Present address: Laboratory for Morphogenetic Signaling Riken Center for Developmental Biology, 2-2-3, Minatojima-minamimachi, Chuo-ku, Kobe, 650-0047, Japan

Summary. Meinhardt [15] proposed the boundary model (BM) to explain pattern formation in developmental subfields. One of applications of his model was to explain formation of supernumerary legs after amputating a leg and reimplanting it onto a contralateral stump. The boundary model was cemented by Campbell and Tomlinson [6]. Their model (CTBM) postulates that key genes during leg development such as *hedgehog* (*hh*), *wingless* (*wg*), and *decapentaplegic* (*dpp*) are also involved in leg regeneration and that the formation of the proximodistal axis of a regenerating leg is triggered at a site where ventral *wg*-expressing cells abut dorsal *dpp*-expressing cells in the anteroposterior (AP) boundary. To verify this model, we experimentally examined whether the model would fit our results obtained during the leg regeneration of a cricket *Gryllus bimaculatus*. Since the expression patterns of the three genes can be predicted in a regenerating leg by the model, we observed corresponding expression patterns in supernumerary legs formed when the distal part of an amputated leg was grafted onto the contralateral leg stump so as to reverse the AP polarity. Since our observations were essentially consistent with the CTBM, we concluded that we were able to verify the BM by our regeneration experiments.

9.1 Introduction

In a developmental system, a signaling and signal-receiving mechanism exists which enables a cell to develop in a manner appropriate to its position [15]. The necessity of such a signaling system is easily demonstrated by the phenomenon of regeneration. The removal of a part of an organism can elicit a

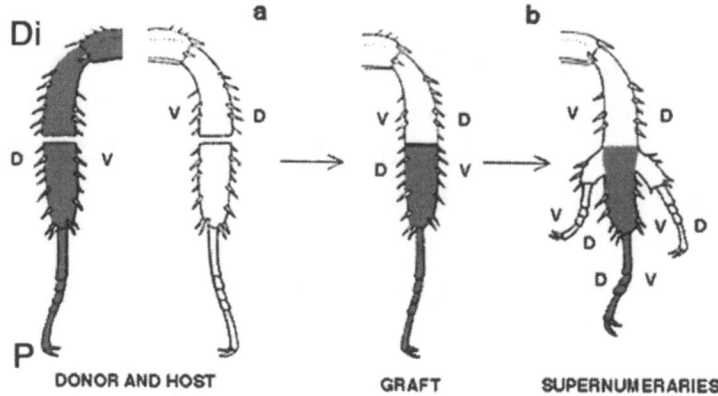

Fig. 9.1. Formation of supernumerary legs (b) of the cockroach after amputating left and right legs at the same level and reimplanting the left distal donor to the proximal host (a). P, proximal; Di, distal; D, dorsal; V, ventral.

signal in the remaining parts which initiates development through a different pathway. Meinhardt [15] proposed the boundary model (BM) to explain such regeneration phenomena under only a few assumptions. He assumed that the borders between compartments are used to create new coordinate systems for regeneration and predicted that a morphogenetic substance is produced by the cooperation of compartments, which provides positional information on the distance of the cells from the border(s). According to his model, one can predict that changes of the geometrical arrangement of the compartments, caused either by surgical interference or by a cell-internal switch in the compartmental specification, can lead to new intersections of borders and therefore to the formation of additional structures [15]. One of the interesting phenomena explained by his model is formation of supernumerary legs after amputating a leg and reimplanting it onto a contralateral stump [2, 3, 4, 5, 11] (see Fig. 9.1).

In 1995, Campbell and Tomlinson [6] proposed their boundary model (CTBM), based upon the relevant molecular biology of cell signaling in the *Drosophila* leg imaginal disc. In the leg disc, the cells of the posterior compartment are defined by the expression of the Engrailed (En) homeodomain protein and they secrete a signaling peptide, Hedgehog (Hh). The dorsal and ventral cells in the anterior compartment close to the AP boundary express *decapentaplegic* (*dpp*) and *wingless* (*wg*), respectively. A single site where *dpp*-expressing cells abut *wg*-expressing cells becomes the presumptive distal tip. With the CTBM, they could explain grafting experiments with the three signaling molecules.

Recently, we demonstrated with the cricket that *hh*, *wg*, and *dpp* are involved in the initiation of proximodistal axis formation during the regeneration of cricket legs [16]. Here, we introduce our experiments for a verification of the boundary model for formation of supernumerary legs and discuss their significance in the field of developmental biology.

9.2 Experimental results

Although insect leg regeneration has been intensively studied with cockroaches since it was first reported by Bodenstein [1], only a few studies have been carried out from a molecular biological point of view. As observed in cockroaches and other hemimetabolous insects, cricket legs develop during embryogenesis without formation of the leg imaginal disc so that the nymph possesses functional legs and the ability to regenerate these legs. Thus, we have been working on limb development [17, 18] and regeneration with the cricket, *Gryllus bimaculatus* [16]. Three major axes of the cricket leg are the anteroposterior (AP), dorsoventral (DV), and proximodistal (PD) axes. The PD axis relates to the distance from the body trunk, while the AP and DV axes unite to form the single circumferential axis (Fig. 9.1).

When a metathoracic leg of a *Gryllus* nymph in the third instar is amputated at the tibia, the distal missing part is completely recovered after about 30-35 days through four molts subsequent to the amputation. We observed expression patterns of *hh*, *wg* and *dpp* during regeneration and found that they are very similar to expression patterns of the three genes in the *Drosophila* leg imaginal disc [16].

The ability to regenerate distal structures was demonstrated more dramatically by grafting experiments that resulted in the development of supernumerary legs [2, 3, 5, 11]. We performed one of grafting experiments with the cricket as shown in Fig. 9.2B, as reported by French [12]: In the third instar nymph, the distal part of the amputated right mesothoracic leg (graft) was grafted onto the stump of the amputated left metathoracic leg (host) at the mid tibia level with a 180° rotation to keep the DV axis in register but misaligning the AP axis (Fig. 9.2A). After two weeks or so, two supernumerary legs were induced at the positions shown in Fig. 9.2A.

We examined whether the expression of *wg* or *dpp* can be ectopically induced in the blastema of supernumerary legs. Our results, obtained by whole-mount in situ hybridization (WISH), are shown in Fig. 9.3. In the host/graft boundary, *hh* is expressed in the original posterior regions of the host and graft epidermis (Fig. 9.3A). On the whole, the expression patterns of *hh* (En) are illustrated in Fig. 9.3D-F.

On the other hand, an ectopic signal for *wg* can be observed intensely at the ventral region of each supernumerary blastema (Fig. 9.3G-I), whereas a normal signal can be observed in the ventral region of the host and graft legs (Fig. 9.3G). Our deduced relationship between the *hh* and *wg* expression

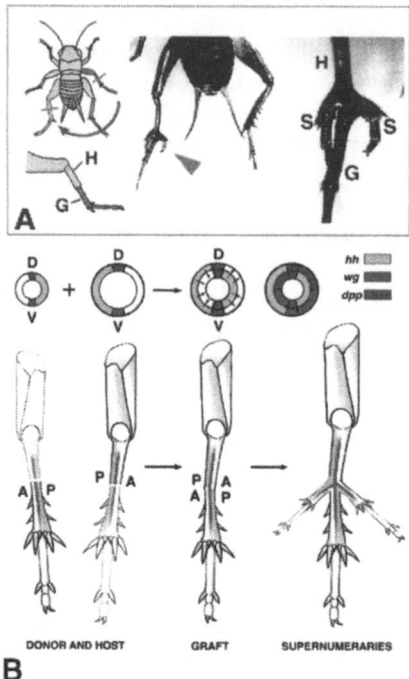

Fig. 9.2. Formation of supernumerary legs by grafting operation in the *Gryllus* nymph and prediction of expression patterns of *hh* (green), *wg* (red) and *dpp* (blue) during regeneration by the CTBM. (A) The distal part of the right mesothoracic leg (graft: G) was amputated and grafted onto the stump of the left metathoracic leg (host: H) in the third instar of the cricket. This operation resulted in the inversion of the anteroposterior (AP) polarity of the graft to the host, leading to the formation of two supernumerary (S) legs, one indicated by the red arrowhead in the center panel and the other by S in the right panel. (B) Prediction of the expression patterns of the three genes *hh* (green), *wg* (red), and *dpp* (blue) by the CTBM during formation of the supernumerary legs shown in (A). In the upper series, the left small and adjacent large circles represent cross-sections of the graft and the host, respectively, showing the expression patterns of *hh* (green), *wg* (red) and *dpp* (blue). When grafted so as to reverse the anteroposterior polarity, Hh in the posterior compartment of the host and graft legs induced the expressions of *wg* and *dpp* in the ventral side and dorsal side of the anterior compartment (double circles). The site where the *wg*-expression cells abut the *dpp*-expressing cells became an organizer that induced the formation of the proximodistal axis. A, anterior; P, posterior; D, dorsal; V, ventral. From Fig. 9.3 of [16].

patterns is illustrated in Fig. 9.3J-L. A weak signal for the *dpp* expression can be detected in the dorsal side of each supernumerary blastema. A typical expression pattern of *dpp* revealed by WISH is shown in Fig. 9.3M (dorsal view) (n = 3 out of 8), while dorsal localization of signals can be confirmed on oblique sections of WISH samples (Fig. 9.3N, O). Our deduced relationship among the *hh*, *wg*, and *dpp* expression patterns is shown in Fig. 9.3P and Q. The ectopic induction of the *wg* and *dpp* expressions, probably induced by Hh, was observed at two sites of the host/graft junction. Although the expression pattern of *dpp* differed from the prediction, the essential features of the expression patterns of the three genes indicated that the initiation of the PD axis formation occurs at a site where *wg*-expressing cells abut *dpp*-expressing cells.

9.3 Discussion

9.3.1 Molecular interpretation of supernumerary leg formation by surgical experiments

Since the expression patterns of the above three genes during regeneration are essentially consistent with the predictions from the CTBM, we can now conclude that the CTBM has been verified by our study [16]. The next question is whether the supernumerary formation induced by various surgical experiments can be interpreted with the CTBM and expression patterns of the three genes. Here, it might be worthwhile to revisit the several studies on supernumerary leg formation reported by Bohn [3]. According to Bohn [3], there are at least five types of surgical experiments capable of inducing supernumerary leg formation, as shown in Fig. 9.4.: (1) by transplanting a left distal graft onto the right proximal stump with a 180° rotation to keep the DV axis in register but misaligning the AP axis, which is shown in the present paper (Fig. 9.4A); (2) by transplanting a left distal graft onto the right proximal stump while keeping the AP axis in register but misaligning D/V (Fig. 9.4B); (3) by transplanting the ventral half of a left tibia in place of the dorsal half of a right tibia so as to form a double-ventral surface in the tibia, and then amputating the distal part (Fig. 9.4C); (4) by cutting a notch into the inner ventral site (Fig. 9.4D); and (5) by transplanting a V-shaped sector graft cut from the dorsal side onto a notch formed in the ventral side with a 180° rotation to keep the AP axis in register but misaligning the DV axis (Fig. 9.4E). These supernumerary legs have been called Bohn's legs. Presumptive expression patterns of *hh*, *wg* and *dpp* are illustrated in the original figures reported by Bohn [3] in Fig 9.4, assuming that *dpp* and *wg* are expressed or inducible in the dorsal side and ventral side, respectively, in the operated legs. In the cases of type (1), (2), (3) and (5), every supernumerary leg is induced at a site where the *dpp*-expressing cells abut the *wg*-expressing cells, as predicted by the CTBM. In the case of type (4), it is likely that dorsal cells in

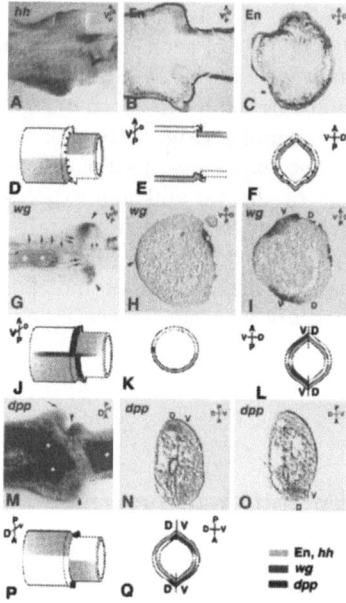

Fig. 9.3. Expression patterns of En, *hh*, *wg*, and *dpp* in supernumerary leg regeneration. (A) Expression of *hh*, as revealed by whole-mount in situ hybridization. (B, C) Localization of En in cryosections, as revealed by immunostaining with anti-Engrailed antibody (mAb4D9). Sagittal (B) and transverse (C) sections of a blastema. Cells expressing En are localized in each posterior half of the host and graft-derived epidermis. (D, E, F) Schematic drawings of *hh* (D) and En (E, F) expression patterns. To distinguish the host epidermis from the graft epidermis, outer and inner circles are used. In the corresponding actual transverse sections, the graft- and the host-derived epidermis are not distinguishable, but identified by En-stained cells. The Hh signals indicated by the arrows possibly induce expressions of *wg* and *dpp* in the adjacent cells. (G) Expression of *wg*, as revealed by whole-mount in situ hybridization. A distinctive expression of *wg* can be observed in the ventral region for both supernumerary blastemata indicated by the large arrowheads. The arrows indicate signals in the ventral side of the host. The small arrowheads indicate weak signals of the graft. Nonspecific staining in the trachea is indicated by the asterisks. (H, I) The expression of *wg* in transverse sections of G at the line passing both the blastemata (I) and at a more proximal position (H). The asterisks indicate nonspecific staining in residuary cuticles. (J, K, L) Schematic drawings of deduced relationship between *wg* (red) and *hh* (En) (green) expression domains. (M) Dorsal view of the expression pattern of *dpp*. Significant signals for *dpp* can be observed on the dorsal side of both blastemata. The sections at the positions indicated by the arrowheads are shown in (N) and (O). Nonspecific staining in the trachea is indicated by the asterisks. An arrow indicates a tibial spur. (N. O) Expression of *dpp* in oblique sections of the posterior (N) and anterior (O) blastemata shown in (M). (P, Q) Schematic drawings of deduced relationships among expression domains of *hh* (En) (green), *wg* (red), and *dpp* (blue). Expression signals of *dpp* in the dorsal side of the host and the graft could not be detected under the present conditions. The anteroposterior and dorsoventral polarities in each panel indicate those in the host leg. From Fig. 6 [16].

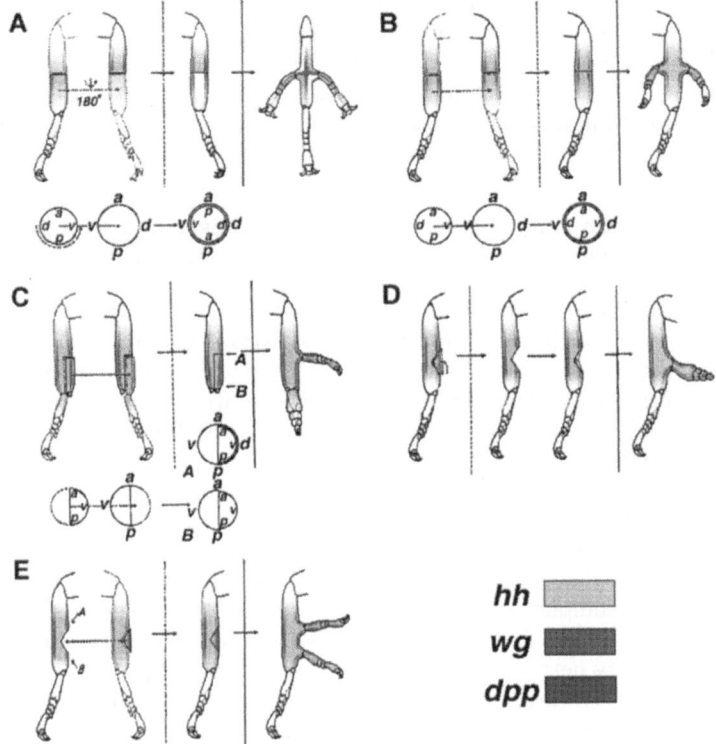

Fig. 9.4. A molecular interpretation of Bohn's legs. Bohn [3] reported that the amputation/transplantation of cockroach legs can induce supernumerary legs (Bohn's legs). There are five types of surgical experiments to induce Bohn's legs as shown in A-E. Presumptive expression patterns of *hh* (green), *wg* (red) and *dpp* (blue) deduced from the present study are painted in the original figures reported in [3]. In the cases of B, C and E, the expression patterns clearly indicate that Bohn's legs are formed at sites where *wg*-expressing cells abut *dpp*-expressing cells. In (A), as shown in the present paper, Hh induces expressions of both *wg* and *dpp* in the anterior compartment, resulting in the formation of the site where *wg*-expressing cells abut *dpp*-expressing cells. In (D), the dorsal anterior cells migrate to contact the posterior cells during wound healing, inducing the expression of *dpp* in the anterior cells. As a result, the *dpp*-expressing cells abut the *wg*-expressing cells in the ventral region. Accordingly, all experimental results can be interpreted with the CTBM. (A) Transplanting a right distal graft onto the left proximal stump with a 180° rotation to keep the DV axis in register but misaligning the AP axis. (B) Transplanting a right distal graft onto the left proximal stump while keeping the AP axis in register but misaligning the DV axis. (C) Transplanting the ventral half of a left tibia in place of the dorsal half of a right tibia so as to form a double-ventral surface in the tibia, and then amputating the distal part. (D) Cutting a notch into the inner ventral site. (E) Transplanting a V-shaped sector graft cut from the dorsal side onto a notch formed in the ventral side (A and B with arrows) with a 180° rotation to keep the AP axis in register but misaligning the DV axis. Supernumerary legs are formed at two sites (A and B). a, anterior; p, posterior; v, ventral; d, dorsal.

the anterior compartment migrate during wound healing and contact cells in the posterior compartment, resulting in induction of *dpp* expression. Consequently, *dpp*-expressing cells can abut *wg*-expressing cells in the ventral side of the leg. Thus, it is likely that supernumerary leg formation is in general due to the confrontation of the *wg*- and *dpp*-expressing cells as predicted by the BM.

9.3.2 The mechanism of pattern formation described by the BM may be conserved in both invertebrates and vertebrates

Recently, Campbell [7] and Galindo et al. [13] demonstrated independently that epidermal growth factor (EGF) is activated by Dpp and Wg at the distal tip during PD pattern formation in the *Drosophila* leg. EGF could be a morphogenetic substance produced by the cooperation of compartments, as predicted by Meinhardt [15]. We also found that the EGF receptor gene (*Egfr*) is expressed in the cricket leg bud (unpublished data). Thus, EGF may act as a morphogen in insect leg development and also in leg regeneration.

Gierer and Meinhardt [14] have shown that short-range autocatalysis coupled with long-range inhibition can give rise to spatial pattern. They also pointed out that the sharp boundaries between the patches of differently determined cells (compartments) open a new possibility for the generation of positional information. In the case of leg pattern formation, Hh/Shh generates the boundary between the anterior and posterior compartments, the two factors of Wg/Wnt and Dpp/BMP induce the distal organizer region which then produces a morphogen, EGF/FGF [8, 10, 19].

We speculate that a similar molecular mechanism may be generally used in biological pattern formation. Since the Toolkit genes are likely to be involved in pattern formation of various biological systems [9], biological patterns may be generated with those Toolkit proteins which play essential roles in the reaction-diffusion mechanism.

Acknowledgements

We would like to thank K. Kitagawa and M. Saito for their technical assistance. This work was supported by grants from the Japan Society for the Promotion of Science (Research for the Future) and from the Ministry of Education, Sports, Culture, Science and Technology to S.N.

References

1. Bodenstein, D. (1933). Beintransplantationen an Lepidopterenraupen. II. Zur Analyses der Regeneration der Brustbeine von Vanessa urticae-Raupen. *Wilhelm Roux Arch. EntwMech. Org.* **130**:747-770.

2. Bodenstein, D. (1937). Beintransplantationen an Lepidopterenraupen. IV. Zur Analyse experimentell erzeugter Beinmehrfachbildungen. *Wilhelm Roux Arch. EntwMech. Org.* **136**:745-785.

3. Bohn, H. (1965). Analyse der Regenerationsfahigkeit der Insekten-extremitat durch Amputations-und Transplationsbersuche an Larven der afrikanschen Schabe Leucophaea maderae Fabr. (Blattaria). II. Mitt. Achsendetermination. *Wilhelm Roux Arch. EntwMech. Org.* **156**:449-503.

4. Bryant, S.V. and Iten, L.E. (1976). Supernumerary limbs in amphibians: experimental production in Notophthalmus viridescens and a new interpretation of their formation. *Dev. Biol.* **50**:212-234.

5. Bulliere, D. (1970). Interpretation des regenerats multiples chez les insectes. *J. Embryol. Exp. Morph.* **23**:337-357.

6. Campbell, G. and Tomlinson, A. (1995). Initiation of the proximodistal axis in insect legs. *Development* **121**:619-628.

7. Campbell, G. (2002). Distalization of the *Drosophila* leg by graded EGF-receptor activity. *Nature* **418**:781-785.

8. Capdevila, J. and Izpisua Belmonte, J.C. (2001). Patterning mechanisms controlling vertebrate limb development. *Annu. Rev. Cell Dev. Biol.* **17**:87-132. Review.

9. Carroll, S.B., Grenier, J.K. and Weatherbee, S.D. (2001). *From DNA to Diversity: Molecular Genetics and the Evolution of Animal Design.* Massachusetts, USA, Blackwell Science.

10. Drossopoulou, G., Lewis, K.E., Sanz-Ezquerro, J.J., Nikbakht, N., McMahon, A.P., Hofmann, C. and Tickle, C. (2000). A model for anteroposterior patterning of the vertebrate limb based on sequential long- and short-range Shh signalling and Bmp signalling. *Development* **127**:1337-1348.

11. French, V. (1976). Leg regeneration in the cockroach, Blatella germanica. II. Regeneration from a non-congruent tibial graft/host junction. *J. Embryol. Exp. Morphol.* **35**:267-301.

12. French, V. (1984). The structure of supernumerary leg regenerates in the cricket. *J. Embryol. Exp. Morphol.* **81**:185-209.

13. Galindo, MI., Bishop, SA., Greig, S. and Couso, JP. (2002). Leg patterning driven by proximal-distal interactions and EGFR signaling. *Science* **297**:256-259.

14. Gierer, A. and Meinhardt, H. (1972). A theory of biological pattern formation. *Kybernetik* **12**:30-39.

15. Meinhardt, H. (1982). *Models of Biological Pattern Formation.* London: Academic Press.

16. Mito, T., Inoue, Y., Kimura, S., Miyawaki, K., Niwa, N., Shinmyo, Y., Ohuchi, H. and Noji, S. (2002). Involvement of *hedgehog, wingless,* and *dpp* in the initiation of proximodistal axis formation during the regeneration of insect legs, a verification of the modified boundary model. *Mech Dev.* **114**:27-35.

17. Niwa, N., Saitoh, M., Ohuchi, H., Yoshioka, H., and Noji, S. (1997). Correlation between Distal-less expression patterns and structures of appendages in development of the two-spotted cricket, *Gryllus bimaculatus. Zool. Sci.* **14**:115-125.

18. Niwa, N., Inoue, Y., Nozawa, A., Saito, M., Misumi, Y., Ohuchi, H., Yoshioka, H., and Noji, S. (2000). Correlation of diversity of leg morphology in *Gryllus bimaculatus* (cricket) with divergence in *dpp* expression pattern during leg development. *Development* **127**:4373-4381.

19. Ohuchi, H. and Noji, S. (1999). Fibroblast-growth-factor-induced additional limbs in the study of initiation of limb formation, limb identity, myogenesis, and innervation. *Cell Tissue Res.* **296**:45-56. Review.

The Heterotopic Shift in Developmental Patterns and Evolution of the Jaw in Vertebrates

Shigeru Kuratani

Laboratory for Evolutionary Morphology, Center for Developmental Biology,
RIKEN, 2-2-3 Minatojima-minami, Chuo-ku, Kobe, Hyogo 650-0047, JAPAN
E-mail: saizo@cdb.riken.go.jp
Tel & Fax: +81-78-306-3064

10.1 Introduction

What is evolutionary novelty? Generally, two distinct types of morphological changes are recognized in animal evolution. Firstly, the shape of a given structure can change to serve a new function, although the basic architecture of the derived structure is retained, and the homologous relationship is maintained through evolution. For example, the wing of the bat has been derived from the arm of the ancestral mammal, and all the anatomical components of the mammalian arm, such as the bones, muscles, and nerve branches, can be found in the bat wing, arranged in the same order and with the same topographical patterns. In other words, bat wings develop under the strong constraints that constitute the mammalian body plan. Wagner and Müller [15] have called such changes "adaptation" and regard them as not fundamentally new or innovative, but as evolutionary changes. Most evolutionary changes belong to this category, at least to the extent that specific fields of evolutionary biology, such as comparative embryology and morphology, can establish.

The other type of evolutionary change, however, involves the fundamental rearrangement of a developmental pattern that causes a disruption of morphological homology. In this case, the developmental constraints themselves are altered, or even discarded, and thus a truly innovative pattern can be achieved. This situation has been defined as "novelty" [11, 15]. One possible example is found in the turtle shell. The carapace of the turtle develops from the ribs, which are secondarily shifted to the superficial layer of the body (reviewed by [2, 4]). This change also involves alteration of the topographical relationships between the ribs and the scapula. As a result, the anatomical pattern of the trunk of this animal has diversified greatly, compared with the standard amniote pattern.

The jaw of vertebrates is usually regarded as an evolutionary invention that arose in the early phase of their history. One of the rostral gill arches of an ancestral vertebrate divided dorsoventrally and enlarged to function as a biting jaw. We investigated the developmental pattern of the oral region in a jawless vertebrate, the Japanese lamprey (*Lampetra japonica*), to determine if the jaw is simply an adaptation or an innovation involving a rearrangement of the ontogenetic process. As noted above, this evaluation must be based upon morphological homology and comparison of the developmental patterning between the lamprey and gnathostomes.

10.2 Embryonic development of the lamprey and the mandibular arch

We have previously shown that the lamprey embryonic head does not constitute an intermediate state between the amphioxus and shark, but shares the same basic morphological pattern as that of the gnathostomes: the paraxial mesoderm is segmented only postotically, and the distribution pattern of the cephalic crest cells is almost identical to that found in amniote embryos. The latter phenomenon is associated with the conservative patterning and behavior of the cephalic neural crest in vertebrates. Before emigration, the neural crest is largely specified along the anteroposterior axis of the embryo. In terms of molecular level developmental patterning, homologous regulatory genes are also expressed in the same pattern as found in gnathostome embryos. In the context of jaw evolution, a large part of the jaw-forming ectomesenchyme (the mandibular arch crest cells) in gnathostomes originates from the mesencephalic neural crest, where the homeobox gene *Otx2* is expressed. Expression of this gene is also a prerequisite for normal patterning of the murine lower jaw ([10]; reviewed by [7]). In the lamprey embryo, the apparent lamprey cognate of *Otx2* (*LjOtxA*) is similarly expressed rostral to the mid-hindbrain boundary, and the crest cells derived from that region migrate into the mandibular arch ([5]; Fig. 10.1). Therefore, a similar crest-cell origin and the same gene expression pattern occur in the mandibular arch of the lamprey embryo as in the gnathostome.

10.3 Epigenetic interactions and the patterning of the jaw

Other downstream developmental patterning of the mandibular arch also demonstrates the similarities between these animal groups. The patterning of the proximal–distal (P–D) polarity of the gnathostome jaw, for example, arises via epithelial–mesenchymal tissue interactions involving growth factors and target homeobox genes [1, 12, 14]. In gnathostomes, fibroblast growth

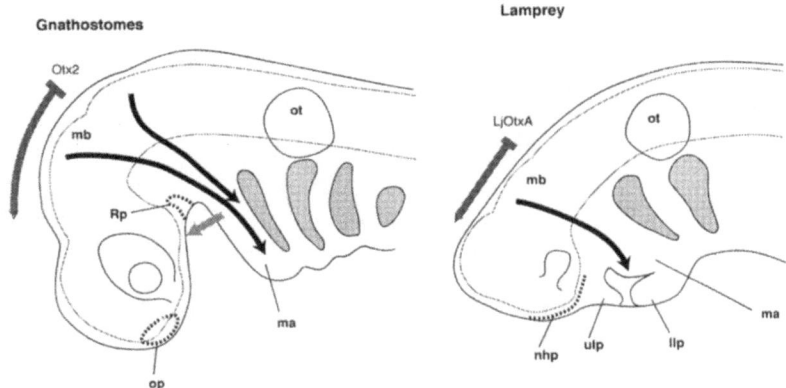

Fig. 10.1. Comparison of the gnathostome and lamprey embryonic heads. In both embryos, *Otx* gene cognates are expressed rostral to the mid-hindbrain boundary, and the crest cells derived from the Otx-positive neuroectoderm migrate into the mandibular arch (black arrows). In the gnathostome embryo, the upper jaw grows secondarily from the dorsal part of the mandibular arch (gray arrow). *Abbreviations: llp*, lower lip; *ma*, mandibular arch; *mb*, midbrain; *nhp*, nasohypophysial plate; *op*, olfactory placode; *ot*, otocyst; *Rp*, Rathke's pouch; *ulp*, upper lip.

factor 8 (FGF8) is derived from the central part of the ectoderm covering the mandibular arch, and upregulates its target homeobox gene, *Dlx*1, in the underlying ectomesenchyme. On the other hand, the distal ectoderm of the upper and lower jaws produces another type of growth factor, bone morphogenetic growth factor 4 (BMP4), and similarly induces its target homeobox gene, *Msx*1, in the distal ectomesenchyme (Fig. 10.2). Therefore, local expression of homeobox genes is established to define the P–D polarity of the jaw in gnathostomes, and the cause of this patterning can be found in the ectodermal distribution of upstream factors that form a prepattern. To determine whether it is possible to define the P–D patterning in the lamprey oral region, genes encoding the above molecules have been isolated and their spatial expression patterns observed [13]. The larval oral apparatus of the lamprey also consists of upper and lower components, the upper and lower lips. As in the gnathostome embryo, lamprey cognate proteins of the corresponding growth-factor-encoding genes, *LjFgf8/17* and *LjBmp2/4a* (*Lj* stands for *Lampetra japonica*), are expressed in the perioral ectoderm and the distal tip ectoderm, respectively. Lamprey proteins encoded by the homeobox genes *LjDlx1/6* and *LjMsxA* are consistently expressed in the perioral and distal ectomesenchyme, respectively. Moreover, by implanting beads soaked with mammalian growth-factor proteins into the lamprey embryonic head, it

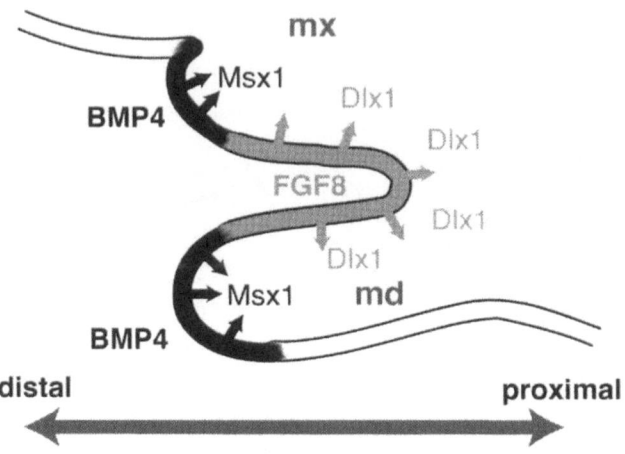

Fig. 10.2. Proximal and distal patterning of the gnathostome mandibular arch and epigenetic tissue interactions. BMP4 in the distal ectoderm of the mandibular arch induces its target homeobox gene, *Msx1*, and FGF8 in the proximal ectoderm induces another homeobox gene, *Dlx1*. Such localized expression patterns of homeobox genes define the proximal and distal polarity of the jaw in later development. *Abbreviations: md*, mandibular or lower jaw region; *mx*, maxillar or upper jaw region.

was possible to upregulate ectopically the endogenous putative target genes, implying that the apparently similar expression patterns of these oral patterning genes are based on conserved gene cascades that have remained unchanged through the evolutionary history of the vertebrates.

10.4 Heterotopy and homology

The observations described above provide possible evidence of the homology between larval lamprey lips and gnathostome jaws, which has been assumed by previous comparative morphologists. Careful examination of the mesenchymal morphology, however, revealed that the upper lip-forming crest cells do not correspond to the upper jaw mesenchyme with respect to their topographical relationships with other embryonic structures. In gnathostome embryos, for example, both the upper and lower jaw mesenchymes derive from the crest cells in the mandibular arch proper. However, in the lamprey, the upper lip ectomesenchyme originates from the cells surrounding the premandibular mesoderm, not from the mandibular arch domain but from the premandibular

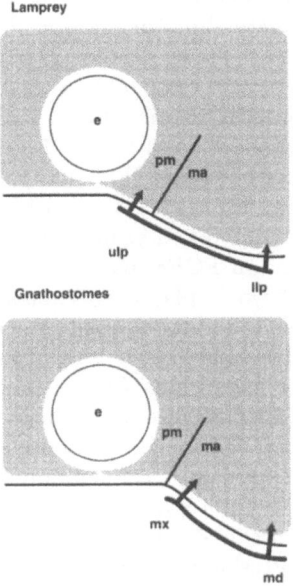

Fig. 10.3. Heterotopic shift of the epigenetic interactions involved in jaw evolution. In the lamprey, the epigenetic interaction that produces the proximal–distal polarity of the oral apparatus involves both the mandibular (ma) and premandibular (pm) ectomesenchyme, whereas in the gnathostomes, the shared molecular cascades are restricted posteriorly to act only on the mandibular arch ectomesenchyme. *Abbreviations: e, eye; llp, lower lip; ma, mandibular ectomesenchyme; md, mandibular process (lower jaw); mx, maxillary process (upper jaw); pm, premandibular ectomesenchyme; ulp, upper lip.*

domain of the head (reviewed by [9]). Therefore, purely morphological comparisons of the oral structures deny any homology between the lips and jaws. They further establish that homologous genes are not expressed in homologous sets of cells or embryonic structures. Strictly speaking, the homology may have been disrupted in the invention of the gnathostome jaw.

What factors could be involved in the evolution of the jaw? In both lampreys and gnathostomes, epigenetic tissue interactions between the ectoderm and mesenchyme established the localized gene expression involved in the patterning of the oral apparatus. The same (or homologous) molecular cascades function in the formation of the dorsal and ventral protrusions around the mouth opening. Paleontological evidence suggests that some Paleozoic agnathans possessed a movable 'oral plate' that may have functioned like our lower jaw [6]. If the same gene cascades function in a different place in the embryo, these genes may always have served to produce such oral protrusions since early in vertebrate history. Haeckel [3] first coined the term 'heterotopy' to describe such evolutionary shifts of place in development, as opposed to the concept of 'heterochrony', which denotes a shift in the timetable of development. In this case, it appears reasonable to assume that the action of the FGF8 molecule was secondarily restricted posteriorly to the mandibular arch domain, leading to an upregulation of the *Dlx1* gene only in the mandibular-arch crest cells in the gnathostomes (Fig. 10.3). To ascertain whether this could actually occur, Shigetani et al. [13] tried to expand the distribution of FGF8 in the early chick embryo and examined the downstream effects. Two days after implantation of an FGF8-soaked bead in the rostral part of the ventral ectoderm, an embryo displayed an expanded expression domain for *Dlx*1, the target gene of FGF8. *Dlx1* was not only expressed in the mandibular ectomesenchyme in this embryo, but its expression was also apparent in the postoptic mesenchyme from which *Dlx1* expression is normally absent. In terms of the morphological pattern, this embryo exhibited a 'lamprey-like' expression domain for *Dlx1*. Although a lamprey-like oral apparatus could not be produced in this experiment, the lamprey-like target-gene expression pattern was achieved by heterotopic shift of the upstream gene expression domain. This phenomenon can also be regarded as an embryonic phenocopy in the sense of a gene expression cascade.

From the above discussion, it seems most likely that a heterotopic shift in ectodermal growth factors or in the site of tissue interactions was a prerequisite for the evolution of the gnathostome jaw. This change probably occurred in concert with the patterning of placodes such as Rathke's pouch or the nasal placodes, which also require tissue interactions based on the specific topographical relationships of cell layers. These placodes in the lamprey develop as single median structures, unlike those of the gnathostome embryo, in which separate placodes develop (reviewed by [9]). Therefore, these shared molecular cascades precipitate site-specific tissue interactions to induce specific structures throughout the vertebrate species. However, topographical rearrangement of embryonic cell populations may alter the topographical and anatomical relationships of organs, leading to the loss of morphological homology. The gnathostome jaw, in this sense, appears to be a true novelty that has not evolved simply through the gradual modification and specialization of one of the rostral gill arches of the ancestral vertebrate.

References

1. Bei, M., and Maas, R. (1998). FGFs and BMP4 induce both *Msx*1-independent and *Msx*1-dependent signaling pathways in early tooth development. *Development* **125**, 4325-4333.
2. Gilbert, S.C., Loredo, G.A., Brukman, A., and Burke, A.C. (2001). Morphogenesis of the turtle shell: the development of a novel structure in tetrapod evolution. *Evol. Devel.* **3**, 47-58.

3. Haeckel, E. (1875). Die Gastrea und die Eifurchung der Thiere. *Jena Z. Natur-wiss.* **9**, 402–508.
4. Hall, B.K. (1998). *Evolutionary Developmental Biology.* 2nd Ed. Chapman & Hall, London.
5. Horigome, N., Myojin, M., Hirano, S., Ueki, T., Aizawa, S., and Kuratani, S. (1999). Development of cephalic neural crest cells in embryos of Lampetra japonica, with special reference to the evolution of the jaw. *Dev. Biol.* **207**, 287–308.
6. Janvier, P. (1996). *Early Vertebrates.* Oxford Scientific Publications, New York.
7. Kuratani, S., Matsuo, I., and Aizawa, S. (1997). Developmental patterning and evolution of the mammalian viscerocranium: genetic insights into comparative morphology. *Dev. Dyn.* **209**, 139–155.
8. Kuratani, S., Horigome, N., and Hirano, S. (1999). Developmental morphology of the cephalic mesoderm and re-evaluation of segmental theories of the vertebrate head: evidence from embryos of an agnathan vertebrate, *Lampetra japonica. Dev. Biol.* **210**, 381–400
9. Kuratani, S., Nobusada, Y., Horigome, N., and Shigetani, Y. (2001). Embryology of the lamprey and evolution of the vertebrate jaw: insights from molecular and developmental perspectives. *Phil. Trans. R. Soc. Lond.* **B 356**, 1615–1632.
10. Matsuo, I., Kuratani, S., Kimura, C., Takeda, N., and Aizawa, S. (1995). Mouse *Otx2* functions in the formation and patterning of rostral head. *Genes Dev.* **9**, 2646–2658.
11. Müller, G.B., and Wagner, G.P. (1991). Novelty in evolution: restructuring the concept. *Annu. Rev. Ecol. Syst.* **22**, 229–256.
12. Shigetani, Y., Nobusada, Y., and Kuratani, S. (2000). Ectodermally-derived FGF8 defines the maxillomandibular region in the early chick embryo: epithelial-mesenchymal interactions in the specification of the craniofacial ectomes-enchyme. *Dev. Biol.* **228**, 73-85.
13. Shigetani, Y., Sugahara, F., Kawakami, Y., Murakami, Y., Hirano, S., and Kuratani, S. (2002). Shape precedes structure: heterotopic shift of epithelial-mesenchymal interactions for vertebrate jaw evolution. *Science* **296**, 1319–1321.
14. Vainio, S., Karavanova, I., Jowett, A., and Thesleff, I. (1993). Identification of BMP-4 as a signal mediating secondary induction between epithelial and mesenchymal tissues during early tooth development. *Cell* **75**, 45-58.
15. Wagner, G.P., and Müller, G.B. (2002). Evolutionary innovations overcome ancestral constraints: a re-examination of character evolution in male sepsid flies (Diptera: Sepsidae). *Evol. Dev.* **4**, 1–6.

Morphogenesis and Pattern Formation in
Plants

The Development of Cell Pattern in the Arabidopsis Root Epidermis

Olga Ortega-Martínez and Liam Dolan

Department of Cell and Developmental Biology, John Innes Centre, Norwich, NR4 7UH, UK
mail: liam.dolan@bbsrc.ac.uk

Summary. The root epidermis of *Arabidopsis thaliana* is composed of files (stripes) of trichoblasts that develop into hair cells, which are separated by 1-3 files of atrichoblasts that develop into hairless epidermal cells. This pattern forms during embryogenesis and is maintained in the post-embryonic seedling root by positional signals. The maintenance of the pattern of alternating files requires the movement of small transcriptional regulators, such as CPC which moves from atrichoblasts to trichoblasts. In trichoblasts CPC represses the transcription of *WER* and *GL2*, transcription factors that promote atrichoblast/non-hair cell development. These data are consistent with a lateral inhibition model of epidermal development. According to this model the atrichoblasts negatively regulate atrichoblast fate in neighbouring cells by promoting their development into trichoblasts.

11.1 Introduction

A central process during the development of multicellular organisms is the formation of patterned arrays of cells in tissues such as bristles on a fly's wing or the guard cells on the surface of the leaf. The flexibility of plant cell fate has led to the suggestion that positional signalling is the major regulator although it has been suggested that cell lineage may be important in the establishment of guard cell fate in the developing stomatal complex. In animals both cell lineage and positional signalling have been shown to be involved in the specification of cell fate and the development of patterned groups of cells. The aim of this review is to summarise what is known about the development of pattern in the root epidermis of *Arabidopsis thaliana*, which has been used as a model system for the molecular dissection of the patterning process in plants.

The *Arabidopsis* root epidermis is composed of two cell types, hair cells (HC) and non-hair cells (NHC) [6, 8, 10, 11]. HC are derived from meristematic trichoblasts and NHC are derived from meristematic atrichoblasts. Trichoblasts are shorter and less vacuolated than atrichoblasts allowing the two cell types to be distinguished early in development. After trichoblasts have elongated in the direction of the long axis of the root, tubular outgrowths emerge from the basal end of a hair cell and elongate by a process called tip-growth, in which cell growth is restricted to the tip of the cell. These projections are called root hairs. These hairs are conduits of ions and water from the soil to the plant and in some species are the sites of secretion of chemicals into the soil environment [5].

The young root of a seedling can be considered to be composed of three developmental regions. At the root tip is the meristem (division zone) which includes the population of dividing cells and the initials. The latter can be considered to be the stem cells of the root. In the elongation zone, cells elongate rapidly in the absence of cell division in a direction parallel to the long axis of the root. Once cells have stopped elongating they undergo further differentiation in the appropriately named differentiation zone [2, 3]. It is in the differentiation zone that root hairs form. The first sign is the formation of a bulge at the end of the cell nearest the root-tip. Then a tip-growing root hair emerges from this bulge. When hairs reach a length between 20 and 40 μm hair growth accelerates achieving rates of growth of approximately 1-2 μm.min-1. In the last phase the hair stops elongating, the tip vacuolates and the characteristic dense mass of secretory vesicles at the apex disperses [11].

11.2 Positional information is involved in fate specification

Epidermal cell types are arranged in files, with atrichoblast files alternating with trichoblast files. The number of atrichoblast files can increase during development so that trichoblast files may be separated by up to three atrichoblast files [11, 4]. The identity of cells is determined by their position relative to the underlying cells of the cortex. HC files are located over the intercellular space (junction) between two underlying cortical cell files. NHC files are located over the outer cortical cell wall, which lies parallel to the surface of the root - the so-called periclinal wall [8, 6, 11] (Fig. 11.1).

Clonal analysis and laser microsurgery has shown that root epidermal cell identity is specified by cell position and is independent of the clonal origin of the cells (lineage). In a clonal analysis the pattern of cell divisions was monitored in developing roots [4]. The number of cells in each file in the epidermis increases by undergoing transverse divisions - divisions where the new wall is orientated perpendicular to the long axis of the root. Occasionally longitudinal divisions or T-divisions occur in trichoblasts producing a clone of two parallel sister files (Fig. 11.2) where one file is located in the HC position and the other is located in the NHC position. The file in the HC position develops a hair, while the file in the NHC position assumes the non-hair identity as shown by the absence of hair formation and the expression of non-hair cell specific marker genes. As both files derived from the same cell, this implies position is the most important determinant of cell fate.

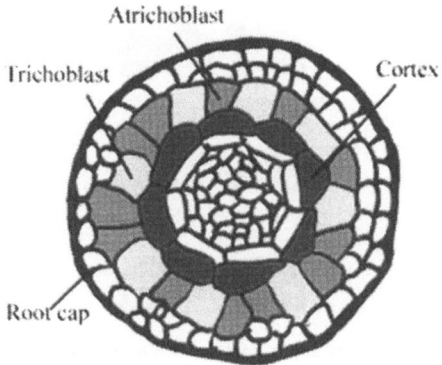

Fig. 11.1. Schematic representation of a transverse section through the meristem of a wild type *Arabidopsis* root, indicating the position of trichoblasts which will develop into root hair cells and atrichoblasts which will develop into non-root hair cells. The trichoblasts (light grey) are located over the junctions between two cortical cells (black). Atrichoblasts (dark grey) are located over single cortical cells (black). This schematic represents the meristematic zone, where the root epidermis is surrounded by a protective cell layer called the root cap (white).

In a second approach, lasers were used to ablate a variety of different cell types in the root (Fig. 11.3). Cell ablation creates a gap, and neighbouring cells can migrate into the vacant space. We can then assay the identity of cells that filled the gap. If lineage were of primary importance in defining cell identity then the invading cells would be expected to maintain their original identity. If position were of primary importance then we would expect that the invading cells would adopt the identify of cells normally found in that new location. Following ablation of an atrichoblast, trichoblasts invaded the gap. These cells then expressed atrichoblast specific marker genes and remained hairless. These observations show that not only is positional information required for the specification of cell identity, but also that cell identity is flexible until relatively late in development (until the last cell division before the beginning of cell elongation) [4].

11.3 Genetic model of epidermal development

In *Arabidopsis* several genes have been identified that play an important role during root epidermal cell differentiation. They were identified in screens for mutants that developed either fewer or more hairs than the wild type. The genes have been cloned and encode putative transcriptional regulators whose pattern of expression is controlled by the positional cues defined in the

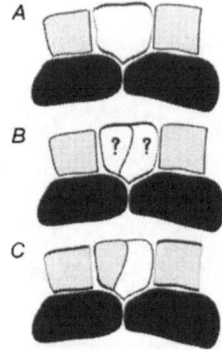

Fig. 11.2. Schematic representation of the clonal analysis that shows the role of positional information in specifying cell identity [4]. (A) Organisation of the cells in the root - white cells are trichoblasts and grey cells are atrichoblasts. Black cells are the underlying cortical cells. (B) A longitudinal division in the trichoblast forms two daughter cells (marked with "?") in the position previously occupied by a single trichoblast. (C) Only the cell in contact with the junction between the two underlying cortical cells develops as a trichoblast (grey).

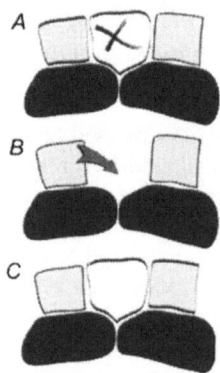

Fig. 11.3. Schematic representation of events that occur upon the laser ablation of a trichoblast [4]. Trichoblasts are in white and atrichoblasts are in grey. (A) The "X" marks the trichoblast that was ablated with a laser. (B) A neighbouring atrichoblast (grey) invades the space left upon ablation of the trichoblast. (C) The invading cell (grey) develops as a trichoblast.

previous section and by each other. *GLABRA2 (GL2), WEREWOLF (WER),* are positive genetic regulators of non-hair cell development. Plants lacking either *WER* or *GL2* function develop hairs on every cell. *CAPRICE (CPC)* and *TRYPTICHON (TRY)* are positive regulators of hair cell development. Plants lacking *CPC* function develop very few root hairs (Fig. 11.4). Plants lacking *TRY* function are wild type but plants lacking both *CPC* and *TRY* develop fewer hairs than the plants lacking *CPC* function alone. This suggests

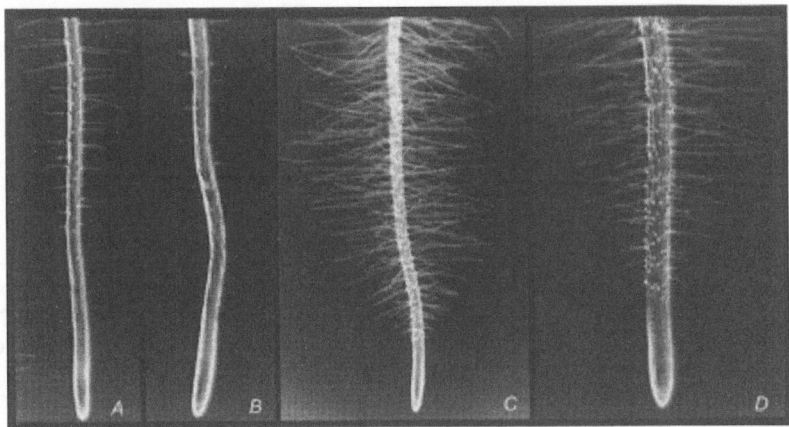

Fig. 11.4. Phenotypes of wild type root and mutants with defective epidermal cell differentiation in the root. (A) Wild type Columbia. (B) *cpc* mutant develops few hair cells. (C) *gl2* mutant is hairy because it develops hairs in all epidermal cells. (D) *wer* mutant also develops hairs on every epidermal cell.

that both CPC and TRY are required for the promotion of hair cell identity in the developing root epidermis.

Both WER and $GL2$ are transcribed in atrichoblasts. WER encodes a MYB-related transcription factor that positively regulates $GL2$ transcription in atrichoblast cells (Fig. 11.5). $GL2$ is a homeodomain-leucine zipper DNA binding protein that is thought to act by promoting the expression of genes required for NHC development and/or repressing the expression of genes required for HC development [9, 17, 20].

CPC and TRY are putative transcriptional regulators that promote root hair cell development. TRY and CPC are small MYB-related proteins that lack a transcriptional activation domain and therefore may act as transcriptional repressors - perhaps negatively regulating atrichoblast-promoting genes. CPC is transcribed in atrichoblasts and the CPC protein moves into neighbouring trichoblasts where it promotes trichoblast cell development [15, 24]. CPC transcription in atrichoblasts is promoted by WER. WER therefore promotes trichoblast development by promoting the expression of CPC, which acts in a non cell-autonomous fashion. Once in the trichoblasts, CPC represses $GL2$ and WER transcription, thus promoting HC fate. CPC transcription occurs in trichoblasts in plants homozygous for the *cpc* mutation. This indicates that CPC protein also represses the transcription of the CPC gene in trichoblast during normal development [15]. TRY promotes trichoblast development but its transcription pattern and protein accumulation pattern has not yet been described. Nevertheless the elevated levels of transcription of a TRY reporter gene observed in *try cpc* double mutant background, suggests that TRY, like CPC, is preferentially transcribed in atrichoblasts. If TRY

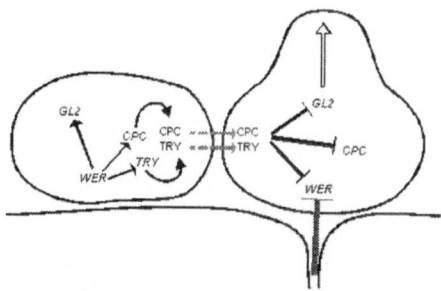

Fig. 11.5. Model of genetic regulation between *GL2*, *WER*, *TRY* and *CPC* in the *Arabidopsis* root epidermis. Positional cues from the cortex start the cascade of events that determine the correct cell fate in the epidermis. High levels of *WER*, *GL2* and *CPC* transcripts accumulate in the atrichoblasts and low levels of *WER*, *GL2* and *CPC* accumulate in the trichoblasts. These levels are maintained by regulatory feedback loops between *WER* and *CPC*. Inhibition of *GL2* expression in trichoblasts by CPC and TRY is central to the model (model is based on [15, 21]).

were transcribed in atrichoblasts it is predicted that the protein will move from the atrichoblast into the trichoblasts where is promotes trichoblast development, as was shown for *CPC* [21].

It has been proposed that epidermal cell identity is regulated by the spatial ratio of WER to CPC protein. A high WER to CPC ratio in NHC results in the activation of *GL2* expression and thus the promotion of NHC fate while a low WER to CPC ratio in HC results in the inhibition of *GL2* transcription, forming a hair cell [14]. High WER in NHC also reinforces HC fate in a non-cell autonomous fashion by positively regulating CPC in NHC.

11.4 Other regulators of hair development

ECTOPIC ROOT HAIR3 (ERH3) encodes a microtubule severing protein called katanin. Katanin proteins are required for the reorganisation of microtubules in animal cells and act by severing microtubules into smaller segments, which are unstable and in turn depolymerise rapidly. *erh3* mutants develop HC in the position of NHC (ectopic hair cells). *erh3* mutants also have a higher frequency of cells in the hair position that do not produce hairs (ectopic non-hair cells) indicating that the microtubule reorganisation is required for the specification of both cell types [22, 27]. In addition, the misexpression of cell type specific markers indicates that ERH3 activity is also required for the specification of cortex and root cap identity. This suggests that microtubules may be involved in the specification of cell identity throughout the plant and not solely in the root epidermis. Support for the role of microtubules in the specification of cell identity comes from the observation that plants expressing an antisense α-tubulin 6 gene have defective microtubule organisation and also develop ectopic root hairs [1]. The mechanism by which microtubules control the development of cell identity is unclear. It is possible that the directional movement of regulatory proteins such as CPC from cell to cell may require microtubules. Analogous roles for microtubules moving proteins required for position dependent differentiation during the development of invertebrate body axes have been described [12, 18, 26].

Ectopic root hairs also develop on roots exposed to high levels of the gaseous growth factor ethylene or its precursor aminocyclopropanecarboxylic acid (ACC). This implies ethylene is a positive regulator of HC development. Blocking ethylene biosynthesis results in the development of hairless roots, which is consistent with ethylene being a positive regulator. CTR1 (CON-STITUTIVE TRIPLE RESPONSE1) is a negative regulator of the ethylene response and acts by blocking the ethylene signal transduction pathway. *ctr1* mutants grown in normal conditions look as though they are being exposed to ethylene. The development of ectopic root hair cells in ctr1 mutants indicates that ethylene promotes root hair development [11]. *ethylene over producer (eto)* mutants over produce ethylene and develop ectopic hairs. This also indicates that ethylene promotes root hair development [7]. The *ethylene resistant 1(etr1)* mutant which has a defective ethylene receptor develops shorter root hairs than the wild type [19]. Together these data indicate that ethylene is a positive regulator of either hair cell fate or hair outgrowth. The growth regulator auxin is also required for correct development regulating both the site of hair initiation and hair elongation. For example *auxin resistant 3 (axr3)* mutants develop fewer root hairs than the wild type [16]. It is likely that it is tip growth rather than cell specification that is defective in these mutants backgrounds. Consistent with this view is the finding that the expression of the atrichoblast specific gene *GL2* is normal in mutants with defective growth factor signalling. Resolution of the precise role of auxin and ethylene requires further study.

11.5 Future prospects

It is clear that the development of the striped pattern of epidermal cell types in the *Arabidopsis* root epidermis involves the movement of small transcriptional regulators from atrichoblast to trichoblasts. This is consistent with a lateral inhibition model, which has been shown to operate in animal epithelia [13]. It remains to be seen if proteins move in the opposite direction, from trichoblast to atrichoblast. In addition, most studies to date have focussed on the maintenance of the pattern in the mature seedling and have not examined the expression of the genes during pattern formation in the embryo. The exception is *GL2*, which is expressed throughout the outer layers of the future root meristem in the heart stage embryo and is restricted to atrichoblasts in mature embryos. The examination of the expression patterns of the other regulatory genes during embryogenesis will be instructive as to the precise mechanism of pattern formation as distinct from its maintenance.

Acknowledgements

O.O-M is funded by the John Innes Foundation. The John Innes Centre, the European Union and the BBSRC fund LD's laboratory. We thank Eoin Ryan for many comments that improved this manuscript.

References

1. Bao, Y., Kost, B., and Chua, N.H. (2001). Reduced expression of α-tubulin genes in *Arabidopsis thaliana* specifically affects root growth and morphology, root hair development and root gravitropism. *Plant J*, **28**, 145-157.

2. Berger, F., Hung, C.Y., Dolan, L., and Schiefelbein, J. (1998). Control of cell division in the root epidermis of *Arabidopsis thaliana*. *Dev Biol*, **194**, 235-245.

3. Berger, F., Linstead, P., Dolan, L., and Haseloff, J. (1998). Stomata patterning on the hypocotyl of *Arabidopsis thaliana* is controlled by genes involved in the control of root epidermis patterning. *Dev Biol*, **194**, 226-234.

4. Berger, F., Haseloff, J., Schiefelbein, J., and Dolan, L. (1998). Positional information in root epidermis is defined during embryogenesis and acts in domains with strict boundaries. *Curr Biol*, **8**, 421-430.

5. Brigham, L.A., Michaels, P.J., and Flores, H.E. (1999). Cell-specific production and antimicrobial activity of naphthoquinones in roots of *Lithospermum erythrorhizon*. *Plant Physiol*, **119**, 417-428.

6. Bünning, E. (1951). Über die Differenzierungsvorgange in der Cruciferenwurzel. *Planta*, **36**, 126-153.

7. Cao, X.F., Linstead, P., Berger, F., Kieber, J., and Dolan, L. (1999). Differential ethylene sensitivity of epidermal cells is involved in the establishment of cell pattern in the *Arabidopsis* root. *Physiol Plant*, **106**, 311-317.

8. Cormack, R. (1935). Investigations on the development of root hairs. *New Phytol*, **34**, 30-54.

9. Di, Cristina, M., Sessa, G., Dolan, L., Linstead, P., Baima, S., Ruberti I, and Morelli, G. (1996). The *Arabidopsis* Athb-10 (GLABRA2) is a HD-Zip protein required for regulation of root hair development. *Plant J*, **10**, 393-402.

10. Dolan, L., Janmaat, K., Willemsen, V., Linstead, P., Poethig, S., Roberts, K., and Scheres, B. (1993). Cellular organisation of the *Arabidopsis thaliana* root. *Development*, 119, 71-84.

11. Dolan, L., Duckett, MC., Grierson, C.S., Linstead, P., Schneider, K., Lawson, E., Dean, C., Poethig, S., and Roberts, K. (1994). Clonal relationships and cell patterning in the root epidermis of *Arabidopsis*. *Development*, **120**, 2465-2474.

12. Gotta, M., and Ahringer, J. (2001). Axis determination in *C. elegans*: initiating and transducing polarity. *Curr Opin Genet Dev*, **11**, 367-373.

13. Heitzler, P., and Simpson, P. (1991). The choice of cell fate in the epidermis of *Drosophila*. *Cell*, **64**, 1083-1092.

14. Lee, M.M., and Schiefelbein, J. (1999). WEREWOLF a MYB-related protein in *Arabidopsis*, is a position-dependent regulator of epidermal cell patterning. *Cell*, **99**, 473-483.

15. Lee, M.M., and Schiefelbein, J. (2002). Cell pattern in the *Arabidopsis* root epidermis determined by lateral inhibition with feedback. *Plant Cell*, **14**, 611-618.

16. Leyser, H.M., Pickett, F.B., Dharmasiri, S., and Estelle, M. (1996). Mutations in the *AXR3* gene of *Arabidopsis* result in altered auxin response including ectopic expression from the SAUR-AC1 promoter. *Plant J*, **10**, 403-413.

17. Masucci, J.D., Rerie, W.G., Foreman, D.R., Zhang, M., Galway, M.E., Marks, M.D., and Schiefelbein, J.W. (1996). The homeobox gene *GLABRA2* is required for position-dependent cell differentiation in the root epidermis of *Arabidopsis thaliana*. *Development*, **122**, 1253-1260.

18. O'Connell, K.F., Maxwell, K.N., and White, J.G. (2000). The *spd-2* gene is required for polarization of the anteroposterior axis and formation of the sperm asters in the *Caenorhabditis elegans* zygote. *Dev Biol*, **222**, 55-70.

19. Pitts, R.J., Cernac, A., and Estelle, M. (1998). Auxin and ethylene promote root hair elongation in *Arabidopsis*. *Plant J*, **16**, 553-560.

20. Rerie, W.G., Feldmann, K.A., and Marks, M.D. (1994). The *GLABRA2* gene encodes a homeo domain protein required for normal trichome development in *Arabidopsis*. *Genes Dev*, **8**, 1388-1399.

21. Schellmann, S., Schnittger, A., Kirik, V., Wada, T., Okada, K., Beermann, A., Thumfahrt, J., Jurgens, G., and Hulskamp, M. (2002). *TRYPTYCHON* and *CAPRICE* mediate lateral inhibition during trichome and root hair patterning in *Arabidopsis*. *EMBO J*, **21**, 5036-5046.

22. Schneider, K., Mathur, J., Boudonck, K., Wells, B., Dolan, L., and Roberts, K. (1998). The *ROOT HAIRLESS 1* gene encodes a nuclear protein required for root hair initiation in *Arabidopsis*. *Genes Dev*, **12**, 2013-2021.

23. Tanimoto, M., Roberts, K., and Dolan, L. (1995). Ethylene is a positive regulator of root hair development in *Arabidopsis thaliana*. *Plant J*, **8**, 943-948.

24. Wada, T., Kurata, T., Tominaga, R., Koshino-Kimura, Y., Tachibana, T., Goto, K., Marks, M.D., Shimura, Y., and Okada, K. (2002). Role of a positive regulator of root hair development, *CAPRICE*, in *Arabidopsis* root epidermal cell differentiation. *Development*, **129**, 5409-5419.

25. Wada, T., Tachibana, T., Shimura, Y., and Okada, K. (1997). Epidermal cell differentiation in *Arabidopsis* determined by a Myb homolog, *CPC*. *Science*, **277**, 1113-1116.

26. Wallenfang, M.R., and Seydoux, G. (2000). Polarization of the anterior-posterior axis of *C. elegans* is a microtubule-directed process. *Nature*, **408**, 89-92.

27. Webb, M., Jouannic, S., Foreman, J., Linstead, P., and Dolan, L. (2002). Cell specification in the *Arabidopsis* root epidermis requires the activity of EC-TOPIC ROOT HAIR 3 - a katanin-p60 protein. *Development*, **129**, 123-131.

Pattern Formation during Dicotyledonous Plant Embryogenesis

Masahiko Furutani[1,3], Mitsuhiro Aida[2], and Masao Tasaka[1,4]

[1]Graduate School of Biological Sciences, Nara Institute of Sciences and Technology,
  Ikoma, Nara, 630-0101 Japan
  Tel. +81-743-72-5487
  Fax. +81-743-72-5489
[2]Department of Molecular Cell Biology, University of Utrecht, Padualaan 8,
  3584CH Utrecht, The Netherlands
[3]e-mail: ma-furut@bs.aist-nara.ac.jp
[4]e-mail: m-tasaka@bs.aist-nara.ac.jp

Summary. A basic body plan consisting of two axes, apical-basal and radial (central-peripheral), is established during the embryogenesis of higher plants. The embryo forms the shoot and root meristems, which are essential for postembryonic development, at the opposite ends of the apical-basal axis. Recently, a molecular genetic approach using the model plant *Arabidopsis thaliana* has provided insight into the establishment of the basic body plan and the molecular mechanisms regulating the formation of the apical meristems. The phytohormone auxin in particular has been shown to be involved in pattern formation during the early stages of embryogenesis. In this review, we focus on several recent studies of the establishment of the body plan and pattern formation.

Keywords: *Arabidopsis thaliana*, auxin, bilateral symmetry, embryogenesis, *PIN1*.

12.1 Introduction

The embryos of angiosperms develop in the maternal tissues of the ovule (Fig. 12.1). Following fertilization, the zygote develops into a seedling, a juvenile form of the plant. The seedling is a relatively simple structure which lacks most adult structures (Fig. 12.2A). During postembryonic development, the seedling starts to produce adult organs, the shoot and root meristems, from specialized groups of cells at both ends. In dicotyledonous plants, including a widely studied model species, *Arabidopsis thaliana*, the embryo develops two cotyledons from its apical region, located in bilaterally symmetrical positions [1, 2, 3, 8, 32].

The patterning processes that occur during embryogenesis have been studied in some angiosperms physiologically or genetically. For example, treatment

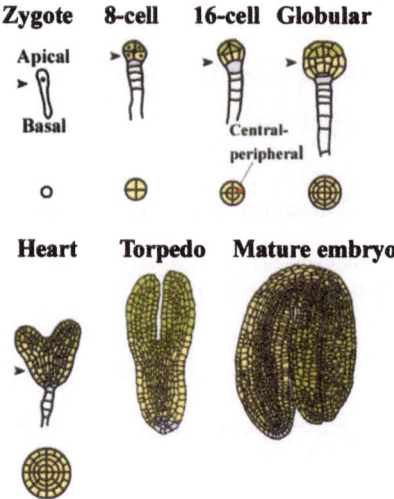

Fig. 12.1. *Arabidopsis* embryos at various stages of development. Embryos are illustrated in longitudinal sections above, with transverse sections taken at the positions of the arrowheads below. Following fertilization, the zygote divides asymmetrically into a small apical cell and a large basal cell. After several rounds of cell division, the apical cell gives rise to the embryo proper, which develops into the seedling, a juvenile form of the plant. The basal cell mainly forms the suspensor, proposed to be a supporting organ, through which nutrients and growth factors can pass to the developing embryo and which degenerates during embryogenesis. The hypophysis is generated from the basal cell, leading to the formation of the quiescent center and central collumera. The cell division at the 4-cell stage splits the embryo proper into the apical and basal region. Derivatives of the apical and basal regions and the hypophysis are shown in green, yellow, and pink, respectively. The apical region of the embryo proper gives rise to apical structures of the seedling, such as the shoot meristem and most of the cotyledons, while the basal region forms the hypocotyl, radicle, and the root meristem.

of *Brassica juncea* or wheat embryos with inhibitors of polar auxin transport or high concentrations of auxin severely disturbs patterning [16, 20, 31]. Molecular genetic studies using *Arabidopsis* also have identified genes involved in pattern formation of the embryo. This article reviews recent studies on pattern formation during the early stages of embryogenesis.

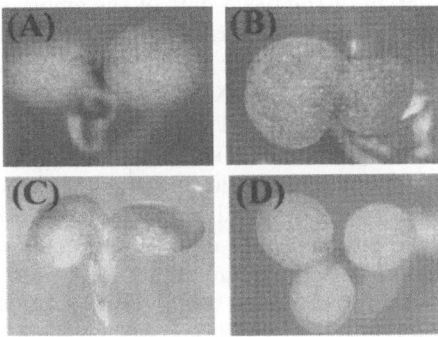

Fig. 12.2. Phenotypes of auxin-related *Arabidopsis* mutants. Wild-type (A), *pin1* (B), *mp* (C) and *pid* (D) seedlings.

12.2 Axis

12.2.1 Apical-basal axis

During embryogenesis, the shoot meristem, cotyledons, hypocotyl, radicle, and root meristem are formed along the apical-basal axis. Several *Arabidopsis* mutants have been isolated by screening seedlings that display abnormal patterns in any of these. The *gnom/embryo-defective30* (*gn/emb30*) mutation causes variable defects in apical-basal polarity, and in the most severe case, results in a ball-shaped seedling without any polarity [35]. The *GN/EMB30* gene encodes a guanine-nucleotide exchange factor for the ADP-ribosylation factor G protein (ARF-GEF) and is suggested to be involved in membrane vesicle trafficking [35, 45, 50]. Polar localization of the PIN-FORMED1 (PIN1) protein, a member of a family of putative auxin efflux carriers on the plasma membrane, is not coordinated in *gn/emb30* embryos, raising the possibility that *gn/emb30* embryos are defective in polar auxin transport, which would cause the phenotype [53]. In fact, the gn/emb30 phenotype can be mimicked by treatment of globular-stage embryos with a high concentration of auxin or an inhibitor of polar auxin transport in *Brassica juncea*, a species closely related to *Arabidopsis* [31]. Therefore, *GN/EMB30* may be involved in the establishment of apical-basal polarity partially through regulating PIN1 protein localization.

Seedlings of the *monopteros* (*mp*) mutant completely lack the basal structure of radicle and root meristem, and show abnormal patterning of the apical region (Fig. 12.2C; [7]). The *MP* gene encodes AUXIN RESPONSE FACTOR 5 (ARF5), a transcription factor that binds to auxin response elements (AuxREs) [7, 23, 56]. AuxREs are *cis* regulatory elements found in the promoters of several auxin responsive genes. These results suggest that auxin is involved in the formation of the basal part of the embryo. In fact, some mutants with reduced auxin sensitivity such as *bodenlos* (*bdl*) and *auxin-resistant6* (*axr6*) show phenotypes that are very similar to *mp* [21, 26]. The *BDL* gene encodes IAA12, a member of the AUX/IAA protein family [22]. Members of this family are thought to repress ARF protein activity through dimerization with ARF, and their degradation is stimulated by auxin [29, 57]. The *bdl* phenotype is caused by an amino acid exchange in the conserved degradation domain of IAA12. The corresponding mutations in other members of the AUX/IAA family have been shown to inhibit their auxin-dependent degradation, raising the possibility that the mutated IAA12 protein is stabilized and inhibits MP-mediated transcription in the *bdl* mutant [19, 43, 46, 59, 61]. Consistent with this idea, the BDL (IAA12) and MP proteins interact in the yeast two-hybrid system and the expression patterns of each gene overlap almost completely during early embryogenesis [22]. Therefore, BDL may prevent MP from binding to the AuxREs of target genes involved in root initiation.

The *fackle/hydra2* (*fk/hyd2*) and *hydra1* (*hyd1*) mutants show pleiotropic phenotypes, including stunted hypocotyls and multiplication of the shoot meristem and cotyledons [27, 49, 52]. The *FK/HYD2* and *HYD1* genes encode a sterol C-14 reductase and a Delta8-Delta7 sterol isomerase, both of which function in the biosynthetic pathway of sterols. Seedlings mutant for each gene similarly have abnormal concentrations of the three major *Arabidopsis* sterols. These results suggest that sterols are important for correct pattern formation during embryogenesis. The phenotypes of the mutants are partially rescued by the inhibition of auxin and ethylene signaling but not by exogenous sterols, suggesting that altered auxin and ethylene signaling are partially responsible for the observed phenotype.

The *topless* (*tpl*) mutant shows a loss of apical structures or the transformation of the apical portion into basal structures [33]. The *tpl* mutant is temperature sensitive and develops seedlings with no shoot meristem and with cotyledons fused to various extents when embryos are grown at the standard temperature. At high temperature, by contrast, a second root develops at the top instead of a shoot meristem and cotyledons. Marker genes, such as *SHOOT MERISTEMLESS* (*STM*) and *UNUSUAL FLORAL ORGANS* (*UFO*) for the shoot meristem, and *KNAT1* for the hypocotyls, are not expressed in developing *tpl* embryos. Instead, marker genes for the basal region of the embryo are expanded. These results indicate that the *tpl* mutation transforms structures that normally have an apical fate into structures with a basal fate during embryogenesis. Since only a single mutant allele has been reported and this allele shows a weak semi-dominance, the role of the wild-type *TPL* gene in apical-basal patterning remains unclear. Nevertheless, the phenotype of the mutant indicates that a developmental switch that determines the fate of the apical region exists during plant embryogenesis and that this switch is compromised in the *tpl* allele.

12.2.2 Radial (central-peripheral) axis

The first radial patterning emerges at the transition from the 8- to the 16-cell stage in *Arabidopsis* (Fig. 12.1; [36]). At the 8-cell stage, each cell of the embryo proper divides periclinally (new cell walls parallel to the surface) and gives rise to the inner and outer cells, the latter of which form the protoderm. Subsequently, a series of periclinal divisions subdivides the basal part of the embryo to form the hypocotyl and root with concentric layers of different cell types, such as the stele, endodermis, cortex, and epidermis, resulting in radial symmetry in transverse sections.

Several mutants that affect radial patterning have been identified. Among them, the *short-root* (*shr*) mutant fails to undergo the periclinal division that generates the endodermis and cortex layers of the hypocotyl and radicle, resulting in a single layer of cells with cortical identity [17, 25, 48]. Thus, the *SHR* gene is involved in both the periclinal division that leads to the formation of endodermis and cortex and the specification of the endodermis. *SHR* encodes a putative transcription factor of the *GRAS* family. Interestingly, *SHR* mRNA is found only in the stele internal to the adjacent layer that gives rise to endodermis and cortex, while its protein is found both in the stele and the adjacent layer, suggesting that the SHR protein moves from the stele into the adjacent layer, where it promotes the periclinal division and the specification of the endodermis. In the adjacent layer, SHR promotes the periclinal division partially through activating another GRAS gene, *SCARECROW* (*SCR*) [12, 25, 40, 60]. Thus, the movement of the SHR protein could act as a signal that provides the positional information necessary for the correct tissue patterning of the endodermis and cortex.

Another class of genes was initially identified through screening mutants that affect different aspects of development, and was later found to be expressed in a specific domain within the radial axis. Among the genes in this class, the *PHABULOSA* (*PHB*), *PHAVOLUTA* (*PHV*), and *REVOLUTA* (*REV*) genes encode putative transcription factors of the HD-ZIP family and are implicated in establishing the polarity of lateral organs in the shoot, such as leaves and cotyledons [37, 42, 63]. *REV* is expressed in the adaxial portion of the cotyledon primordia, the shoot meristem, and the provascular cells of the hypocotyl and root at the heart stage, marking the central domain within the radial axis [42]. *PHB* mRNA is also detected in the central region within the apical half of the globular embryo [37]. Another gene, *KANADI1* (*KAN1*), which was also initially identified as a gene involved in the polarity of shoot lateral organs, encodes a putative transcription factor of the GARP family [15, 28, 37]. *KAN1* is expressed ubiquitously in the early globular embryo but is soon restricted to peripheral cells in the basal region of the embryo. Moreover, ectopic expression of *KAN1* results in the loss of central structures, including the shoot meristem and the vascular tissue. These data suggest that, besides their roles in lateral organ polarity, *PHB*, *PHV*, *REV*, and *KAN1* are also involved in radial patterning during embryogenesis.

12.2.3 Bilateral symmetry

In dicotyledonous plants, two cotyledon primordia and the shoot meristem develop in the apical region of the embryo after the globular stage [2, 8, 32]. Regions a few cells wide that intervene between the cotyledons become boundaries in which growth is suppressed, so that the cotyledons are separated from each other. Due to the symmetrical positioning of the cotyledons and their boundaries, the apical region of the embryo exhibits bilateral symmetry. Previous studies have shown that the positioning of the cotyledons, their boundaries, and the shoot meristem are affected by the phytohormone auxin [20, 31]. Treatment of *Brassica juncea* embryos with polar auxin transport inhibitors or a high concentration of auxin results in embryos that exhibit a cup-shaped cotyledon without bilateral symmetry. Several *Arabidopsis* genes implicated in auxin transport or signal transduction, such as *PIN1* and *MP* (see above), have also been suggested to be involved in the bilateral symmetry of the embryo (Fig. 12.2B, C; [3, 6, 7, 41]). Mutations in these genes cause various defects in position, number, or separation during cotyledon development resulting in the disruption of bilateral symmetry. Besides cotyledon phenotypes, *mp* mutants also show defects in the establishment of the embryonic axis and the development of vascular tissues, through which auxin is proposed to be transported with polarity. The mutation of another gene, *PINOID* (*PID*), causes a phenotype similar to, but milder than, that of *pin1* mutants (Fig. 12.2D; [6]). The *PID* gene has been cloned and encodes a serine/threonine kinase, whose expression is induced by auxin treatment [4, 11]. Stems of loss-of-function *pid* mutants show a reduction of polar auxin transport as in the *pin1* mutant, while overexpression of *PID* enhances the activity of auxin transport [6]. Furthermore, plants overexpressing *PID* exhibit reduced root growth, a phenotype that is suppressed by treatment with polar auxin transport inhibitors [4]. Therefore, *PID* may function as a positive regulator of polar auxin transport, which is involved in the establishment of bilateral symmetry by affecting the development of the cotyledons and their boundaries.

Expression of marker genes in *pin1* embryos

The addition of the new bilateral pattern to the previously established radial pattern occurs during the initiation of cotyledon primordia. Several *Arabidopsis* genes are expressed with bilaterally symmetrical patterns before morphological changes in symmetry become apparent, and these genes can be used as good molecular markers for bilateral pattern development. Among them, the *CUP-SHAPED COTYLEDON1* (*CUC1*) and *CUC2* genes are expressed in a stripe across the top part of the globular embryo (Fig. 12.3A; [1, 55]). After the initiation of cotyledon primordia, this stripe pattern continues in a region that will give rise to the cotyledon boundaries and shoot meristem. Another gene, *FILAMENTOUS FLOWER* (*FIL*) is expressed in the abaxial side of cotyledon primordia from the late globular stage, thus representing

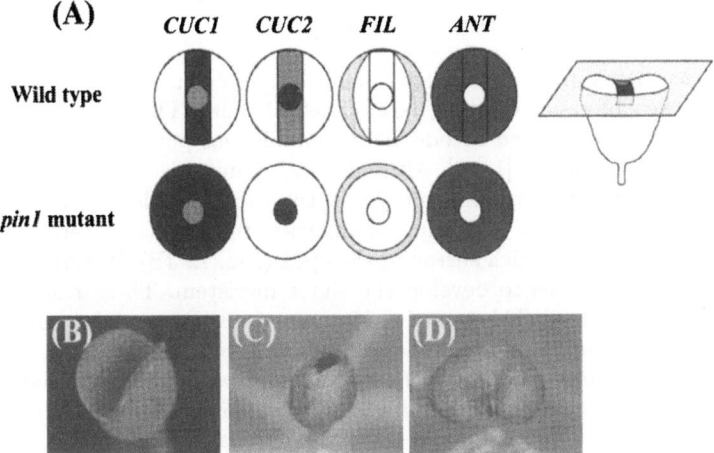

Fig. 12.3. Expression of marker genes during embryogenesis. Schematic diagram of marker-gene expression in wild-type (A, above) and *pin1* (A, below) embryos, viewed from above. Definition of the transverse section plane through the embryo (A, right). The intensity of the color shows the strength of the signal. Seedlings of *cuc1 cuc2* (B), *cuc1 pin1* (C), and *cuc2 pin1* (D).

another marker for bilateral symmetry (Fig. 12.3A; [47, 51]). In contrast, the *AINTEGUMENTA (ANT)* gene is expressed in a ring of cells around the embryonic apex at the globular stage, displaying a radially symmetrical pattern (Fig. 12.3A; [13, 30, 32]). The ring of *ANT* expression continues even after the cotyledon primordia have emerged. Therefore, the embryo seems to maintain positional information that reflects the radial symmetry even after the acquisition of bilateral symmetry.

Phenotypes of pin1 seedlings, such as defects in cotyledon separation and development, can be traced back to embryogenesis. Expression patterns of the above marker genes have been analyzed in pin1 embryos at the early heart stage, shortly after the emergence of cotyledon primordia [3]. The three bilateral markers, *CUC1*, *CUC2*, and *FIL*, show severely altered patterns: *CUC1* expression extends to a large part of the embryonic apex, including the bulging cotyledon primordia, while *CUC2* expression is confined to the center of the apex. *FIL* expression forms a ring that surrounds the apex of the embryo (Fig. 12.3A). In contrast, the radial symmetry of *ANT* expression is not markedly altered in *pin1* (Fig. 12.3A). These results indicate that radial symmetry is established normally in *pin1* embryos while bilateral symmetry is not, thus supporting an important role of *PIN1* in the establishment of bilateral symmetry.

PIN1 may be involved in the establishment of the apical pattern partially through regulating *CUC1* and *CUC2*

Some of the phenotypes found in *pin1* mutant seedlings may be attributed to changes in the expression of other developmental genes. The main candidates are *CUC1* and *CUC2*, which encode highly homologous putative transcription factors of the NAC family [1, 55]. Although single mutants of *cuc1* or *cuc2* do not display obvious phenotypes, embryos of the *cuc1 cuc2* double mutant fail to suppress growth at the cotyledon boundaries, resulting in severely fused, cup-shaped cotyledons which surround the apex (Fig. 12.3B). In addition, the double mutant also fails to develop the shoot meristem. These results suggest that *CUC1* and *CUC2* have redundant functions in promoting cotyledon separation at the boundaries as well as in shoot meristem initiation.

The expression patterns of *CUC1* and *CUC2* are variable and spatially disturbed in *pin1* embryos, consistent with the variable occurrence of cotyledon fusion in the mutant. In *pin1*, expression of *CUC1* is expanded while that of *CUC2* is restricted, suggesting that the activity of *CUC1* remains while *CUC2* activity is reduced. Double-mutant analyses also support this idea. The *cuc1 pin1* double mutant displays a greatly enhanced phenotype compared to the *pin1* single mutant, resulting in a completely fused cup-shaped cotyledon that surrounds the entire seedling apex (Fig. 12.3C; [3]). This result suggests that *CUC1* function is still largely maintained in the *pin1* single-mutant background. On the other hand, the *cuc2* mutation only moderately increases the frequency of the *pin1* phenotype, suggesting that CUC2 activity is reduced in pin1 single mutants (Fig. 12.3D; [3]). Taken together, these data indicate that *PIN1* promotes cotyledon separation by activating *CUC2* expression as well as by affecting the spatial expression pattern of *CUC1*.

Interestingly, the *cuc1 cuc2* double mutant can establish bilateral symmetry in spite of the severe cotyledon fusion phenotype, indicating that *CUC1* and *CUC2* are not required for bilateral symmetry establishment *per se*. Therefore, the bilateral symmetry defect of *pin1* mutants is not likely to be due to changes in *CUC1* or *CUC2* expression.

Auxin distribution and apical pattern formation

As discussed above, genetic and physiological evidence indicates that polar auxin transport is important for apical patterning of the embryo. Polar auxin transport is also important in other developmental processes, such as the induction of the lateral root, tissue patterning in the root meristem, and initiation of organ primordia from the shoot meristem. It has been proposed that polar transport of auxin occurs by the action of specific auxin influx and efflux carriers in each cell. Genetic and molecular approaches indicate that the AUX1 protein is a candidate influx carrier while members of the PIN family, including PIN1, are putative transport proteins of auxin efflux complexes [5, 18, 34, 44, 54].

How does polar auxin transport affect patterning? One can expect that it produces an asymmetric auxin distribution, to which cells can respond differently depending on the auxin concentration. Indeed, many developmental changes caused by the inhibition of polar auxin transport are associated with changes in auxin distribution, and it has been shown that at least some of these responses require auxin perception. Therefore, it is important to know the distribution of auxin during embryogenesis. Auxin levels in plants can be monitored either by direct or indirect methods. Tandem mass spectrometry has been used to directly measure auxin levels, but this method does not provide cellular resolution and is difficult to apply to tiny tissues such as *Arabidopsis* embryos. Another method is to monitor the activity of an auxin-responsive reporter gene construct, such as *DR5::GUS*, which seems to correlate very well with direct auxin measurements and has been successfully used to monitor auxin levels in *Arabidopsis* roots with cellular resolution [56, 58]. However, *DR5::GUS* may not be very informative when applied to processes that occur during early embryogenesis, such as the establishment of bilateral symmetry, because its expression can be detected only after the shift in symmetry has occurred. Other auxin-responsive constructs expressed before or during the process have not yet been reported. In addition, the reporter expression monitors only the endogenous auxin response and not auxin levels *per se*.

Although the exact cellular distribution of auxin in the globular embryo is not known, the expression patterns of auxin transporters or genes involved in auxin biosynthesis may provide some clues about auxin distribution. The *YUCCA* gene encodes a flavin monooxygenase-like enzyme which catalyzes a rate-limiting step in tryptophan-dependent auxin biosynthesis (Fig. 12.4A; [62]). Expression of *YUCCA* starts at the globular stage in the central region of the apical half of the embryo, which is reminiscent of the site of auxin biosynthesis (Fig. 12.4B; Furutani et al. unpublished). On the other hand, *PIN1* mRNA is detected in the provascular cells just below the expression domain of *YUCCA* as well as in the presumptive cotyledon primordia, which flank the *YUCCA* domain (Fig. 12.4C; [3]). This may suggest that auxin is produced in the central region and is transported to the presumptive cotyledon primordia as well as to the basal region by the action of PIN1 (Fig. 12.4D). Expression of *PIN1* in presumptive cotyledon primordia may cause a difference in auxin concentration between the presumptive cotyledon primordia and their boundary, creating a bilaterally symmetrical distribution of auxin.

The expression of *PIN1* is initially detected in all cells but then becomes bilaterally symmetrical in the peripheral region by the late globular stage. How is this shift of the *PIN1* expression pattern regulated? One possibility is that, since *PIN1* is important for bilateral symmetry, *PIN1* itself is involved in the establishment of its expression pattern with bilateral symmetry. It is also possible that other factors regulate the expression of *PIN1*. Analyses of the relationship between *PIN1* and other genes involved in patterning of the

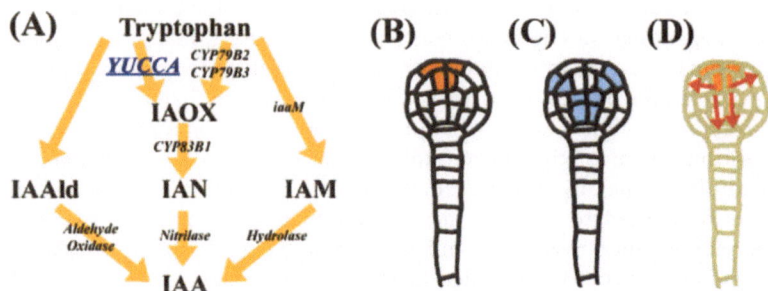

Fig. 12.4. Auxin distribution during early embryogenesis. (A) Tryptophan-dependent IAA pathway. The *YUCCA* gene encodes a flavin monooxygenase-like enzyme that catalyzes an essential step in tryptophan-dependent auxin biosynthesis. (B) Schematic diagram of *YUCCA* expression at the globular stage. The orange region represents the signal of *YUCCA* mRNA. (C) Expression of *PIN1* mRNA at the globular stage is illustrated by the blue region. (D) Schematic diagrams of a model for auxin distribution in the globular stage embryo. Red arrows in (D) show the direction of auxin flow. The orange region shows the site of auxin biosynthesis.

embryonic apex, such as *MP* or *PINOID*, might provide some insights into the mechanism by which embryonic symmetry is established.

12.3 Perspective

Molecular genetic analyses using *Arabidopsis* have identified important factors involved in pattern processes during embryogenesis, such as establishment of the body axes and bilateral symmetry, and have provided us with new insights into these processes. So far, most of the mutants that have been characterized in detail were identified from screenings for abnormal seedling phenotypes. However, mutations that disrupt early patterning processes may result in embryo lethality at early stages of embryogenesis, and this type of mutant would be missed in such screenings. In fact, a large number of embryo-lethal mutants have been isolated in *Arabidopsis* and maize, but few of them have been characterized in detail [10, 14, 24, 38, 39]. Because such embryonic lethality can also be caused by mutations in housekeeping genes, we need to develop efficient techniques to distinguish patterning mutants from those defective in housekeeping genes. Detailed analyses of factors that are linked to auxin, such as *PIN1, YUCCA, PID, or MP*, and their relationships to other developmental genes, would also help to uncover the mechanisms that regulate apical pattern formation.

References

1. Aida, M., Ishida, T., Fukaki, H., Fujisawa, H. and Tasaka, M. (1997). Genes involved in organ separation in Arabidopsis: an analysis of the *cup-shaped cotyledon* mutant. *Plant Cell.*, **9**, 841-857.
2. Aida, M., Ishida, T. and Tasaka, M. (1999). Shoot apical meristem and cotyledon formation during *Arabidopsis* embryogenesis: interaction among the *CUP-SHAPED COTYLEDON* and *SHOOT MERISTEMLESS* genes. *Development*, **126**: 1563-1570.
3. Aida, M., Vernoux, T., Furutani, M., Traas, J. and Tasaka, M. (2002). Roles of *PIN-FORMED1* and *MONOPTEROS* in pattern formation of the apical region of the *Arabidopsis* embryo. *Development*, **129**: 3965-3974.
4. Benjamins, R., Quint, A., Weijers, D., Hooykaas, P. and Offringa, R. (2001). The PINOID protein kinase regulates organ development in *Arabidopsis* by enhancing polar auxin transport. *Development*, **128**: 4057-4067.
5. Bennett, M.J., Marchant, A., Green, H.G., May, S.T., Ward, S.P., Millner, P.A., Walker, A.R., Schulz, B. and Feldmann, K.A. (1996). *Arabidopsis AUX1* gene: a permease-like regulator of root gravitropism. *Science*, **273**: 948-950.
6. Bennett, S.R.M., Alvarez, J., Bossinger, G. and Smyth, D.R. (1995). Morphogenesis in pinoid mutants of *Arabidopsis thaliana*. *Plant J.*, **8**: 505-520.
7. Berleth, T. and Jurgens, G. (1993). The role of the *monopteros* gene in organising the basal body region of the *Arabidopsis* embryo. *Development*, **118**: 575-587.
8. Bowman, JL. and Eshed, Y. (2000). Formation and maintenance of the shoot apical meristem. *Trends Plant Sci.*, **5**: 110-115.
9. Busch, M., Mayer, U. and Jurgens, G. (1996). Molecular analysis of the *Arabidopsis* pattern formation of gene *GNOM*: gene structure and intragenic complementation. *Mol Gen Genet.*, **250**: 681-691.
10. Clark, J.K. and Sheridan, W.F. (1991). Isolation and Characterization of 51 *embryo-specific* Mutations of Maize. *Plant Cell*, **3**: 935-951.
11. Christensen, S.K., Dagenais, N., Chory, J. and Weigel, D. (2000). Regulation of auxin response by the protein kinase *PINOID*. *Cell*, **100**: 469-478.
12. Di, Laurenzio, L., Wysocka-Diller, J., Malamy, J.E., Pysh, L., Helariutta, Y., Freshour, G., Hahn, M.G., Feldmann, K.A. and Benfey, P.N. (1996). The *SCARECROW* gene regulates an asymmetric cell division that is essential for generating the radial organization of the *Arabidopsis* root. *Cell*, **86**: 423-433.
13. Elliott, R.C., Betzner, A.S., Huttner, E., Oakes, M.P., Tucher, W.Q.J., Gerentes, D., Perez, P. and Smyth, D.R. (1996). *AINTEGUMENTA*, an *APETALA2*-like gene of *Arabidopsis* with pleiotropic roles in ovule development and floral organ growth. *Plant Cell*, **8**: 155-168.
14. Errampalli, D., Patton, D., Castle, L., Mickelson, L., Hansen, K., Schnall, J., Feldmann, K. and Meinke, D. (1991). Embryonic Lethals and T-DNA Insertional Mutagenesis in *Arabidopsis*. *Plant Cell*, **3**: 149-157.
15. Eshed, Y., Baum, S.F., Perea, J.V. and Bowman, J.L. (2001). Establishment of polarity in lateral organs of plants. *Curr Biol.*, **11**: 1251-1260.
16. Fischer, C., Speth, V., Fleig-Eberenz, S. and Neuhaus, G. (1997). Induction of zygotic polyembryos in wheat: Influence of auxin polar transport. *Plant Cell*, **9**: 1767-1780.
17. Fukaki, H., Wysocka-Diller, J., Kato, T., Fujisawa, H., Benfey, P.N. and Tasaka, M. (1998). Genetic evidence that the endodermis is essential for shoot gravitropism in *Arabidopsis thaliana*. *Plant J.*, **14**: 425-430.

18. Galweiler, L., Guan, C., Muller, A., Wisman, E., Mendgen, K., Yephremov, A. and Palme, K. (1998). Regulation of polar auxin transport by AtPIN1 in *Arabidopsis* vascular tissue. *Science*, **282**: 2226-2230.
19. Gray, W.M., Kepinski, S., Rouse, D., Leyser, O. and Estelle, M. (2001). Auxin regulates SCFTIR1-dependent degradation of AUX/IAA proteins. *Nature*, **414**: 271-276.
20. Hadfi, K., Speth, V. and Neuhaus, G. (1998). Auxin-induced developmental patterns in *Brassica juncea* embryos. *Development*, **125**: 879-887.
21. Hamann, T., Mayer, U. and Jurgens, G. (1999). The auxin-insensitive *bodenlos* mutation affects primary root formation and apical-basal patterning in the *Arabidopsis* embryo. *Development*, **126**: 1387-1395.
22. Hamann, T., Benkova, E., Baurle, I., Kientz, M. and Jurgens, G. (2002). The *Arabidopsis BODENLOS* gene encodes an auxin response protein inhibiting MONOPTEROS-mediated embryo patterning. *Genes Dev.*, **16**: 1610-1615.
23. Hardtke, C.S. and Berleth, T. (1998). The *Arabidopsis* gene *MONOPTEROS* encodes a transcription factor mediating embryo axis formation and vascular development. *EMBO J.*, **17**: 1405-1411.
24. Heckel, T., Werner, K., Sheridan, W.F., Dumas, C. and Rogowsky, P.M. (1999). Novel phenotypes and developmental arrest in early embryo specific mutants of maize. *Planta*, **210**: 1-8.
25. Helariutta, Y., Fukaki, H., Wysocka-Diller, J., Nakajima, K., Jung, J., Sena, G., Hauser, M.T. and Benfey, P.N. (2000). The *SHORT-ROOT* gene controls radial patterning of the *Arabidopsis* root through radial signaling. *Cell*, **101**: 555-567.
26. Hobbie, L., McGovern, M., Hurwitz, L.R., Pierro, A., Liu, N.Y., Bandyopadhyay, A. and Estelle, M. (2000). The *axr6* mutants of *Arabidopsis* thaliana define a gene involved in auxin response and early development. *Development*, **127**: 23-32.
27. Jang, J.C., Fujioka, S., Tasaka, M., Seto, H., Takatsuto, S., Ishii, A., Aida, M., Yoshida, S. and Sheen, J. (2000). A critical role of sterols in embryonic patterning and meristem programming revealed by the *fackel* mutants of *Arabidopsis thaliana*. *Genes Dev.*, **14**: 1485-1497.
28. Kerstetter, R.A., Bollman, K., Taylor, R.A., Bomblies, K. and Poethig, R.S. (2001). *KANADI* regulates organ polarity in *Arabidopsis*. *Nature*, **411**: 706-709.
29. Kim, J., Harter, K. and Theologis, A. (1997). Protein-protein interactions among the Aux/IAA proteins. *Proc. Natl. Acad. Sci.*, **94**: 11786-11791.
30. Klucher, K.M., Chow, H., Reiser, L. and Fischer, R.L. (1996). The *AINTEGUMENTA* gene of *Arabidopsis* required for ovule and female gametophyte development is related to the floral homeotic gene, *APETALA2*. *Plant Cell*, **8**: 137-153.
31. Liu, C.M., Xu, Z.H. and Chua, N-H. (1993). Auxin polar transport is essential for the establishment of bilateral symmetry during early plant embryogenesis. *Plant Cell*, **5**: 621-630.
32. Long, J.A. and Barton, M.K. (1998). The development of apical embryonic pattern in *Arabidopsis*. *Development*, **125**: 3027-3035.
33. Long, J.A., Woody, S., Poethig, S., Meyerowitz, E.M. and Barton, M.K. (2002). Transformation of shoots into roots in *Arabidopsis* embryos mutant at the *TOPLESS* locus. *Development*, **129**: 2297-2306.
34. Marchant, A., Kargul, J., May, S.T., Muller, P., Delbarre, A., Perrot-Rechenmann, C. and Bennett, M.J. (1999). AUX1 regulates root gravitropism

in *Arabidopsis* by facilitating uptake within root apical tissue. *EMBO J.*, **18**: 2066-2073.

35. Mayer, U., Buttner, G. and Jurgens, G. (1993). Apical-basal pattern formation in the *Arabidopsis* embryo: studies on the role of the *gnom* gene. *Development*, **117**: 149-162.

36. Mayer, U., Torres-Ruiz, R.A., Beleth, T., Misera, S. and Jurgens, G. (1991). Mutations affecting body organization in the *Arabidopsis* embryo. *Nature*, **353**: 402-407.

37. McConnell, R.J., Emery, J., Eshed, Y., Bao, N., Bowman, J. and Barton, K.M. (2001). Role of *PHABULOSA* and *PHAVOLUTA* in determining radial patterning in shoots. *Nature*, **411**: 709-713.

38. Meinke, D.W. and Sussex, I.M. (1979). Embryo-lethal mutants of *Arabidopsis thaliana*. A model system for genetic analysis of plant embryo development. *Dev Biol.*, **72**: 50-61.

39. Meinke, D.W. and Sussex, I.M. (1979). Isolation and characterization of six embryo-lethal mutants of *Arabidopsis thaliana*. *Dev Biol.*, **72**: 62-72.

40. Nakajima, K., Sena, G., Nawy, T. and Benfey, P.N. (2001). Intercellular movement of the putative transcription factor SHR in root patterning. *Nature*, **413**: 307-311.

41. Okada, K., Ueda, J., Komaki, M.K., Bell, C.J. and Shimura, Y. (1991). Requirement of the auxin polar transport system in early stages of *Arabidopsis* floral bud formation. *Plant Cell*, **3**: 677-684.

42. Otsuga, D., DeGuzman, B., Prigge, M.J., Drews, G.N. and Clark, S.E. (2001). *REVOLUTA* regulates meristem initiation at lateral positions. *Plant J.*, **25**: 223-236.

43. Ouellet, F., Overvoorde, P.J. and Theologis, A. (2001). IAA17/AXR3. Biochemical insight into an auxin mutant phenotype. *Plant Cell*, **13**: 829-842.

44. Palme, K. and Galweiler, L. (1999). PIN-pointing the molecular basis of auxin transport. *Curr. Opin. Plant Biol.*, **2**: 375-381.

45. Peyroche, A., Paris, S. and Jackson, C.L. (1996). Nucleotide exchange on ARF mediated by yeast Geal protein. *Nature*, **384**: 479-481.

46. Ramos, J.A., Zenser, N., Leyser, O. and Callis, J. (2001). Rapid degradation of auxin/indoleacetic acid proteins requires conserved amino acids of domain U and is proteasome dependent. *Plant Cell*, **13**: 2349-2360.

47. Sawa, S., Watanabe, K., Goto, K., Kanaya, E., Morita, E.H. and Okada, K. (1999). *FILAMENTOUS FLOWER*, a meristem and organ identity gene of *Arabidopsis*, encodes a protein with a zinc finger and HMG-related domains. *Genes Dev.*, **13**: 1079-1088.

48. Scheres, B., Di, Laurenzio, L., Willemsen, V., Hauser, M.T., Janmaat, K., Weisbeek, P. and Benfey, P.N. (1995). Mutations affecting the radial organisation of the *Arabidopsis* root display specific defects throughout the embryonic axis. *Development*, **121**: 53-62.

49. Schrick, K., Mayer, U., Horrichs, A., Kuhnt, C., Bellini, C., Dangl, J., Schmidt, J. and Jurgens, G. (2000). FACKEL is a sterol C-14 reductase required for organized cell division and expansion in *Arabidopsis* embryogenesis. *Genes Dev.*, **14**: 1471-1484.

50. Shevell, D.E., Leu, W.M., Gillmor, C.S., Xia, G., Feldmann, K.A. and Chua, N-H. (1994). *EMB30* is essential for normal cell division, cell expansion and cell adhesion in *Arabidopsis* and encodes a protein that has similarity to Sec7. *Cell*, **77**: 1051-1062.

51. Siegfried, K.R., Eshed, Y., Baum, S.F., Otsuga, D., Drews, G.N. and Bowman, J.L. (1999). Members of the *YABBY* gene family specify abaxial cell fate in *Arabidopsis*. *Development*, **126**: 4117-4128.

52. Souter, M., Topping, J., Pullen, M., Friml, J., Palme, K., Hackett, R., Grierson, D. and Lindsey, K. (2002). The *hydra1* mutants of *Arabidopsis* are defective in sterol profiles and auxin and ethylene signaling. *Plant Cell*, **14**: 1017-1031.

53. Steinmann, T., Geldner, N., Grebe, M., Mangold, S., Jackson, C.L., Paris, S., Galweiler, L., Palme, K. and Jurgens, G. (1999). Coordinated polar localization of auxin efflux carrier PIN1 by GNOM ARF GEF. *Science*, **286**: 316-318.

54. Swarup, R., Friml, J., Marchant, A., Ljung, K., Sandberg, G., Palme, K. and Bennett, M. (2001). Localization of the auxin permease AUX1 suggests two functionally distinct hormone transport pathways operate in the Arabidopsis root apex. *Genes Dev*, **15**: 2648-2653.

55. Takada, S., Hibara, K., Ishida, T. and Tasaka, M. (2001). The *CUP-SHAPED COTYLEDON1* gene of *Arabidopsis* regulates shoot apical meristem formation. *Development*, **128**: 1127-1135.

56. Ulmasov, T., Hagen, G. and Guilfoyle, T.J. (1997). ARF1, a transcription factor that binds to auxin response elements. Science, **276**: 1865-1868.

57. Ulmasov, T., Hagen, G. and Guilfoyle, T.J. (1999). Dimerization and DNA binding of auxin response factors. *Plant J.*, **19**: 309-319.

58. Ulmasov, T., Murfett, J., Hagen, G. and Guilfoyle, T.J. (1997). Aux/IAA proteins repress expression of reporter genes containing natural and highly active synthetic auxin response elements. *Plant Cell*, **9**: 1963-1971.

59. Worley, C.K., Zenser, N., Ramos, J., Rouse, D., Leyser, O., Theologis, A. and Callis, J. (2000). Degradation of Aux/IAA proteins is essential for normal auxin signalling. *Plant J.*, **21**: 553-562.

60. Wysocka-Diller, J.W., Helariutta, Y., Fukaki, H., Malamy, J.E. and Benfey, P.N. (2000). Molecular analysis of SCARECROW function reveals a radial patterning mechanism common to root and shoot. *Development*, **127**: 595-603.

61. Zenser, N., Ellsmore, A., Leasure, C. and Callis, J. (2001). Auxin modulates the degradation rate of Aux/IAA proteins. *Proc. Natl. Acad. Sci.*, **98**: 11795-11800.

62. Zhao, Y., Christensen, S.K., Fankhauser, C., Cashman, J.R., Cohen, J.D., Weigel, D. and Chory, J. (2001). A role for flavin monooxygenase-like enzymes in auxin biosynthesis. *Science*, **291**: 306-309.

63. Zhong, R. and Ye, Z.H. (2001). Alteration of auxin polar transport in the Arabidopsis *ifl1* mutants. *Plant Physiol.*, **126**: 549-563.

13

Regulation of Inflorescence Architecture and Organ Shape by the *ERECTA* Gene in Arabidopsis

Keiko U. Torii*, Laurel A. Hanson, Caroline A.B. Josefsson and Elena D. Shpak

Department of Biology, University of Washington, Seattle, WA 98195-5325 USA and *CREST, Japan Science and Technology Corporation
E-mail:ktorii@u.washington.edu telephone:+1-206-221-5701
fax:+1-206-685-1728

Summary. The architecture of higher plants is largely determined by the size, shape, and arrangement of the shoot organs that are formed in a reiterative manner by the shoot apical meristem. Immense variations in plant architecture, due to altered shape, size, and position of the individual shoot unit, has significance in adaptation as well as domestication of crop plants. The Arabidopsis *erecta* mutant displays a dramatic alteration in inflorescence architecture and organ shape. Morphometric analysis of representative *erecta* alleles with different severities revealed that *ERECTA* regulates pedicel length and plant size in a quantitative manner but has complex effects on floral organ size. The organs of *erecta* mutants contain a lesser number of larger, and isotropically expanded cortex cells, suggesting that *ERECTA* is required for a coordinated cell proliferation or cell expansion within the same tissue layer (i.e. cortex). The molecular identity of *ERECTA* as a leucine-rich repeat receptor-like kinase (LRR-RLK) is consistent with its predicted role in cell-cell coordination.

13.1 Introduction

The aboveground body of higher plants is made up of repeating units of the shoot, which is constituted by the stem (internode) and leaf (lateral organ at the node). Modification in the position, size, and shape of the individual shoot unit thus provides immense variation in plant architecture. Such diversity in plant form has significance in adaptation as well as domestication of the plant species. For instance, the Maize *teosinte branched1* (*tb1*) gene regulates the branching pattern, and alteration in the tb1 expression level played a pivotal role in the domestication of Maize (*Zea mays* spp. mays) from its wild ancestor, Teosinte (*Zea mays* spp. parviglumis), by reducing the lateral branching [4].

We study genetic regulation of inflorescence architecture using *Arabidopsis thaliana* as a model system. As a typical rosette plant, Arabidopsis produces rosette leaves at the vegetative stage with no apparent internodal elongation. As the SAM enters the reproductive state, the primary inflorescence stem rapidly elongates while the SAM generates in a rhythmical manner a few shoots that contain cauline leaves and axillary meristems and, subsequently, multiple floral meristems. The SAM itself maintains its indeterminacy as an inflorescence meristem. Each floral meristem differentiates a flower at the tip and a pedicel at the base.

Several genes controlling inflorescence patterning have been identified. They can be categorized in three major classes: (1) genes specifying the meristem identity (e.g. *TERMINAL FLOWER* and *LEAFY* [2, 24]); (2) genes regulating the shoot positioning (e.g. *PIN-FROMED1, CLAVATA*, and *FASCIATA* [3, 6, 9, 14, 16]); and (3) genes controlling the shape and elongation patterns of the shoot (e.g. *ERECTA* [21]). For instance, both loss-of-function mutations in the *TERMINAL FLOWER* (*TFL*) gene and constitutive over-expression of *LEAFY* (*LFY*) convert Arabidopsis inflorescence from indeterminate to determinate (i.e. terminal flowers are formed at the top of the inflorescence) [2, 24]. *35S::LFY* Arabidopsis plants produce single flowers at the axils of rosette leaves. Interestingly, *Jonopsidium acaule* (Violet cress) has morphology strikingly similar to the *35S::LFY* Arabidopsis plants, suggesting that altered expression pattern of *LFY* may account for the evolution of inflorescence architecture [18].

Disruption of polar auxin transport (PAT) by a mutation in the auxin efflux carrier encoded by *PIN-FROMED1* (*PIN1*) or application of PAT inhibitor prevents the emergence of lateral organs, resulting in the formation of pin-like naked inflorescence [6, 16]. Micro application of auxin droplets at the periphery of the pin inflorescence meristem induced lateral organ formation [17], indicating that local hot spots of auxin trigger initiation of organ primordia and therefore impact the positioning of organs, or inflorescence phyllotaxis. Three *CLAVATA* (*CLV*) loci and two *FASCIATAs* are required for proper maintenance of the stem cell populations at the SAM, and their loss-of-function mutations lead to a stochastic loss of phyllotaxis and disorganized inflorescence architecture [3, 9, 14].

Loss-of-function mutations in the *ERECTA* gene dramatically alter inflorescence architecture by modifying the shape of the shoot unit and internodal elongation pattern [10, 21]. While the inflorescence of wild type Arabidopsis undergoes elongation of the internodes between individual flowers, displaying a typical racemose inflorescence, the tip of the *erecta* inflorescence clusters takes the form of a head due to congestion of flowers caused by significant reduction in internodal elongation. As a consequence, the surface of the young inflorescence tip becomes flattened, disk-shaped, and somewhat umbellate. At a later stage in development, *erecta* forms a compact inflorescence with short and thick internodes, pedicels, and siliques. The *ERECTA* gene encodes a leucine-rich repeat receptor-like kinase (LRR-RLK), one of the prominent classes of receptors that perceive and transduce the signals from the extracellular environment to modulate cellular proliferation [21]. Thus proper cell-cell signaling mediated by *ERECTA* is required to manifest the inflorescence patterning in Arabidopsis.

As a first step to characterize the severity of *erecta* alleles, we identified the molecular lesion in a null allele *erecta-105*. Our detailed morphometric analysis of four representative *erecta* alleles with different severities revealed that *ERECTA* promotes plant height, pedicel length, and inflorescence architecture in a quantitative manner, while it exerts more complex effects on specifying floral organ shape. Underlying cellular defects of *erecta* plants were revealed by histological analysis.

13.2 Materials and Methods

13.2.1 Plant material, growth conditions, and morphometric analysis

All Arabidopsis plants are in the ecotype Columbia (Col). *erecta-103* and *erecta-105* mutants were backcrossed four times to the wild-type Col to clean up the background mutations [21]. *erecta-116* was backcrossed three times to the wild-type Col [12, 13]. Plants were grown on soil (Sunshine Mix4:Vermiculite:Perlite=2:1:1) in a long day cycle (18 hour light/6 hour dark) at 21 °C. Morphometric measurements for plant height, pedicel length, and fruit length were carried out using a high precision caliper (Mitutoyo, Japan). To accurately measure the length and width of floral organs (sepal, petal, long and short stamens, and pistils), flowers at stage 13 [19] were dissected under the microscope (Olympus SZX12), photographed, and enlarged by printing.

13.2.2 Molecular cloning and DNA sequencing of *erecta-105*

Genomic DNA was isolated from rosette leaves of *erecta-105* using a PhytoPure plant DNA extraction kit (Amersham Pharmacia). The genomic DNA fragment covering the DNA rearrangement in *erecta-105* was amplified by using a primer pair ERg1761 (GTA TAT CTA AAA ACG CAG TCG) and ERg4056rc (TGA ACC AGT CAG CTT GTT ACT GTG). The amplified 6kb fragment was cloned into pCR2.1-TOPO vector (Invitrogen). The insert was sequenced by using various *ERECTA*-specific primers to determine the junction of gene rearrangement.

13.2.3 Light Microscopy

Tissue samples were fixed overnight in 4% (v/v) paraformaldehyde (pH7.0) at 4 °C, dehydrated with a graded series of ethanol, and infiltrated with the polymethacryl resin Technovit 7100 (Kulzer; Energy Beam Science) followed by embedding and polymerization. Nine *mm* sections were prepared using

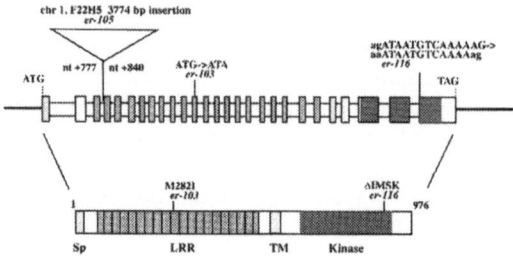

Fig. 13.1. Molecular lesions in representative *erecta* alleles with different severity. Shown is a structure of the *ERECTA* locus and the predicted *ERECTA* protein. *erecta-116*, *erecta-103* and *erecta-105* represent weak, intermediate, and null allele, respectively. Locations of the mutations and the predicted consequences of each lesion are indicated. Sp, signal sequence; LRR, LRR-domain; TM, transmembrane domain; and Kinase, cytoplasmic serine-threonine kinase domain.

the Leica RM-6145 microtome. The tissue sections were stained with 0.1% toluidine blue in 0.1M NaPO4 buffer (pH7.0) and observed under bright field illumination.

13.3 Results

13.3.1 Molecular basis of allele severity in *erecta* mutants

To understand how reduction in the *ERECTA* pathway affects Arabidopsis growth and development, we chose four alleles with different severity among 25 available alleles: a wild-type allele *ERECTA*, a weak allele *erecta-116*, an intermediate allele *erecta-103*, and a null allele *erecta-105*. The severity of the alleles has been determined based upon plant height and pedicel length [12, 13, 21]. *erecta-116* has a G to A substitution that destroys the splicing acceptor site at intron 26. Sequencing of *erecta-116* cDNA revealed that this substitution results in the missplicing of intron 26 by usage of the immediate downstream AG as a cryptic splicing acceptor site, which leads to an in-frame deletion of twelve nucleotides coding for Ile-Met-Ser-Lys within a non-conserved region of the kinase subdomain X (Fig. 13.1) [12]. The subtle phenotype of *erecta-116* plants suggests that the erecta-116 protein retains substantial biological activity despite the deletion of four amino acids.

erecta-103 carries a Met-to-Ile (ATG to ATA) missense mutation within the 10th repeat of the LRR domain (Fig. 13.1) [21]. The mutation falls outside of the LRR hypervariable region, which is predicted to form a β-sheet/β-turn structure important for ligand-receptor interaction [7]. Therefore, *erecta-103* mutations may affect receptor structure in a way that leads to partial loss of function, rather than altering the ligand-recognition site itself.

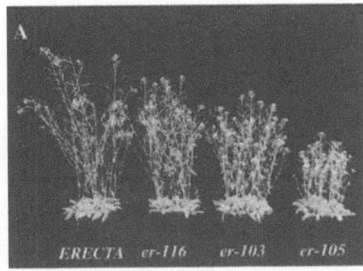

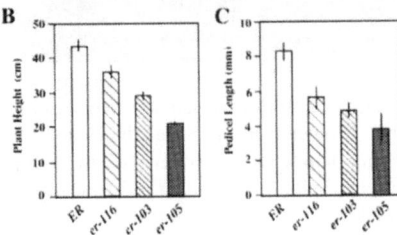

Fig. 13.2. *erecta* mutations alter the plant size and pedicel length in a quantitative manner. (A) Six-week-old wild-type *ERECTA*, *erecta-116* (weak), *erecta-103* (intermediate), and *erecta-105* (null) plants. (B). Height of fully-grown wild-type *ERECTA*, *erecta-116* (weak), *erecta-103* (intermediate), and *erecta-105* (null) plants. Bar=mean; Error bars=SD; n=8. (C) Length of the mature pedicels bearing on the main inflorescence stem of wild-type *ERECTA*, *erecta-116* (weak), *erecta-103* (intermediate), and *erecta-105* (null) plants. Bar= mean; Error bars= SD; n= 40.

The null allele *erecta-105* was induced by fast-neutron irradiation, which led to an insertion of 4kb DNA of unknown origin somewhere between nucleotide +5 and +1056 from the translational initiation codon ATG [21]. We investigated the precise location of the insertion and identity of the inserted DNA by sequencing the PCR fragment that covers the entire insertion. We found that a 3774 bp fragment corresponding to part of Chromosome I (BAC F22H5, at nucleotide position 27138 to 30912) is inserted into the *ERECTA* locus between nucleotide position +777 and +840 from the initiation codon (Fig. 13.1). A deletion of 63 bp nucleotides, corresponding to a part of the third intron and fourth exon, was associated with the insertion. Since *erecta-105* does not produce any *ERECTA* transcripts [21], this insertion likely leads to reduced stability of the *erecta-105* messenger RNA.

13.3.2 *ERECTA* regulates plant size and pedicel length in a quantitative manner

Shown in Fig. 13.2(A) are fully-grown Arabidopsis plants that each carry one of four representative *erecta* alleles. As evident from the photograph as

Fig. 13.3. *ERECTA* regulates inflorescence architecture. (A) Inflorescence top view of wild type at bolting. Flower buds are tightly closed and elongated pedicels are evident. (B) Inflorescence top view of *erecta-105* at bolting. The inflorescence tip forms a characteristic flat, disk-shape due to reduced internodal/pedicel elongation. (C) Inflorescence top view of *erecta-116* at bolting. Flower buds are closed, but pedicels are not as elongated as those of the wild type. (D) Inflorescence top view of *erecta-103* at bolting. The inflorescence tip forms a flat, disk-shape like *erecta-105*, but the pedicels are longer. Bars=5 mm.

well as the statistical data (Fig. 13.2(B)), the severity of the *erecta* mutation directly reflects reduction in plant height in a quantitative manner. Similarly, elongation of pedicels, the basal stem of flowers, is reduced in a quantitative manner as the *erecta* mutation becomes more severe (Fig. 13.2(C)).

13.3.3 *ERECTA* regulates inflorescence architecture

The inflorescence of Arabidopsis wild type is slender with dispersed flowers and tightly closed flower buds due to elongated internodes and pedicels (Fig. 13.3(A)). In contrast, the *erecta-105* mutation confers characteristic compact flower bud clusters that have an open, flat surface due to short internodes at bolting (Fig. 13.3(B)). Flower buds of *erecta-116* align nearly parallel to the main axis, thus resembling the wild-type inflorescence architecture (Fig. 13.3(C)). However, pedicel elongation is significantly inhibited in *erecta-116* (Fig. 13.3(C)). The inflorescence architecture of *erecta-103* is similar to that of *erecta-105*, as both form a flat, disk-like surface, but is less compact, perhaps due to slightly longer pedicels (Fig. 13.3(D)). Therefore, the loss of *ERECTA* function is sufficient to alter inflorescence architecture in an allele-specific manner.

13.3.4 *ERECTA* regulates the shape of floral organs

The effects of *erecta* mutations on floral organ size have not been previously reported. We therefore analyzed the short (lateral) stamen length (Fig. 13.4(A)), long (medial) stamen length (Fig. 13.4(B)), pistil length (Fig. 13.4(C)), sepal shape (Fig. 13.4(D)), and petal shape (Fig. 13.4(E)) of the flowers, which had just reached anthesis (stage 13) [19]. The analysis revealed that pistils and stamens get shorter as the severity of the *erecta* mutation increases, except that *erecta-103* develops longer long (medial) stamens than *erecta-116*. Analysis of petal shape revealed that petals are grouped into two populations of different sizes (Fig. 13.4(E)). These populations may represent perianth organs at the medial and lateral positions, respectively. The sepals and petals of wild-type *ERECTA* are slender (i.e. longer and narrower), while those of null allele *erecta-105* are round (i.e. shorter and wider). The petal size of *erecta-103* is intermediate, but its sepal size is close to that of the wild type (Fig. 13.4(D) and (E)). On the other hand, both sepals and petals of *erecta-116* are consistently smaller than those of wild type and *erecta-103* (Fig. 13.4(D) and (E)). These results indicate that, while it is consistent that a complete loss of *ERECTA* function leads to shorter and wider floral organs, partial loss-of-function mutations confer complex effects on the shape of sepals and petals, as well as elongation of the stamens.

13.3.5 Cellular defects of *erecta* mutants suggest that *ERECTA* regulates cell proliferation

As a first step to understanding how *ERECTA* controls organ elongation at a cellular level, we examined the pedicel tissues of wild type and the null allele *erecta-105* plants. Fig. 13.5 shows sections of developing pedicels at flowering (stage 14) [19]. At this stage, pedicels of wild-type plants are approximately three times longer than those of the null allele *erecta-105* (data not shown). In spite of this, cells in the cortex and pith of *erecta-105* pedicels are notably expanded and larger, suggesting that the short pedicel phenotype is due to fewer numbers of cells. The thicker pedicel phenotype, on the other hand, seems mainly due to expanded cortex cells (Fig. 13.5(C) and (E)). Unlike cortex cells, the epidermal cells are greatly reduced in length (Fig. 13.5(E)). Quantitative measurements of pedicel/stem cells reported essentially the same results [10]. Our observations and those in [10] indicate that organ elongation defects in *erecta* are not caused by reduced cell size. It appears that *ERECTA* promotes proliferation of cells in inner layers, notably the cortex, or, alternatively, it may participate in coordinated growth between the epidermal and inner layers.

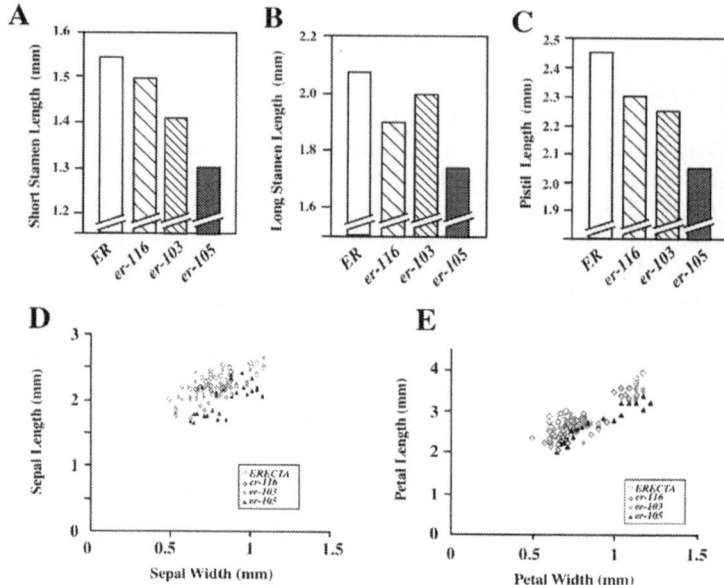

Fig. 13.4. *erecta* mutations affect size and shape of floral organs in a complex manner. (A) Short stamen length. 12 short stamens from 6 flowers at stage 13 were measured for each allele. Bar = mean. (B) Long stamen length. 28 long stamens from 7 flowers at stage 13 were measured for each allele. Bar = mean. (C) Pistil length. 10 pistils at stage 13 were measured for each allele. Bar = mean. (D) Scatter plots of the sepal shape. For each *erecta* allele, length and width of 32 sepals from 8 flowers at stage 13 were measured. Open circle, *ERECTA*; Light-shaded diamond, *erecta-116*; Shaded square, *erecta-103*; Closed triangle, *erecta-105*. (E) Scatter plots of the petal shape. For each *erecta* allele, length and width of 32 petals from 8 flowers at stage 13 were measured. Open circle, *ERECTA*; Light-shaded diamond, *erecta-116*; Shaded square, *erecta-103*; Closed triangle, *erecta-105*.

13.4 Discussion

Congestion of flowers and flower buds at the tip of the inflorescence creates the diversity in the architecture of racemose inflorescence. In some extreme cases, such as in clovers (Trifolium), the entire flowers and flower buds cluster to form a ball-like umbellate cluster [23]. Unlike some other aspects of inflorescence architecture, such as determinacy and branching patterns, the genetic basis of inflorescence clustering has not been heavily investigated. Other than *ERECTA*, two loci, *CORYMBOSA2* (*CRM2*) and *BREVIPEDI-CELLUS* (*BP*), are known to affect clustering of Arabidopsis flower buds ([5, 20, 21, 22], K.U. Torii, unpublished). *CRM2* encodes a novel protein of

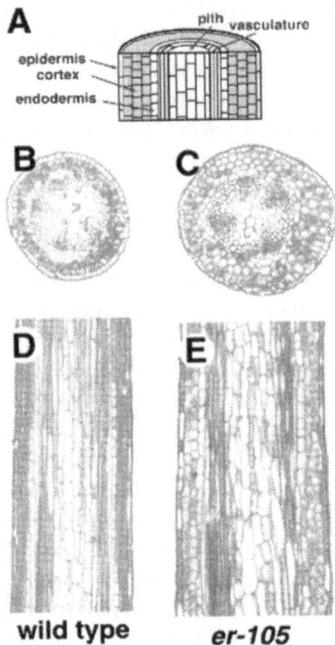

Fig. 13.5. Cellular defects of *erecta. erecta* pedicel has larger, expanded cortex/pith cells. (A) Schematic diagram of the Arabidopsis pedicel. Transverse (B and C) and longitudinal (D and E) sections of the wild-type (B and D) and *erecta-105* (C and E) pedicels at flowering stage 14. Plastic sections were cut at 5μm thickness. Photographs are taken at the same magnifications.

unknown function and *BP* encodes KNAT1, a member of the class I KNOX proteins, many of which specify the undifferentiated state of the shoot meristem [5, 20, 22]. Inflorescence phenotypes of both *crm2* and *bp* are severely enhanced by the loss-of-function of *ERECTA*, suggesting that *ERECTA* may share downstream targets with *CRM2* and *BP* or, alternatively, they may act in partially redundant, independent pathways ([5, 20], K.U. Torii, unpublished).

Our finding that *ERECTA* regulates elongation of floral organs is consistent with a study by [8], who analyzed quantitative traits loci (QTL) for Arabidopsis floral morphology. They reported that the major QTL for floral length characters, QTLFII-2, was mapped at 36 cM of Chromosome II, near the *ERECTA* locus. Indeed, QTLFII-2 explains 77%, 30%, 59%, 33%, and 66% variance in sepal length, petal length, long stamen length, short stamen length, and pistil length, respectively [8]. Because the QTL analysis was done using recombinant inbred lines generated from crosses between Arabidopsis ecotypes Col and Landsberg *erecta* (L*er*), it is unclear whether the *ERECTA* locus affects a majority of the floral length characters in other ecotypes of Arabidopsis as well as related Brassica species. L*er* has a missense mutation within the *ERECTA* gene that alters a conserved Ile to Lys in the kinase domain, and the phenotype is severe [21]. Our morphometric analysis of four *erecta* alleles with different severities in the same ecotypic background (Col) revealed that *ERECTA* controls size and shape of floral organs, notably sepals and petals, in a complex manner. For instance, the null allele *erecta-105* confers sepals shorter and wider, while weak allele *erecta-116* produces smaller sepals (Fig. 13.3 and 13.4). Therefore, it would be of great interest to see whether subtle alterations in *ERECTA* expression and function accounts for a natural variation in size and shape of floral organs.

The degree of *erecta* mutation directly reflects the plant size (Fig. 13.2). Our findings thus place *ERECTA* as an intrinsic size regulator in Arabidopsis (Fig. 13.2). How does *ERECTA* regulate organ size and internodal elongation? Our histological analysis revealed that *erecta* confers a reduced number of larger, expanded cells mainly in the cortex (Fig. 13.5), suggesting that *ERECTA* is required for proper cell proliferation and/or polarity. Because *ERECTA* codes for an LRR-RLK, an exciting scenario would be that *ERECTA* perceives cell-cell signals that coordinate proliferation of cells in the same tissue type (e.g. cortex). Alternatively, signals may emanate from different tissue layers (e.g. epidermis), and in such a case, *ERECTA* acts to coordinate proliferation of the different tissue types during organogenesis. Distinguishing these two hypotheses relies on the future identification of the corresponding *ERECTA* ligand(s). The important question that remains to be addressed is whether the elongation defects in other organs, such as petals and pistils, are due to the same cellular defects as in the case of pedicels. For instance, unlike pedicels, petals are made with fewer cell types and do not have radial organization [1]. Does *ERECTA* control the number and/or size of the petal cells? If so, which cell types? Allele-specific effects of *ERECTA* on floral organ size (Fig. 13.4) may have cellular and molecular significance, such that different floral organs possess organ-specific components of the *ERECTA* signaling pathway. This hypothesis is supported by a recent identification of *ELK4*, which encodes the β-subunit of trimeric G protein, as a silique-specific interactor of the erecta phenotype [13].

Genetic control of plant organ size is just beginning to be understood. The Arabidopsis *AINTEGUMENTA* (*ANT*) gene is necessary and sufficient to promote organ size [11, 15]. The loss-of-function *ant* mutant produces smaller organs consisting of fewer but larger cells [15]. The cellular phenotypes in ant somewhat resembles those in *erecta*, except that the ant mutation does not alter inflorescence architecture. The ectopic, overexpression of *ANT* leads to an organ enlargement, or hyperplasia [11, 15]. It has been proposed [15] that *ANT* promotes organ size by maintaining meristematic competence of cells during organogenesis, based upon the observation that *CaMV35S::ANT* displays prolonged organ growth. It is not clear at this point whether *ERECTA* promotes organ growth by increasing the rate of proliferation or, as is the case of *ANT*, by promoting the duration of cell proliferation. Alternatively, *ERECTA* may directly regulate cell differentiation or cell shape, and the observed difference in cell numbers in *erecta* pedicels may simply be a secondary consequence. Our gain-of-function approach by overexpressing *ERECTA* was hampered by post-transcriptional downregulation of the *ERECTA* protein (E.D. Shpak and K.U. Torii, unpublished). An alternative approach to activate the *ERECTA* signal transduction pathway, such as expressing a ligand-independent, constitutively-active form of *ERECTA* may address the primary role of *ERECTA* in regulating organ shape, plant size, and inflorescence architecture.

Acknowledgements

We thank Emi Hill and Chris Berthiaume for commenting on the manuscript. This work was supported in part by the start-up fund and the RRF grant No. 2499 from the Univ. of Washington, and by the CREST award from the Japan Science and Technology Corporation (JST) to K.U.T.

References

1. Bowman, J.L. (1993). *Arabidopsis: An Atlas of Morphology and Development*. *New York*: Springer-Verlag.
2. Bradley, D., Ratcliffe, O., Vincent, C., Carpenter, R. and Coen, E. (1997). Inflorescence commitment and architecture in Arabidopsis. *Science*, **275**, 80-83.
3. Clark, S.E., Williams, R.W. and Meyerowitz, E.M. (1997). The CLAVATA1 gene encodes a putative receptor kinase that controls shoot and floral meristem size in Arabidopsis. *Cell*, **89**, 575-585.
4. Doebley, J., Stec, A. and Hubbard, L. (1997). The evolution of apical diminance in maize. *Nature*, **386**, 485-488.
5. Douglas, S.J., Chuck, G., Dengler, R.E., Pelecanda, L. and Riggs, C.D. (2002). KNAT1 and ERECTA regulate inflorescence architecture in Arabidopsis. *Plant Cell*, **14**, 547-558.
6. Galweiler, L., Guan, C., Muller, A., Wisman, E., Mendgen, K., Yephremov, A. and Palme, K. (1998). Regulation of polar auxin transport by AtPIN1 in Arabidopsis vascular tissue. *Science*, **282**, 2226-2230.
7. Jones, D.A. and Jones, J.D.G. (1997). The role of Leucine-rich repeats in plant defences. *Advances in Botanical Research incorporating Advances in Plant Pathology*, **24**, 90-167.
8. Juenger, T., Purugganan, M. and Mackay, T.F. (2000). Quantitative trait loci for floral morphology in Arabidopsis thaliana. *Genetics*, **156**, 1379-1392.
9. Kaya, H., Shibahara, K.I., Taoka, K.I., Iwabuchi, M., Stillman, B. and Araki, T. (2001). FASCIATA genes for chromatin assembly factor-1 in arabidopsis maintain the cellular organization of apical meristems. *Cell*, **104**, 131-142.

10. Komeda, Y., Takahashi, T. and Hanzawa, Y. (1998). Development of infloresc-neces in Arabidopsis thaliana. *J.Plant Res*, **111**, 283-288.
11. Krizek, B.A. (1999). Ectopic expression AINTEGUMENTA in Arabidopsis plants results in increased growth of floral organs. *Developmental Genetics*, **25**, 224-236.
12. Lease, K.A., Lau, N.Y., Schuster, R.A., Torii, K.U. and Walker, J.C. (2001). Receptor serine/threonine protein kinases in signaling: analysis of the Erecta receptor-like kinase of Arabidopsis thaliana. *New Phytol*, **151**, 133-144.
13. Lease, K.A., Wen, J., Li, J., Doke, J.T., Liscum, E. and Walker, J.C. (2001). A mutant Arabidopsis heterotrimeric G-protein beta subunit affects leaf, flower, and fruit development. *Plant Cell*, **13**, 2631-2641.
14. Leyser, H. and Furner, I. (1992). Characterization of 3 shoot apical meristem mutants of Arabidopsis thaliana. *Development*, **116**, 397-403.
15. Mizukami, Y. and Fischer, R. (2000). Plant organ size control: AINTEGU-MENTA regulates growth and cell numbers during organogenesis. *Proc Natl Acad Sci USA*, **97**, 942-947.
16. Okada, K., Ueda, J., Komaki, M.K., Bell, C.J. and Shimura, Y. (1991). Requirement of the Auxin Polar Transport System in Early Stages of Arabidopsis Floral Bud Formation. *Plant Cell*, **3**, 677-684.
17. Reinhardt, D., Mandel, T. and Kuhlemeier, C. (2000). Auxin regulates the initiation and radial position of plant lateral organs. *Plant Cell*, **12**, 507-518.
18. Shu, G., Amaral, W., Hileman, L.C. and Baum, D.A. (2000). LEAFY and the evolution of rosette flowering in violet cress (Jonopsidium acaule, Brassicaceae). *Am J Bot*, **87**, 634-641.
19. Smyth, D.R., Bowman, J.L. and Meyerowitz, E.M. (1990). Early flower development in Arabidopsis. *Plant Cell*, **2**, 755-767.
20. Suzuki, M., Takahashi, T. and Komeda, Y. (2002). Formation of corymb-like inflorescences due to delay in bolting and flower development in the corymbosa2 mutant of Arabidopsis. *Plant Cell Physiol*, **43**, 298-306.
21. Torii, K.U., Mitsukawa, N., Oosumi, T., Matsuura, Y., Yokoyama, R., Whittier, R.F. and Komeda, Y. (1996). The Arabidopsis ERECTA gene encodes a putative receptor protein kinase with extracellular leucine-rich repeats. *Plant Cell*, **8**, 735-746.
22. Venglat, S.P., Dumonceaux, T., Rozwadowski, K., Parnell, L., Babic, V., Keller, W., Martienssen, R., Selvaraj, G. and Datla, R. (2002). The homeobox gene BREVIPEDICELLUS is a key regulator of inflorescence architecture in Arabidopsis. *Proc Natl Acad Sci U S A*, **99**, 4730-4735.
23. Weberling, F. (1989). *Morphology of Flowers and Inflorescences*. Cambridge: Cambridge University Press.
24. Weigel, D. and Nilsson, O. (1995). A developmental switch sufficient for flower initiation in diverse plants. *Nature*, **377**, 495-500.

Axis-dependent Regulation of Lateral Organ Development in Plants

Keiro Watanabe[1], Noritaka Matsumoto[1,2], Shunji Funaki[1], Ryuji Tsugeki[1], and Kiyotaka Okada[1,3]

[1] Department of Botany, Graduate School of Science, Kyoto University, Kyoto 606-8502, Japan.
[2] present address: Developmental, Cell and Molecular Biology Group, Duke University, Durham, NC 27708, USA.
[3] email: `kiyo@ok-lab.bot.kyoto-u.ac.jp`

Summary. Leaves and floral organs are known as lateral organs formed from meristem at the top of shoots. Distinct from other plant organs, such as stems or roots, the lateral organs are flat with two faces, the adaxial side and the abaxial side. The structural principle of the lateral organs suggests that their development is dependent on three crossing axes, the apical-basal axis, the adaxial-abaxial axis and the lateral axis, although the molecular nature of the axes is not known. Recent studies of *Arabidopsis* mutant show a couple of examples that putative axes control the expression pattern of genes working in the spatial specification of lateral organs. One of the genes, *FILAMENTOUS FLOWER* (*FIL*), a member of the *YABBY/FIL* gene family encoding a protein with a zinc finger and HMG-related domains, is involved in the specification of the abaxial side of lateral organs. *FIL* gene expression was restricted at the abaxial side of the lateral organ primordia, suggesting that *FIL* gene expression is under the control of the putative adaxial-abaxial axis in the lateral organ primordia. Another gene, *PRESSED FLOWER* (*PRS*), is a member of the homeobox gene family. Loss of function mutant of *PRS* lacks two sepals at lateral positions. Two sepals at the adaxial and the abaxial positions are present, but the marginal cell files of the remaining sepals are missing. *PRS* expression is restricted at the lateral regions of flower primordia and of lateral organs. The expression patterns of *PRS* strongly suggest that it is controlled by the putative lateral axis formed in the primordia. During development of floral meristems, the axis-dependent expression of FIL and PRS are transiently reduced at stage 2, a stage just before the floral organ primordia appear. Their expression recovers after stage 3, but the expression pattern is different before and after the tentative reduction, suggesting that the center of axes formed in the primordia shifted from the inflorescence meristem to the floral meristem. This shift of the axis center suggests the timing when the floral meristem acquires independence from the inflorescence meristem.

14.1 Introduction

Different from the situation of animal systems, maintenance of the meristem activity throughout life is key for the patterning of the plant body. The apical meristem formed during embryonic development is kept active after germination, and continues to produce leaves during vegetative growth. Leaf primordia are generated at the peripheral zone of the vegetative meristem (VM), but after entering the reproductive phase, the VM shifts to an inflorescence meristem (IM), which produces lateral floral meristems (FM) from its flank. Floral meristems generate floral organs, sepals, petals, stamens, and carpels in a manner similar to leaf generation. Leaves, floral organs, and the arrangement of floral organs have symmetric structures, suggesting that an axis-dependent development mechanism is working.

As shown in Fig. 14.1, the leaf structure indicates the presence of three axes, basal - apical, right - left, and adaxial - abaxial. Considering the process of leaf development, the relative orientation of the axes could be determined in relation to the position of the meristem from which the leaf primordia are formed. For example, the adaxial and abaxial sides could be determined as the sides close to and side far from the meristem, respectively. It is well known that cell shape and structure are different in the adaxial and abaxial sides of leaves. The adaxial side shows large cells with the shape of a jigsaw puzzle, whereas cells at the abaxial side are smaller and their shape is simpler than those of the adaxial side. The trichomes are more abundantly distributed on the adaxial side, but stomatas are found on the abaxial side rather than on the adaxial side. In addition to the epidermal cells, organization of inner tissue is decided along the abaxial-adaxial axis of leaves. The palisade and spongy tissues are located at the adaxial and at the abaxial sides, respectively. The xylem and phloem are located on the adaxial and abaxial sides in the vascular bundles, respectively.

In addition to the leaf structure, we can speculate the presence of similar axes in other lateral organs. The symmetrical arrangement of floral organs is fixed against the position of the inflorescence axis. In an *Arabidopsis* flower, two short stamens are at the lateral sides, whereas four long stamens are at the abaxial and adaxial sides of the carpels. The fixed position and size of the floral organs could be understood by assuming that the adaxial - abaxial axis and the lateral axis pivoted at the inflorescence meristem (Fig. 14.1, bottom center) and that the axes control floral organ development. When looking closer at the structure of each floral organ, we notice that the floral organs have similar axis-dependent structure as leaves. In this case, the adaxial - abaxial axis and the lateral axis assumed in the floral organs are pivoted at the floral meristem, not at the inflorescence meristem (Fig. 14.1, bottom right). Therefore, we could distinguish three meristem systems in the aerial part of plants. Each system has lateral organs, and the basic structure of the lateral organs could be based on adaxial - abaxial axis based at the position of the meristem.

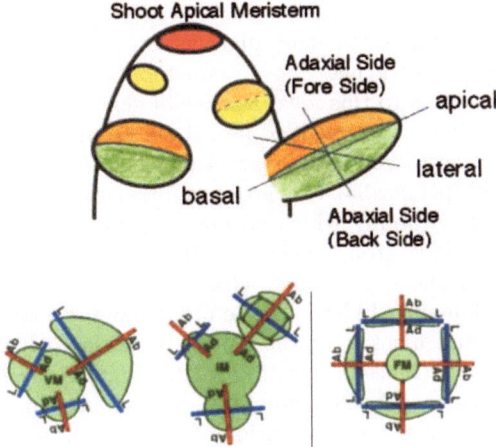

Fig. 14.1. Axes postulated in the lateral organ. Top: Primordium of lateral organs (painted in yellow) is formed at the shoot apical meristem (painted in red). The primordium develops according to the three crossing axes, apical-basal, adaxial-abaxial, and lateral axes. The abaxial side (fore side) and the abaxial side (back side) of the organ are painted in orange and green, respectively. Bottom: Top views of the three shoot meristems. Left: vegetative meristem (VM) forms leaves. Center: Inflorescence meristem (IM) generates floral primordia, which develop into floral meristems. Right: Floral meristem (FM) forms floral organs: sepals, petals, stamens and carpels. Red bar and blue bar indicate the adaxial-abaxial axis and the lateral axis, respectively.

Considering the process of flower development, we can speculate that flower development is guided by two meristem-based axis systems: the inflorescent meristem-based system at early stages and the floral meristem-based system at late stages. The former axis system might work for determining the symmetrical positioning of floral organs, and the latter system for side specific tissue development of the floral organs. It would be interesting to examine whether the two axis systems are established and maintained by the same molecular mechanism(s), and whether the two systems stand at the same time or are shifted from one to the other.

In order to understand the axis-dependent development of lateral organs, we isolated *Arabidopsis thaliana* mutants showing developmental abnormalities in the organs, and examined the expression pattern of the genes.

14.2 Genes expressed in an axis-dependent manner

One of the studies to examine the molecular system of axis-dependent development is to search for the genes whose expression is under control of the conceptual axes. According to this idea, a series of genes have been identified (Table 1) [2]. The genes can be classified into three groups according to their expression pattern. Group 1 genes are expressed at the abaxial side of the lateral organ and/or their primordia. This group is made up of two gene families. One is the *YABBY* gene family which encodes a protein with a zinc finger and a HMG domain. This group includes *FIL, CRC, INO, YABBY2* and *YABBY3*. The other is the *KAN1* gene, which encodes a plant-specific transcription activation domain, GARP. On the other hand, group 2 genes show expression pattern complementary to the group 1 genes, namely at the adaxial side of the lateral organs. Genes in this group, *PHABULOSA, PHABOLUTA*, and *REVOLUTA*, encode a homeodomain and a leucine zipper. Dominant mutations of the genes cause transformation of abaxial leaf fates into adaxial leaf fates, showing that these genes have a role to specify the adaxial side of leaves. Group 3 is a gene expressed according to the lateral axis. *PRESSED FLOWER (PRS)* is expressed at the lateral margins of primordia and young organ of leaves and floral organs.

Table 14.1. Arabidopsis Genes expressed in lateral organs in an axis-dependent expression pattern

Gene	protein structure	ref.
1. GENES EXPRESSED AT THE ABAXIAL SIDE		
FILAMENTOUS FLOWER (FIL)	Zinc finger and HMG domain	[13-15]
CRABS CLAW (CRC)	Zinc finger and HMG domain	[1, 3]
INNER NO OUTER (INO)	Zinc finger and HMG domain	[17]
YABBY2	Zinc finger and HMG domain	[15]
YABBY3	Zinc finger and HMG domain	[15]
KANADI1 (KAN1)	GARP domain	[6, 15]
2. GENES EXPRESSED AT THE ADAXIAL SIDE		
PINHEAD (PID)	elF2-homolog	[8, 9]
PHABULOSA (PHB)	homeodomain and ZIP	[11, 12]
PHAVOLUTA (PHV)	homeodomain and ZIP	[12]
REVOLUTA (REV)	homeodomain and ZIP	[4, 16]
3. GENES EXPRESSED AT THE LATERAL ENDS		
PRESSED FLOWER (PRS)	homeodomain-like	[10]

14.3 Expression pattern of *FIL*

We reported the phenotype of a loss-of-function mutant of *FIL* gene (former name: *Fl-54*) [7, 13]. The mutant shows a pleiotropic phenotype in the inflorescence with two types of flower-like structures, flowers with floral organs of altered number and shape (type A) and filamentous structures with no appendages (type B). Type A flowers often show fewer petals and stamens. Type B filaments were interpreted to be underdeveloped flowers that failed to form receptacles and floral organs. Interestingly, type A and type B structures are formed as a cluster on an inflorescence axis. First, a cluster of type A flowers is formed, and then, a cluster of type B filaments and a cluster of both types follow after bolting. In contrast to the flowers, the mutant does not show structural abnormalities in other organs including leaves and roots. *FIL* was cloned using the map-based chromosome walking procedure and found to encode a protein of 229 amino acid residues with two known domains, a Cys 2-Cys 2 type zinc-finger domain and a HMG domain [5, 14]. A zinc-finger domain is one of the well-known motifs characteristic to transcriptional regulatory factors, and a HMG domain is known to bind double-stranded DNA and make it bend [18], suggesting that the FIL protein may work as a transcription controlling protein. This hypothesis was supported by an observation that the FIL protein was localized in the nucleus [14].

Expression pattern of *FIL* was analyzed by *in situ* hybridization and by promoter::GFP (green fluorescent protein) transgenic plants. *FIL* expression was observed at the lateral organ primordia, and localized in tissues at the abaxial sides of cotyledons, leaf primordia, young leaves, young floral primordia, and primordia of floral organs (Fig. 14.2). This unique expression pattern is consistent to the function of the *FIL* gene, namely initiating and supporting development of abaxial side differentiation. This conclusion was obtained from the 35S promoter :: *FIL* transgenic plants [14]. Leaves of plants expressing *FIL* at both abaxial and adaxial sides of leaf primordia show abnormal adaxial surface interpreted as a chimera of the adaxial and abaxial surfaces. The vascular bundles in the leaves often lacked the ab-ad side specificity. Leaves of the fil mutant look to be normal, but leaves of double mutant line lacking both *FIL* and *YABBY3* genes converted to filamentous structure which lacked the characteristics of abaxial surface [15]. This observation indicates that *FIL* and *YABBY3* genes have redundant function to support abaxialization of lateral organs.

14.4 *PRS* gene and its expression pattern

An *Arabidopsis* flower has four sepals of similar size and shape. However, a flower of the *prs* mutant has two sepals of normal size at the adaxial and the abaxial positions. Two sepals at the lateral positions are, however, missed or underdeveloped. Cell file at the margin of remaining sepals is missing. Leaves

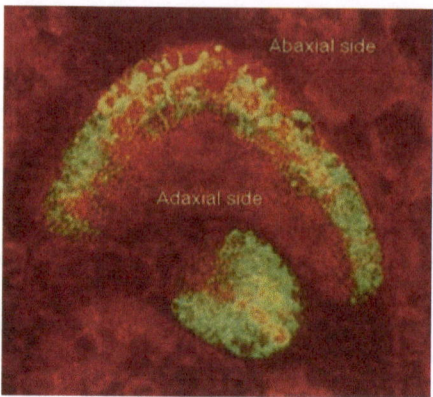

Fig. 14.2. Cross section of a vegetative meristem of a *FIL::GFP* transgenic plant. *FIL* gene is specifically expressed at the abaxial side tissue of young leaves (green area).

and other lateral organs seem to be normal. The *PRS* gene encodes a protein of 244 amino acid residues which carry a homeobox-related domain at the center [10]. Ectopic cell proliferation was observed in the 35S::*PRS* transgenic plants. Overgrowth of epidermal cells on sepals mimics the cells at the margin of sepals [10]. The effect of ectopic expression of *PRS* and the phenotype of *prs* mutant indicates that *PRS* promotes cell proliferation. Expression of *PRS* is unique. In situ hybridization and promoter::*GFP* transgenic plants showed *PRS* is expressed in cells located at the lateral margin of lateral organs (Fig. 14.3).

14.5 Model of the spatio-specific expression

As shown above, *FIL*, *PRS*, and other genes are expressed in a temporally and spatially restricted manner. It strongly indicates that the spatial pattern of gene expression is under the control of some basic system that determines the topological structure of lateral organs and flower primordia.

How could the orientation of axes speculated in the lateral organ promordia be fixed? Our model is shown in Fig. 14.4. A primordium will be trained the relative position of itself against the meristem by flow of positional information originating from the meristem. The trained primordium may fix a boundary separating the abaxial and adaxial regions. Lateral ends of the primordium would be determined at the position where the boundary meets the margin of the primordium. Axis-dependent genes are expressed according to the regions fixed in the primordium. Region-specific expression of *FIL* and *PRS* would be

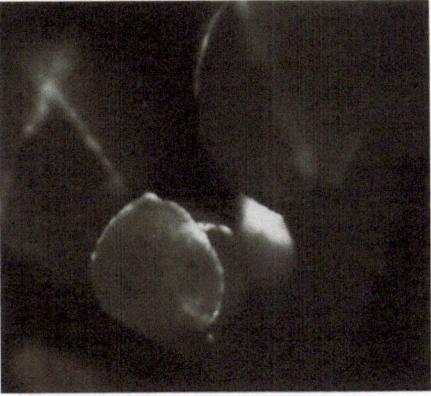

Fig. 14.3. Floral buds of a *PRS::GFP* transgenic plant. *PRS* is expressed at the marginal cells located at the lateral ends of the sepals.

regulated in this way. This process would be repeated as new lateral organ primodia are formed at the peripheral zone of the meristem.

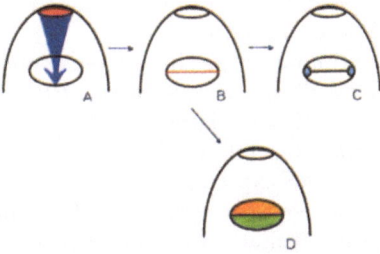

Fig. 14.4. A model showing how the axis-dependent expression is induced. A: Flow of positional information from the apical meristem arrives at the lateral organ primordia. B: The information fixes a boundary separating the adaxial and the abaxial regions. C: Regions at the lateral ends are fixed in the primordia. Expression of lateral region-specific gene (*PRS*) is induced in the regions. D: Expression of abaxial region-specific genes (*FIL* and others) and of adaxial region-specific genes is induced.

This model leads to further questions. What molecule(s) work as the positional information? How is the information conveyed and perceived? How does the information work to make a clear boundary? How are the regions fixed? How are the specific genes expressed?

Although the molecular nature of the positional information is totally un-known, it could be argued that some small peptide or small RNA molecule may work as the signal. Considering that a number of receptor kinases are involved in several developmental processes at the meristem, we can imagine that small diffusible molecules act as ligand(s) for the receptor molecules and carry the positional information. Alternatively, the molecules may be trans-ported from cell to cell through plasmodesmata, which are holes connecting the neighboring plant cells. The boundary separating the gene expression do-mains in the primordium could be formed by a mechanism similar to that known in the imaginal discs of Drosophila, where gradients of diffusible fac-tors form boundaries and successive gene expression.

14.6 Transient reduction of the expression in the flower primordium development

The process of flower development can be divided into two phases. In the first phase, a flower primordium is generated from the peripheral zone of an iflores-cence meristem in a spiral manner similar to leaf formation from a vegetative meristem. In the second phase, the flower primordium generates a floral meris-tem which sets a whorled arrangement of floral organ primordia; the center of the organogenesis is the floral meristem, not the inflorescence meristem. Therefore, a flower primordium could be considered to have acquired inde-pendence from the inflorescence meristem.

We noticed that the timing of the shift of the center of organogenesis co-incides with the temporal disappearance of the *PRS* expression in a flower primordium (Fig. 14.5). *PRS* was expressed first in a flower primordium of early stage 1 at the lateral sides relative to the inflorescence meristem. Obvi-ously, the position of the *PRS*-expressing cells is controlled by the positional relationship with the inflorescence meristem. *PRS* expression disappears at late stage 1 but reappears at stage 3 at four regions where sepal primordia are formed. The expression continues at the lateral marginal regions of the de-veloping sepal primordia. Apart from the carpels, a similar expression pattern is observed in other floral organs [10]. The lateral position in the floral organs can be determined against the center of the floral meristem. It would be rea-sonable to assume that the pivoting center determining the regions of *PRS* expression has shifted from the inflorescence meristem to the floral meristem after the temporal disappearance (Fig. 14.6). Therefore, it is strongly sug-gested that the floral meristem acquires independence at least partially from the inflorescence meristem between early stage 1 and late stage 2 and that the positional signals from the inflorescence meristem are replaced by the signals from the developing floral meristem after late stage 2.

Expression of *FIL* at early stages of flower development was also examined using FIL promoter::*GFP* plants. As shown in Fig. 14.7, the abaxial side-specific expression was observed in the stage 1 floral primordium, but the

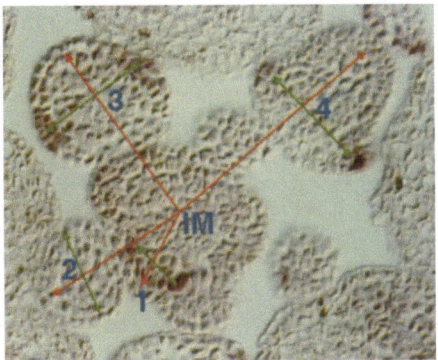

Fig. 14.5. *In situ* hybridization against a cross section of an inflorescence meristem with a probe of *PRS*. *PRS* is expressed at the lateral ends of flower primordia of stage 1, but not of stage 2. The expression reappears in a floral meristem of stage 3 and continues at stage 4 and later. The red line and the green line show the adaxial-abaxial axis and the lateral axis against the center of inflorescence meristem, respectively. IM: inflorescence meristem. Number: stage of floral meristem.

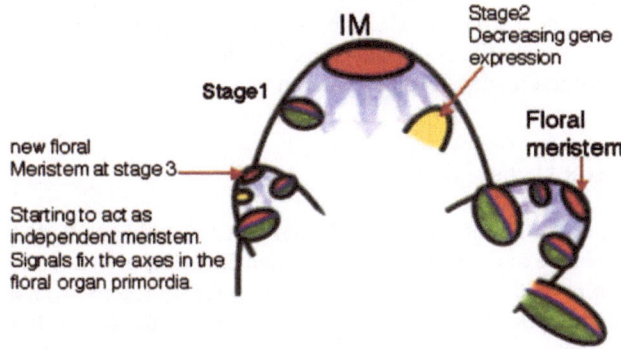

Fig. 14.6. A model showing the structural relationship between the inflorescence meristem and floral meristems of different stage.

Fig. 14.7. A cross section of an inflorescence meristem of a *FIL* promoter::*GFP* transgenic plant. *FIL* is expressed at the abaxial region of young floral primordia of stages 0 and 1 (inner three circles), but the expression is not detected in a primordium of stage 2. The expression recovers at the abaxial region of sepal primordia formed in floral meristems after stage 3 (outer three circles).

expression was missed in flowers of stage 2, and was recovered after stage 3. The timing of disappearance and reappearance of the *FIL* expression was just the same as that of *PRS* expression [unpublished results]. This observation indicates that stage 2 is a critical timing in the development of floral meristem. It is also suggested that the adaxial-abaxial axis and the lateral axis could be formed by a common signaling mechanism.

14.7 Perspectives

Axis formation and maintenance are important for making plant organs with "normal" structure and function. Recent identification of genes showing axis-dependent expression offers a good way to investigate the molecular nature of the hypothetical axes. It would be interesting to examine the mechanism of the unique gene expression of *FIL* and *PRS* by using promoter-deletion, 1-hybrid analysis, and screening mutants showing aberrant expression patterns. There is a discussion that the structural and metabolic integrity of an individual is different in plants and animals. The integrity looks to be looser in plants than animals. Further understanding of the "independence" of the meristem would help to understand the structural basis of an individual plant.

Acknowledgements

This work is supported by grants from the Japanese Ministry of Education, Culture, Sports, Science and Technology (Grant-in-Aid for Scientific Research on Priority Areas number 10182101, 10182103, 1436220, Joint Studies Program for Advanced Studies, and the 21st Century COE Program), Japan

Science and Technology Corporation (CREST program), and the Mitsubishi Foundation.

References

[1] Bowman, J.L. and Smyth, D.R. (1999). *CRABS CRAW*, a gene that regulates carpel and nectary development in *Arabidopsis*, encodes a novel protein with zinc finger and helix-loop-helix domains. *Development* **126**: 2387-2396.

[2] Bowman, J.L., Eshed, Y. and Baum, S.F. (2002). Establishment of polarity in angiosperm lateral organs. *Trends in Genetics* **18**: 134-141.

[3] Eshed, Y., Baum, S.F. and Bowman, J.L. (1999). Distinct mechanisms promote polarity establishment in carpels of *Arabidopsis*. *Cell* **99**: 199-209.

[4] Eshed, Y., Baum, S.F., Perea, J.V. and Bowman, J.L. (2001). Establishment of polarity in lateral organs of plants. *Current Biology* **11**: 1251-1260.

[5] Kanaya, E., Watanabe, K., Nakajima, N., KMorikawa, K., Okada, K. and Shimura, Y. (2001). Zinc release from the CH2C6 zinc finger domain of FILA-MENTOUS FLOWER protein form *Arabidopsis thaliana* induces self-assembly. *J. Biol. Chem.* **276**: 7383-7390.

[6] Kerstetter, R.A., Bollman, K., Taylor, R.A., Bomblies, K. and Poethig, R.S. (2001). *KANADI* controls organ polariry in *Arabidopsis*. *Nature* **411**: 706-709.

[7] Komaki, M.K., Okada, K., Nishino, E. and Shimura, Y. (1988). Isolation and characterization of novel mutants of *Arabidopsis thaliana* defective in flower development. *Development* **104**: 195-203.

[8] Lynn, K., Fernandez, A., Aida, M., Sedbrook, J., Tasaka, M., Masson, P. and Narton, M.K. (1999). The *PINHEAD/TWILLE* gene acts pleiotropically in *Arabidopsis* development and has overlapping functions with the *ARGONAUTE 1* gene. *Development* **126**: 469-481.

[9] Lynn, Newman, K., Fernadez, A.G. and Barton, M.K. (2002). Regulation of axis determinacy by the *Arabidopsis PINHEAD* gene. *Plant Cell* **14**: 3029-3042.

[10] Matsumoto, N. and Okada, K. (2001). A homeobox gene, *PRESSED FLOWER*, regulates lateral axis-dependent development of *Arabidopsis* flowers. *Genes and Development* **15**: 3335-3364.

[11] McConnell, J.R. and Barton, M.K. (1998). Leaf polarity and meristem formation in *Arabidopsis*. *Development* **125**: 2635-2942.

[12] McConnell, J.R., Emery, J., Eshed, Y., Bao, N., Bowman, J. and Barton, M.K. (2001). Role of *PHABULOSA* and *PHAVOLUTA* in determining radial patterning in shoots. *Nature* **411**: 709-713.

[13] Sawa, S., Ito, T., Shimura, Y. and Okada, K. (1999). *FILAMENTOUS FLOWER* controls the formation and development of *Arabidopsis* inflorescences and floral meristems. *Plant Cell* **11**: 69-86.

[14] Sawa, S., Watanabe, K., Goto, K., Liu, Y-G., Shibata, D., Kanaya, E., Morita, E.H. and Okada, K. (1999). *FILAMENTOUS FLOWER*, a meristem and organ identity gene of *Arabidopsis*, encodes a protein with a zinc finger and HMG-related domains. *Genes and Development* **13**: 1079-1088.

[15] Siegfried, K.R., Eshed, Y., Baum, S.F., Otsuga, D., Drews, G.N. and Bowman, J.L. (1999). Members of the *YABBY* gene family specify abaxial cell fate in *Arabidopsis*. *Development* **126**: 4117-4128.

[16] Talbert, P.B., Alder, H.T., Parks, D.W. and Komai, L. (1995). The *REVO-LUTA* gene is necessary for apical meristem development and for limiting cell divisions in the leaves and stems of *Arabidopsis thaliana*. *Development* **121**: 2723-2735.

[17] Villanueva, J.M., Broadhvest, J., Hauser, B.A., Meister, R.J., Schneitz, K. and Gasser, C.S. (1999). *INNER NO OUTER* regulates abaxial-adaxial patterning in *Arabidopsis* ovules. *Genes and Development* **13**: 3160-3169.

[18] Werner, M.H., Huth, J.R., Gronenborn, A.M. and Clore, G.M. (1995). Molecular basis of human 46X, Y sex reversal revealed from the three-dimensional solution structure of the human SRY-DNA complex. *Cell* **81**: 705-714.

15

Formation of a Symmetric Flat Leaf Lamina in Arabidopsis

Chiyoko Machida[1,5], Hidekazu Iwakawa[2], Yoshihisa Ueno[2], Endang Semiarti[3], Hirokazu Tsukaya[4], Mitsuyasu Hasebe[4], Shoko Kojima[1], and Yasunori Machida[2]

[1] College of Bioscience and Biotechnology, Chubu University and CREST, Japan. Science and Technology Corporation, 1200 Matsumoto-cho, Kasugai, Aichi 487-8501, Japan.
[2] Division of Biological Science, Graduate School of Science, Nagoya University, Furo-cho, Chikusa-ku, Nagoya 464-8602, Japan.
[3] Current address: Faculty of Biology, Gadjah Mada University, Sekip Utara, Yogyakarta 55281, Indonesia.
[4] National Institute for Basic Biology, 38 Nishigounaka, Myo-daiji-cho, Okazaki 444-8585, Japan.
[5] Corresponding author Chiyoko Machida. e-mail: cmachida@isc.chubu.ac.jp, phone: +81-568-51-6276, fax: +81-568-51-6276

Summary. The *ASYMMETRIC LEAVES1* (*AS1*) and *ASYMMETRIC LEAVES2* (*AS2*) genes of *Arabidopsis thaliana* are involved in the establishment of the leaf venation system, which includes the prominent midvein, as well as in the development of a symmetric lamina. The gene product also represses the expression of class 1 *knox* homeobox genes in leaves. We have characterized the *AS2* gene, which appears to encode a novel protein with cysteine repeats (designated the C-motif), conserved glycine, and a leucine-zipper-like sequence in the amino-terminal half of the primary sequence. The *Arabidopsis* genome contains 42 putative genes that potentially encode proteins with conserved amino acid sequences in the amino-terminal half. Thus, the AS2 protein belongs to a novel family of proteins that we have designated the AS2 family. Members of this family except AS2 also have been designated ASLs (AS2-like proteins). Overexpression of AS2 cDNA in transgenic *Arabidopsis* plants resulted in upwardly curled leaves, which differed markedly from the downwardly curled leaves generated by loss-of-function mutation of *AS2*. Our results suggest that AS2 functions in the transcription of a certain gene(s) in plant nuclei and thereby controls the formation of a symmetric flat leaf lamina and the establishment of a prominent midvein and other patterns of venation.

15.1 Introduction

Leaves of angiosperms, which are relatively flat organs, exhibit remarkable diversity in terms of their shape and complexity. Nonetheless, the basic structure of each leaf can generally be described in terms of three axes: the proximal-distal, medial-lateral and adaxial-abaxial axes [7, 16, 32, 38]. Thus, plants appear to exploit common mechanisms that are responsible for the establishment of these axes during leaf development.

Leaves develop as lateral organs from a shoot apical meristem (SAM). Various mutants have been isolated with alterations in leaf morphology that are related to the development of shape along each of three axes, to adaxial-abaxial identity, and to the overall shapes of leaves. Some genes responsible for the mutant phenotypes have been cloned and characterized [4, 7, 11, 15, 20, 29, 36, 37]. *PHANTASTICA* (*PHAN*) of *Antirrhinum majus* [38], *PHABULOSA* (*PHB*) and *PHAVOLUTA* (*PHV*) of *Arabidopsis* [22] are involved in adaxial cell fate. *FILAMENTOUS FLOWER* (*FIL*; [27]), *YABBY3* (*YAB3*; [30]), *CRABS CLAW* (*CRC*; [3]) and *KANADI* (*KAN*; [19]) are involved in the specification of abaxial cell fate in the leaf lamina. Mutations in these genes convert flat expanded leaves to filamentous and rod-shaped structures and they also distort adaxial-abaxial identity. Thus, it is likely that the mediolateral development of the leaf lamina might be coupled with the determination by these genes of adaxial-abaxial identity.

With regard to leaf shape along the medial-lateral axis, the leaves of many plant species commonly exhibit obvious but approximate left-right symmetry, with the rachis as the axis [12, 13, 23, 28, 31], even though exceptions have been reported [8, 21, 40]. Such symmetry is independent of the complexity of leaf shape (e.g., simple or compound).

To clarify the mechanisms responsible for the development of symmetrical leaves, several groups including our own have taken advantage of the *as1* and *as2* mutants of *Arabidopsis*. The phenotypes of such mutants are very similar in terms of the asymmetric shape of the lamina and a malformed vein system but the abnormalities are not absolutely identical [2, 4, 6, 24, 28, 34]. *AS1* and *AS2* might be involved in the establishment of the entire venation system, which includes the prominent midvein as the structural axis of left-right symmetry of the leaf, as well as in the development of lamina symmetry [28]. They might also function in maintaining leaf cells in a developmentally determinate state, probably by repressing expression of class 1 *knox* genes. Although the roles of these genes in the establishment of the venation system might be tightly correlated with their roles in maintaining the determinate state of leaf cells, such correlations remain to be investigated. The similarities among abnormalities in *as1*, *as2*, and *as1 as2* double-mutant plants have led to the proposal that *AS1* and *AS2* might somehow interact genetically [5, 28]. To obtain a further insight into the mechanisms whereby *AS1* and *AS2* control lamina formation, it is obviously important to characterize both the *AS1* and the *AS2* genes.

In this study we characterized the *AS2* gene and analyzed the phenotype of plants that overproduce *AS2* and plants of the *as2* mutant. Our results suggest that *AS2* might control the formation of a symmetric and flat leaf lamina.

15.2 Results

15.2.1 Structural characteristics of AS2 and related proteins

We isolated and characterized the *AS2* gene. This gene encodes a putative novel protein that contains the cysteine repeats, designated the C-motif, and a leucine-zipper-like motif. A database search revealed that AS2 belongs to a large family of proteins that we designated the AS2 family [17].

The predicted AS2 proteins contained the leucine-zipper-like sequence, from residue 81 to residue 109, which included five repetitions of hydrophobic amino acid residues, such as valine, isoleucine and leucine, with six-residue intervals (Fig. 15.1B). A search of databases revealed that the N-terminal half of AS2 was similar, in terms of amino acid sequence, to the N-terminal halves of a number of putative proteins encoded by hypothetical genes and ESTs of *Arabidopsis* and other plant species (*Oryza sativa*, *Glycine max*, *Lycopersicon esculentum*, etc.; Fig. 15.1B). The public databases of sequences of the *Arabidopsis* genome include at least 41 ORFs that are predicted to encode proteins with N-terminal halves that are related to that of AS2. Of these proteins, several appear to be only distantly related to AS2 (Fig. 15.1B). As shown in Fig. 15.1B, comparative analysis of the putative AS2-like proteins revealed that the Cx2Cx6Cx3C sequence (where x is an unconserved residue; designated the C-motif) was completely conserved in the N-terminal halves of all 41 predicted proteins identified in the database search and AS2 (positions from 10 to 24 in AS2). The glycine residue at position 46 is conserved in all members of the AS2 family (designated conserved G). In addition, the leucine-zipper-like sequence was also strongly conserved in most of the 41 proteins. More than half of the proteins, including AS2, had additional conserved sequences, as follows: PCAACKFLRRKCxxxCVFAPYFP in and around the C-motif; FAxVHKVFGASNVxKLL between the C-motif and the leucine-zipper-like sequence; and RxxAVxSLxYEAxARxRDPVYGCVGx-ISxLQxQL(V or I)xxLQxxLxxxxxxL(V or I) in and around the leucine-zipper-like sequence (Fig. 15.1B). Despite the strong conservation of amino acid sequence at the N-terminus of AS2, the amino acid sequence of the C-terminal region of AS2 was unlike those of the other AS2-like proteins and other proteins in the database (Fig. 15.1A). In AS2, the C-motif, conserved G, and the leucine-zipper-like sequence characterize the N-terminal sequence. We propose that this characteristic region be designated the AS2 domain and that proteins that include these domains be designated members of the AS2 family. According to this designation, the genome of *Arabidopsis* contains 42 ORFs that potentially encode proteins that belong to the AS2 family. As shown in Fig. 15.1B, we chose the designation *AS2-like* genes (*ASLs*) for the genes or putative genes for the 41 proteins that resembled AS2 and numbered these genes 1 to 41, respectively. Nucleotide sequences of some cDNAs of these genes have been submitted to the GenBank database and named LOB and LBD by Shuai et al. [33].

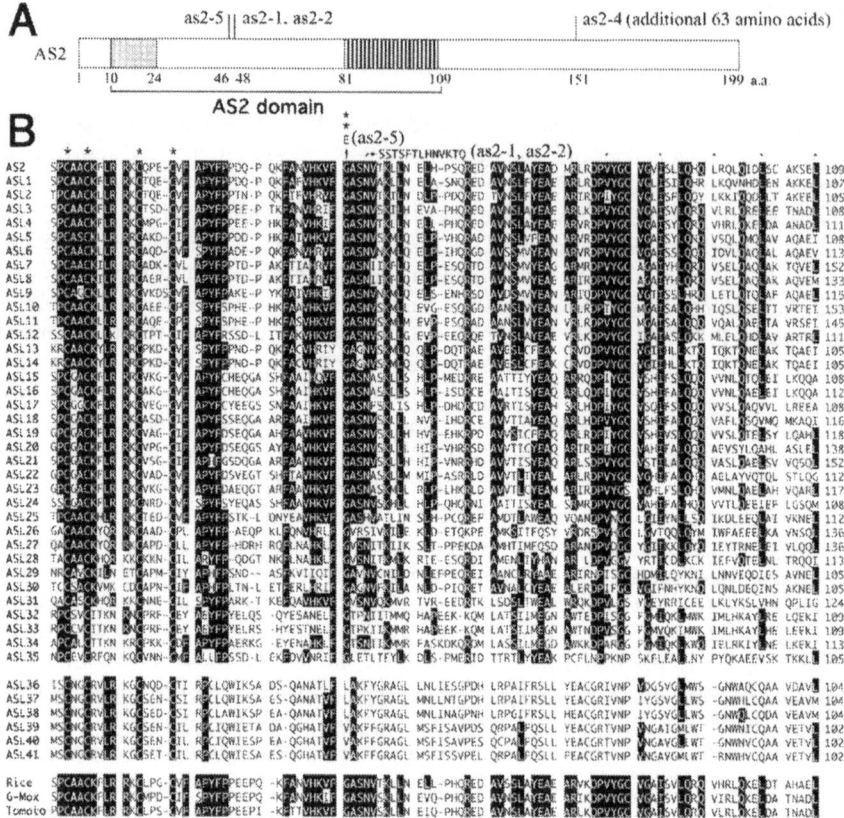

Fig. 15.1. (A) Domain organization and characteristic features of AS2. The region indicated by a bracket below the box is the AS2 domain. The shaded box and the striped box represent the C-motif and the leucine-zipper-like sequence, respectively. Sites of *as2-1*, *as2-2*, *as2-4*, and *as2-5* mutations are indicated. Numbers below the box indicate positions of amino acid residues. (B) Comparison of the amino acid sequences of the AS2 domains in members of the AS2 family. The sequence from residue 8 to residue 109 of AS2 is aligned with sequences from corresponding regions of members (designated ASL; see text) of the AS2 family. White characters on a black background indicate amino acid residues conserved in more than 20 members. Asterisks and dots indicate the consensus sequences of the C-motif and hydrophobic residues in the leucine-zipper-like sequence, respectively. Two asterisks mark the glycine residue that is conserved in all members of the family and mutated in the *as2-5* allele.

15.2.2 The AS2 family consists of at least two major classes of proteins in Arabidopsis

Members of the AS2 family can be divided into at least two classes, class I and class II (Fig. 15.2B). Class I consists of AS2 and 35 proteins (from ASL1 through ASL35) that include a C-motif, conserved G, and a leucine-zipper-like sequence and most of the major conserved residues noted above. Class II consists of six proteins (from ASL36 through ASL41). These proteins include a C-motif, conserved G, and an incomplete leucine-zipper-like sequence, in which the fourth hydrophobic residue is missing, and there is weaker sequence conservation in the AS2 domain. Furthermore, most amino acid residues are conserved among all members of class II, with only a few respective residues being different in each protein in this class. As shown in Fig. 15.2, a phylogenetic tree confirmed that class II is only distantly related to class I and suggested that members of class I can be divided further into sub-classes, namely class Ia and class Ib, with weaker sequence conservation but a complete leucine-zipper-like sequence.

15.2.3 Phenotypes of Arabidopsis plants that overproduce AS2

To investigate the function of the *AS2* gene in further detail, we fused AS2 cDNA to the 35S promoter (35S-AS2) and attempted to generate transgenic *Arabidopsis* plants that overexpressed *AS2*. Our data showed that it was difficult to generate transformed shoots [17]. These data suggested that overexpression of AS2 cDNA had an inhibitory effect on cell proliferation in green tissues and/or the regeneration of shoots. We obtained two transgenic lines that overexpressed AS2 cDNA. We confirmed that *AS2* transcripts that were derived from the cDNA were accumulated in one of the transgenic lines that exhibited the severer phenotype. One line of transgenic plants had a mildly dwarf phenotype and generated leaves that curled upward (data not shown). The other line of transgenic plants, with a severely abnormal phenotype, generated leaves with a very narrow lamina at the early stages of plant growth (Fig. 15.3C). The leaf lamina failed to expand and generated upward curling and rod-like leaves (Fig. 15.3C).

15.3 Discussion

15.3.1 Functional domains in the AS2

The *AS2* gene appears to encode a novel protein of 199 amino acid residues that includes a C-motif (defined as Cx2Cx6Cx3C) and a leucine-zipper-like sequence in its N-terminal half and an apparently unique sequence in its C-terminal half (Fig. 15.1). It is generally accepted that a leucine-zipper sequence is involved in protein-protein interactions [9] and it seems likely that the

Fig. 15.2. An unrooted maximum-likelihood tree for 42 members of the AS2 family of proteins from *Arabidopsis*, as generated by a local rearrangement search. Numbers on branches represent local bootstrap values, which were calculated with the ProtML program. The length of each horizontal branch is proportional to the estimated evolutionary distance. The brackets on the right indicate the classification of members of the AS2 family.

leucine-zipper-like sequence of AS2 might also play a role in the association of AS2 with some other protein(s). Such interaction(s) might be essential for actions of AS2. The C-motif was identified for the first time in this study but it is rather similar to a zinc finger, which generally has the consensus sequence Cx2-3CxnCx2-3 C (n is more than 12) and which functions in interactions with other macromolecules [25, 39]. It is likely that the C-motif is also involved in association(s) with other macromolecules, such as DNA and proteins, and with AS1, in particular [5, 28, our unpublished data]. The identification and characterization of the molecules with which AS2 interacts will be critical to the elucidation of the functions of AS2. The sequence of the C-terminal half

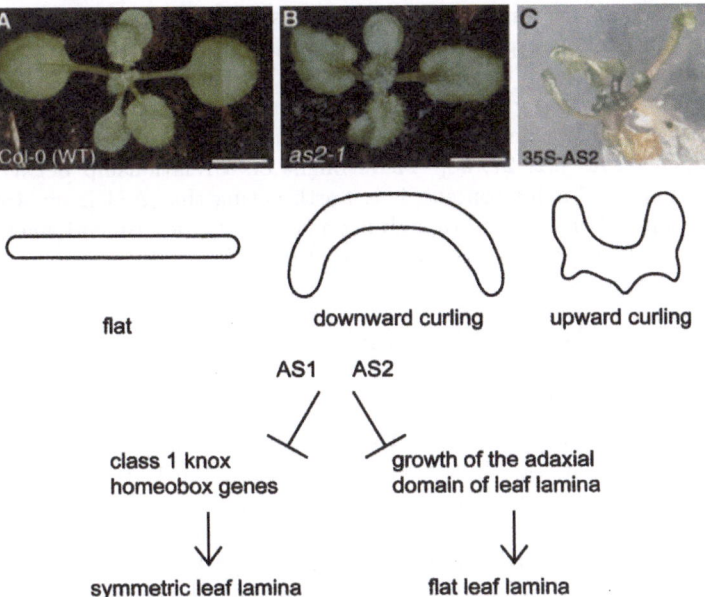

Fig. 15.3. Phenotypes of transgenic *Arabidopsis* plant that overexpressed AS2 cDNA (a) wild type plant; (b) *as2-1* plant; (c) a transgenic plant transformed with pSK35S-AS2. Scale bars: 10 mm.

of AS2 is unlike that of any protein in the databases, including known motifs (Fig. 15.1A). Since the *as2-4* allele had a frame-shift mutation in this region and the phenotype due to this allele is similar to those due to other *as2* alleles [28], it seems that the C-terminal half is essential for the functions of AS2, in addition to the N-terminal half, which contains the newly characterized AS2 domain.

15.3.2 AS2 is involved in the development of a flat leaf lamina

Overexpression of AS2 cDNA induced the upward curling of leaves and cotyledons (Fig. 15.3). By contrast, loss-of-function mutations in *AS2* result in the downward curling of leaves and cotyledons [28]. Therefore, it seems plausible that AS2 might be involved in suppression of the growth of the adaxial domain of the leaf lamina or in stimulation of the growth of the abaxial domain. Considering the growth-inhibitory effects that we observed upon transformation and the expression of *AS2* in the adaxial domain in the cotyledonary primordia in embryos [17], we can speculate that AS2 might function in suppression of the adaxial domain. It will be of interest to determine whether such suppression is achieved via the inhibition of the proliferation or the expansion of cells. Whatever the details of the activity of AS2 at the molecular level, the

present study of overproducers and previous analysis of *as2* mutant leaves
indicate that AS2 is responsible for the expansion of a flat leaf lamina, as well
as for leaf symmetry (Fig. 15.3). As described in the Introduction, mutations
in genes that are responsible for the adaxial-abaxial fates of leaf domains interfere with the fate determination of leaves and often induce formation of
rod-shaped leaves [22, 27, 38]. There might be a relationship between *AS2*
and these genes. In this context, it is worth noting that AS1 is an *Arabidopsis* homolog of PHAN, even though it remains to be determined whether the
functions of AS1 are analogous to those of PHAN.

15.4 Materials and Methods

15.4.1 Phylogenetic analysis

Arabidopsis genes that resemble *AS2* were obtained from the AGI data set
(Proteins from AGI, Total Genome) at TAIR using the TAIR BLAST version
2.0 program (http://arabidopsis.org/Blast/index.html). The *AS2-like* genes
of other plant species were obtained from the nr data sets at GenomeNet
using the GenomeNet BLAST2 program (http://www.blast.genome.ad.jp/).
The sequences of AS2 domains of 42 members of the AS2 family from
Arabidopsis were aligned using CLUSTAL W, version 1.8 [35]. For construction of the maximum-likelihood (ML) tree, we used a neighbor-joining
(NJ) tree as the start tree for a local rearrangement search. We used the
NJdist and ProtML programs in the MOLPHY, version 2.3b3, package
(http://www.ism.ac.jp/software/ismlib/softother.html; [1]). The NJ tree was
obtained with NJdist and the ML tree was obtained with ProtML. The local bootstrap probability of each branch was estimated using the ProtML
program [14, 26].

15.4.2 Transformation of Arabidopsis

Plasmids pSK35S-AS2 and pSK1 were used to transform *Agrobacterium tume-
faciens* strain GV3101 [10, 18]. Whole plants were then transformed by vac-
uum infiltration. Transgenic plants were selected on MS medium contained 15
mg/l hygromycin B and 300 mg/l carbenicillin.

15.4.3 GenBank accession numbers

The GenBank accession number for AS2 is AB080802.

Acknowledgments

This work was supported in part by grants from the Research for the Future Program of the Japan Society for the Promotion of Science (JSPS-
RFTF00L01603 to C. M.), and by Grants-in-Aid for Scientific Research on

Priority Areas (no. 10182101 to Y. M. and no. 13044003 to C. M.) from the Ministry of Education, Science, and Culture and Sports of Japan, and by Core Research For Evolutional Science and Technology (CREST) of the Japan Science and Technology Corporation.

References

[1] Adachi, J. and Hasegawa, M. (1996). MOLPHY version 2.3: programs for molecular hylopgenetics based on maximum likelihood. *Comput. Sci. Monogr.* **28**:1-150.

[2] Berna, G., Robles, P. and Micol, J.L. (1999). A mutational analysis of leaf morphogenesis in Arabidopsis thaliana. *Genetics* **152**: 729-742.

[3] Bowman, J.L. and Smyth, D.R. (1999). CRABS CLAW, a gene that regulates carpel and nectary development in Arabidopsis, encodes a novel protein with zinc finger and helix-loop-helix domains. *Development* **126**: 2387-2396.

[4] Byrne, M.E., Barley, R., Curtis, M., Arroyo, J.M., Dunham, M., Hudson, A. and Martienssen, R.A. (2000). Asymmetric leaves1 mediates leaf patterning and stem cell function in Arabidopsis. *Nature* **408**: 967-971.

[5] Byrne, M.E., Simorowski, J. and Martienssen, R.A. (2002). ASYMMETRIC LEAVES1 reveals knox gene redundancy in Arabidopsis. *Development* **129**:1957-1965.

[6] Byrne, M., Timmermans, M., Kinder, C. and Martienssen, R. (2001). Development of leaf shape. *Curr. Opin. Plant Biol.* **4**: 38-43.

[7] Conway, L.J. and Poethig, R.S. (1997). Mutations of Arabidopsis thaliana that transform leaves into cotyledons. *Proc. Natl. Acad. Sci. U.S.A.* **94**: 10209-10214.

[8] Dengler, N.G. (1999). Anisophylly and dorsoventral shoot symmetry. *Int. J. Plant Sci.* **160**, S67-S80.

[9] Ellenberger, T.E., Brandl, C.J., Struhl, K. and Harrison, S.C. (1992). The GCN4 basic region leucine zipper binds DNA as a dimer of uninterrupted alpha helices: crystal structure of the protein-DNA complex. *Cell* **71**: 1223-1237.

[10] Galbiati, M., Moreno, M. A., Nadzan, G., Zourelidou, M. and Dellporta, S. L. (2000). Large-scale T-DNA mutagenesis in Arabidopsis for functional genomic analysis. Funct. Integr. *Genomics* **1**: 25-34.

[11] Hake, S., Vollbrecht, E. and Freeling, M. (1989). Cloning Knotted, the dominant morphological mutant in maize using Ds2 as a transposon tag. *EMBO J.* **8**: 15-22.

[12] Hickey, L.J. (1973). Classification of the architecture of dicotyledonous leaves. *Am. J. Bot.* **60**: 17-33.

[13] Hickey, L.J. (1979). A revised classification of the architecture of dicotyledonous leaves. In *Anatomy of the Dicotyledons*, edited by Metcalfe, C.R. and Chalk, L.: pp. 25-39 Oxford University Press, New York.

[14] Himi, S., Sano, R., Nishiyama, T., Tanahashi, T., Kato, M., Ueda, K. and Hasebe, M. (2001). Evolution of MADS-box gene induction by FLO/LFY genes. *J. Mol. Evol.* **53**: 387-393.

[15] Hofer, J., Turner, L., Hellens, R., Ambrose, M., Matthews, P., Michael, A. and Ellis, N. (1997). UNIFOLIATA regulates leaf and flower morphogenesis in pea. *Curr. Biol.* **7**: 581-587.

[16] Hudson, A. (2000). Development of symmetry in plants. *Annu. Rev. Plant Physiol. Plant Mol. Biol.* **51**: 349-370.

[17] Iwakawa H., Ueno .Y, Semiarti E., Onouchi H., Kojima S., Tsukaya H., Hasebe Y., Soma T., Ikezaki M., Machida C. and Machida Y. (2002). The ASYMMETRIC LEAVES2 gene of Arabidopsis thaliana, required for formation of a symmetric flat leaf lamina, encodes a member of a novel family of proteins characterized by cysteine repeats and a leucine zipper. *Plant and Cell Physiology* **43**, 467-478

[18] Jefferson, R.A., Kavanagh, T.A. and Bevan, M.W. (1987). GUS fusions: beta-glucuronidase as a sensitive and versatile gene fusion marker in higher plants. *EMBO J.* **6**: 3901-3907.

[19] Kerstetter, R.A., Bollman, K., Taylor, R.A., Bomblies, K. and Poethig, R.S. (2001). KANADI regulates organ polarity in Arabidopsis. *Nature* **411**: 706-709.

[20] Kim, G.T., Tsukaya, H. and Uchimiya, H. (1998). The ROTUNDIFOLIA3 gene of Arabidopsis thaliana encodes a new member of the cytochrome P-450 family that is required for the regulated polar elongation of leaf cells. *Genes Dev.* **12**: 2381-2391.

[21] Lieu, S.M. and Sattler, R. (1976). Leaf development in Begonia hispida var. cucullifera with special reference to vascular organization. *Can. J. Bot.* **54**: 2108-2121. 8675-8680.

[22] McConnell, J.R., Emery, J., Eshed, Y., Bao, N., Bowman, J. and Barton, M.K. (2001). Role of PHABULOSA and PHAVOLUTA in determining radial patterning in shoots. *Nature* **411**: 709-713.

[23] Ogura, Y. (1962). *Plant Anatomy and Morphology.* pp. 102-134 Youkendo, Inc., Tokyo.

[24] Ori, N., Eshed, Y., Chuck, G., Bowman, J.L. and Hake, S. (2000). Mechanisms that control knox gene expression in the Arabidopsis shoot. *Development* **127**: 5523-5532.

[25] Pavletich, N.P. and Pabo, C.O. (1991). Zinc finger-DNA recognition: crystal structure of a Zif268-DNA complex at 2.1 A. *Science* **252**: 809-817.

[26] Sakakibara, K., Nishiyama, T., Kato, M. and Hasebe, M. (2001). Isolation of homeodomain-leucine zipper genes from the moss Physcomitrella patens and the evolution of homeodomain-leucine zipper genes in land plants. *Mol. Biol. Evol.* **18**: 491-502.

[27] Sawa, S., Watanabe, K., Goto, K., Liu, Y.G., Shibata, D., Kanaya, E., Morita, E.H. and Okada, K. (1999). FILAMENTOUS FLOWER, a meristem and organ identity gene of Arabidopsis, encodes a protein with a zinc finger and HMG-related domains. *Genes Dev.* **13**: 1079-1088.

[28] Semiarti, E., Ueno, Y., Tsukaya, H., Iwakawa, H., Machida, C. and Machida, Y. (2001). The ASYMMETRIC LEAVES2 gene of Arabidopsis thaliana regulates formation of a symmetric lamina, establishment of venation and repression of meristem-related homeobox genes in leaves. *Development* **128**: 1771-1783.

[29] Serrano-Cartagena, J., Robles, P., Ponce, M.R. and Micol, J.L. (1999). Genetic analysis of leaf form mutants from the Arabidopsis Information Service collection. *Mol. Gen. Genet.* **261**: 725-739.

[30] Siegfried, K.R., Eshed, Y., Baum, S.F., Otsuga, D., Drews, G.N. and Bowman, J.L. (1999). Members of the YABBY gene family specify abaxial cell fate in Arabidopsis. *Development* **126**: 4117-4128.

[31] Sinha, N.R. (1999). Leaf development in angiosperms. Annu. Rev. Plant Physiol. *Plant Mol. Biol.* **50**: 419-446.

[32] Steeves, T.A. and Sussex, I.M. (1989). *Patterns in Plant Development*. Cambridge University Press., Cambridge.

[33] Shuai, B., Reynaga-Pena, C.G. and Springer, P.S. (2002). The lateral organ boundaries gene defines a novel, plant-specific gene family. *Plant Physiol.* **129**:747-761.

[34] Sun, Y., Zhou, Q., Zhang, W., Fu, Y. and Huang, H. (2002). ASYMMETRIC LEAVES1, an Arabidopsis gene that is involved in the control of cell differentiation in leaves. *Planta* **214**: 694-702.

[35] Thompson, J.D., Higgins, D.G. and Gibson, T.J. (1994). CLUSTAL W: improving the sensitivity of progressive multiple sequence alignment through sequence weighting, positions-specific gap penalties and weight matrix choice. *Nucl. Acid Res.* **22**:4673-4680

[36] Timmermans, M.C., Hudson, A., Becraft, P.W. and Nelson, T. (1999). ROUGH SHEATH2: a Myb protein that represses knox homeobox genes in maize lateral organ primordia. *Science* **284**: 151-153.

[37] Tsiantis, M., Schneeberger, R., Golz, J.F., Freeling, M. and Langdale, J.A. (1999). The maize rough sheath2 gene and leaf development programs in monocot and dicot plants. *Science* **284**: 154-156.

[38] Waites, R., Selvadurai, H.R., Oliver, I.R. and Hudson, A. (1998). The PHANTASTICA gene encodes a MYB transcription factor involved in growth and dorsoventrality of lateral organs in Antirrhinum. *Cell* **93**: 779-789.

[39] Wang, J.H., Avitahl, N., Cariappa, A., Friedrich, C., Ikeda, T., Renold, A., Andrikopoulos, K., Liang, L., Pillai, S., Morgan, B.A. and Georgopoulos, K. (1998). Aiolos regulates B cell activation and maturation to effector state. *Immunity* **9**: 543-553.

[40] Whaley, W.G. and Whaley, C.Y. (1942). A developmental analysis of inherited leaf patterns in Tropaeolum. *Am. J. Bot.* **29**: 105-194.

Part IV

Morphogenesis and Pattern Formation Viewed
from the Behaviour of Individual Cells

Pattern Formation by Cell Movement in Closely-Packed Tissues

Kei Inouye

Department of Botany, Graduate School of Science, Kyoto University, Kyoto
606-8502, Japan
inoue@cosmos.bot.kyoto-u.ac.jp

Summary. Cell movement in 3-dimensional tissues is important in various biological processes such as cell-sorting, wound healing, and metastasis. Unlike solitary cells attached to a flat surface, motile cells tightly packed in 3-dimensional tissues are under strong mechanical constraints, but nevertheless they can still move actively by getting traction from the neighbouring cells. Here, I will describe some characteristics of cell movements in isolation and within multicellular tissues in the cellular slime mould *Dictyostelium*, and propose a possible mechanism whereby the cells inside a tissue can get traction from the substratum. Based on this hypothesis, models for cell sorting and for the generation of coherent motion of cells in tissues will also be outlined.

16.1 Introduction

Current knowledge of the mechanisms of amoeboid movement is largely based on studies of cells firmly attached to an artificial solid surface. Under such conditions, a typical amoeboid cell would move by extending one or more protrusions called pseudopodia while retracting the other end of its body. The pseudopodium is classified into several types based on its shape and dynamics, which are primarily determined by the localisation and behaviour of actin filaments. The molecular mechanisms regulating actin assembly/disassembly and the association of actin filaments to the cell membrane are the focus of intensive studies in current cell biology [2, 16].

In multicellular organisms, most cells are tightly packed to form tissues. However, they are generally capable of deformation and locomotion when dissociated into single cells, and cell movement in 3-dimensional tissues is indeed common during embryogenesis and also in tumour invasion. For instance, spatial patterns of differentiated cells are often formed by rearrangement of the cells. Also, various kinds of coherent motion of cells are observed in early development and in various *in vitro* experimental systems [1, 10, 26].

The situation of motile cells in a 3-dimensional tissue is quite different from that of cells in isolation in two respects. Firstly, the substratum of movement is their neighbouring cells (directly, or indirectly via a thin layer of extracellular matrix) which may also be moving in the same or different direction, and secondly, there are very strong constraints on the direction of pseudopod extension. These characteristics of cell movements in a tissue provide the following problems which are interesting from both theoretical and experimental points of view. (1) How does a cell located deep in a closely-packed tissue get traction for its movement? (2) Is coherent motion of cells an intrinsic property of a tissue composed of motile cells, or does it require a mechanism to organise cell movements?

Here we will focus on the collective movement of amoeboid cells seen in the development of the cellular slime mould *Dictyostelium discoideum*. This eukaryotic microorganism is particularly suitable for the study of cell movement in 3-dimensional tissues; it has both unicellular and multicellular phases in its development, and in the latter phase it exhibits spatial rearrangement of different cell types as well as coherent movement of cells within 3-dimensional tissues. I will describe some characteristics of cell movements in *Dictyostelium*, first at the unicellular stage then in multicellular tissues, with emphasis on the distribution and dynamics of filamentous actin (F-actin) within the cells revealed by the green fluorescent protein (GFP)-tagged F-actin-binding domain of ABP-120 [19]. The growth stage and early aggregation stage will not be distinguished in most part, since the differences seem relatively unimportant in the present context. Models for force transmission between cells moving in 3-dimensional tissues and for the generation of coherent motion in 2-dimensional cell aggregates will be outlined in later sections.

16.2 Movement of *Dictyostelium* cells in isolation

16.2.1 "Crawling" movement

A *Dictyostelium* cell at the growth phase on a solid surface moves about by extending pseudopodia of various sizes and appearance. These pseudopodia are generally rich in F-actin. Accumulation of F-actin at a limited site of the cell membrane precedes the extension of a pseudopodium from that site, and this accumulation continues during the period of its elongation [36]. It is suggested that in *Dictyostelium* too, elongation of actin filaments by addition of monomeric actin molecules to their plus ends at the leading edge of the pseudopodium is a major source of the force needed for its elongation [4].

A typical *Dictyostelium* cell in a growth medium has several pseudopodia, some extending while others retracting, without showing preferred direction of locomotion, and the rate of movement of its centre of mass (usually lower than ~ 0.3 μm/s) is generally much smaller than the maximum speed of pseudopod extension (*ca.* 0.7 μm/s, [23]). The translocation of such cells seems a consequence of the competition between local pseudopodial activities over the cell membrane. Involvement of a self-organising reaction-diffusion dynamics of actin assembly in these spontaneous protrusive activities has been proposed [30]. For the sake of comparison with cell movement within multicellular tissues where cells normally have fixed polarity, we will focus on *Dictyostelium* cells in growth medium that move in a more directional way by extending pseudopodia on one side of the cell body. Fig. 16.1A shows the distribution of actin filaments in a cell showing persistent movement in one direction. The cell is moving by continuously extending many protrusions enriched with F-actin on one side. The other end of the cell, which might be called the "tail" of the cell, retracts while keeping its round shape with a smooth surface suggestive of the existence of tension over it. Also note that the membrane of the tail region is lined with a thin layer of F-actin. Presumably, these cells differ from non-directional "crawling" cells only in the distribution of the membrane regions high in the protrusive activities, which are clustered and segregated from the region with high cortical tension.

The cortical layer of the tail region of a *Dictyostelium* amoeba is invariably rich in myosin II, the molecule responsible for force generation and contraction of the actin-based cell cortex. By contrast, extending pseudopodia generally lack myosin II, and its accumulation in the apical end of a pseudopodium parallels its retraction [17, 35]. Mutant cells lacking myosin II can still extend pseudopodia and translocate, indicating that actin polymerisation alone can give rise to protrusive activities [6, 14]. However, since the speed of locomotion is reduced to less than half in these cells [31], both actin polymerisation and cortical contraction seem to contribute to the "crawling movement" of *Dictyostelium* cells.

16.2.2 "Gliding" movement

Dictyostelium cells sometimes show a peculiar way of locomotion when transferred from suspension onto solid substratum (unpublished observation). These cells extend a single large pseudopodium with a smooth round end and move at a high speed of 0.5 $\sim$ 1 μm/sec. This type of movement can be seen only in a thin film of liquid just deep enough to accommodate the cells without squashing. Such cells do not seem to adhere to the substratum and appear as if they are freely gliding without much change in their shape. We will call this type of motion "gliding" for the sake of convenience, but it should be noted that its mechanism is, as will be shown below, quite different from the gliding motion of some other microorganisms such as myxobacteria.

In *Dictyostelium*, the "gliding" motion is only transient. A "gliding" cell starts to adhere to the substratum within a few minutes, and it comes to assume a familiar shape of *Dictyostelium* amoeba, with several small pseudopodia with more ragged appearance, which is indistinguishable from other cells that have not shown the "gliding" motion. Therefore the "gliding" motion is unlikely to be a special ability possessed by specialised cells.

If a "gliding" *Dictyostelium* amoeba happens to come across a narrow channel bounded on both sides by other cells firmly attached to the substra-

A   B   C   D

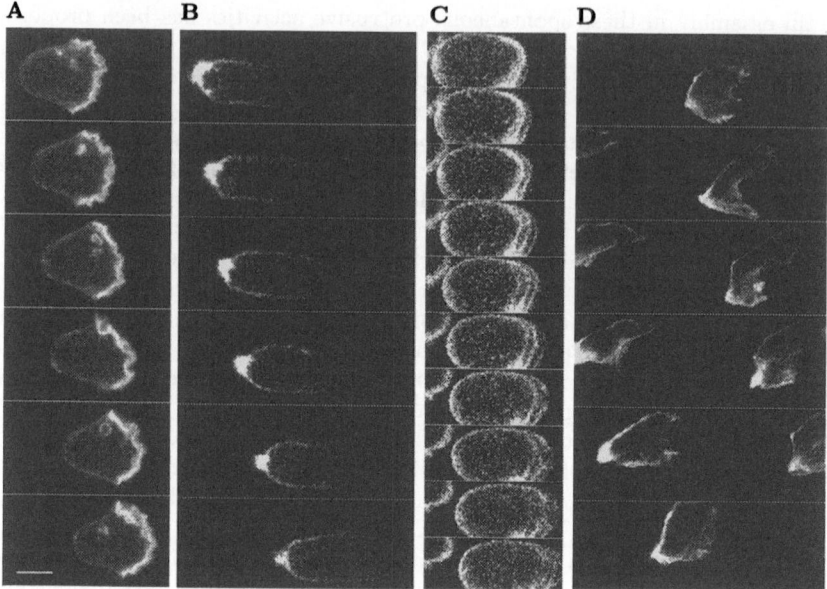

Fig. 16.1. Sequential fluorescence images of *Dictyostelium* cells transformed with the GFP-ABD probe [19] at unicellular and multicellular stages showing unidirectional locomotion. Fluorescence indicates the distribution of F-actin in the cell. Selected frames of the movies of confocal sections are shown at regular intervals as indicated below. Scale bar, 10 μm. All panels at same magnification. **A,** A solitary cell in growth medium. Intervals, 24 seconds. Average velocity, 0.13 μm/s. **B,** A growth phase cell showing "gliding motion" by extending a single large pseudopodium induced by 2 mM quinine. Intervals, 5.6 seconds. Average velocity, 0.90 μm/s. Data by Dr. K. Yoshida. **C,** A "gliding" cell induced by quinine in which layers of F-actin are clearly seen to repeatedly detach from the apical membrane of the pseudopodium. As the pseudopodium extends, the detached F-actin layers move rearwards (for detailed analysis, see [33]). Intervals, 1.8 seconds. Average velocity, 0.56 μm/s. **D,** Two GFP-ABD cells in a chimaeric migrating slug consisting of nonfluorescent wild-type cells and a few percent of GFP-ABD cells. The two cells in the figure are actually surrounded by invisible cells on all sides. Intervals, 24 seconds. Average velocity, 0.54 μm/s (left cell) and 0.42 μm/s (right cell).

tum, its velocity increases when it passes the stretch narrower than its original width. This phenomenon can be best explained if this "gliding" motion of the cell is driven by pressure flow of the cytosol.

Although the "gliding" motion of *Dictyostelium* cells is only rarely encountered under normal conditions, formation of a large cylindrical protrusion can be efficiently induced by various means. Our study of "gliding" motion induced by quinine [33] indicates that in these cells myosin II-dependent contraction of cell cortex provides the propulsive force. The most striking feature of such cells is the absence of the dense accumulation of F-actin in the apical end of the pseudopodium. Instead, F-actin is accumulated more in their tail, and the cortical actin layer on the pseudopod membrane continues to grow from its base towards the apical end during its elongation (Fig. 16.1B). Formation of a cortical layer of F-actin at the front end of the pseudopodium antagonises its elongation, and usually marks the termination of its elongation if the cortical layer is sufficiently rigid and tightly bound to the lipid bilayer of the membrane. However, a pseudopodium lined with a fairly thick cortical actin layer can still elongate by expanding the lipid membrane at its leading front, while leaving behind the cortical actin layer which has been detached from it. The detached actin layer is now pulled backwards until it dissipates because it is connected to the contracting cell cortex (Fig. 16.1C). It is argued that whether the protrusion elongates or retracts depends on the balance between the cortical tension, the hydrostatic pressure of the cytoplasm (which pushes the lipid bilayer of the membrane outwards), and the force needed to detach the cortical cytoskeleton from the lipid bilayer [33]. Posterior distribution of F-actin has also been noted in growth phase cells placed under mechanical constraint [15].

In both "crawling movement" and "gliding movement", there is a continuous circulation of actin molecules within pseudopodia, flowing forwards in the cytosol, most likely in the form of monomeric G-actin, and coming back along the cell membrane in the form of F-actin woven into the cell cortex. If the cortical actin layer is mechanically linked to the substratum of the cell, the outcome would be translocation of the pseudopod tip and, if it is the prevailing pseudopodium of the cell, translocation of the cell.

16.3 Movement of *Dictyostelium* cells in a multicellular environment

Considering the geometrical constraint imposed on *Dictyostelium* cells in closely packed tissues, it might first seem that their movement would be slower than isolated cells on solid substrata. In fact, it is opposite. The speed of cell movement in mounds or large slugs can be several-fold larger than the typical speed of isolated cells. How cells move in 3-dimensional tissues can be observed by mixing a relatively small number of fluorescently-labelled cells into unlabelled cells and allowing them to aggregate together. By observing a slug thus formed under a fluorescent microscope, one can see that *Dictyostelium* cells move very actively in multicellular environments (Fig. 16.1D).

Actively moving cells in slugs have some features common to each other. They have an irregularly shaped front region, often having two or three major protrusions with a pointed end, and a cell body tapering towards its posterior end which has a smooth surface compared with its front (Fig. 16.1D). As in cells moving on solid surface, myosin II accumulates in the rear part of these cells [3, 34], indicating a higher degree of rigidity and cortical tension in this part of the cell membrane. The shape of the cells will therefore be a natural consequence of the difference in the rigidity between the front and rear of the cells (Fig. 16.2A).

The involvement of myosin II in cell movement in multicellular environments has been demonstrated by Knecht and others by comparing the phenotype of mutants defective in myosin II activities. The essential myosin light chain (ELC) is necessary for myosin contractile function but dispensable for the generation of myosin-dependent rigidity of the cortex. Mutant cells lacking ELC move normally when placed on a solid surface but are severely defective in locomotion in migrating slugs [32]. Mutant cells lacking myosin II itself have an additional defect of weakened cortical integrity, and show severe defects in maintaining their shape when placed under mechanical stress in multicellular environments [25]. These results indicate that cell locomotion in slugs requires contraction of the cell cortex, as in isolated cells in "gliding" motion, as well as the mechanical strength of the cells, both of which depend on myosin II.

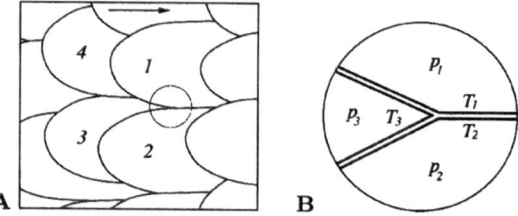

Fig. 16.2. **A,** A schematic diagram showing how the shape of cells in a slug is formed. The shape is more complex in 3-dimensional space. Direction of movement is towards the right. The higher cortical tension in the rear part of the cells helps to maintain the hemi-ellipsoidal shape of that region, while the more flexible front region fills in the clefts formed by the tails of the preceding cells. The cells mentioned in the text are numbered. **B,** Enlargement of the circled region in **A**. Thick lines represent the cell membrane lined with the cortical cytoskeleton. T_i, cortical tension; p_i, pressure, both at the position where the 3 cells meet. T_1, T_2 and p_3 act to break the adhesion between Cell 1 and Cell 2, while p_1, p_2 and T_3 have opposite effects. The curvature of the membranes in the direction perpendicular to the page is assumed to be negligible in the present argument.

In the multicellular environment too, the distribution of F-actin provides an important clue to the mechanism of cell movement. The cell membrane is heavily lined with F-actin from the middle region to the rear of the cell body, whereas the extent of actin accumulation in the front region varies with time and with the local shape of the membrane (Fig. 16.1D). By carefully observing the patterns of inhomogeneity in the thickness of the cortical actin layer, or where possible meshwork patterns of actin filaments, over a period of cell migration, it becomes evident that F-actin in the cortex on the lateral side of the cells is mostly stationary with respect to the substratum (unpublished observations). These observations are unlikely to be artefacts due to possible stabilisation of actin filaments by binding of GFP-ABD because it has sufficiently fast binding kinetics [19]. The following observations provides further support for F-actin being stationary in the lateral cell cortices. The pattern of small indentations and protrusions on the lateral surface of the cells bearing no GFP-tagged actin markers is also stationary with respect to the substratum (unpublished observations). Considering the lack of rigid extracellular matrix structures in *Dictyostelium* slugs, such local and transient geometrical features of the cell surface could only be maintained by structural supports from within the cells, and the actin-based cytoskeleton is the only subcellular structure that could shape the cell membrane in this way.

On the cell front, pointed ends are rich in F-actin, with an appearance similar to the pseudopodia of isolated "crawling" cells, while detachment of actin layers from the front faces of the pseudopodia followed by their rearward movement are also common.

The multicellular structures of *Dictyostelium* can be dissociated into single cells without affecting their pseudopodial activities. Dissociated *Dictyostelium* cells are characteristically non-adhesive to artificial substrata such as plastic dishes or agar, and unable to translocate when placed on a solid substratum and covered with a liquid medium, even though they extend large pseudopodia very actively. However, when confined in a thin layer of liquid, they can translocate very rapidly in a manner very similar to the "gliding" motion of growth phase cells in many respects, including the shape of the cell, speed of motion, and the distributions of F-actin and myosin II, suggesting that the movement of these cells would be predominantly determined by the tail contraction mode [33].

From these results, it can be concluded that *Dictyostelium* cells in a 3-dimensional tissue move by forward flows of the cytosol along with subcellular structures while the cortical actin layer remains stationary with respect to the substratum of the entire tissue. An important corollary of this conclusion is that there is no slip between two neighbouring cells in contact with their lateral surfaces. This is valid irrespective of whether the velocities of the two cells are the same or different. Considering the strong cohesiveness of *Dictyostelium* cells at the multicellular stages, the lack of slip is most likely due to strong attachment between the lateral surfaces of the two adjacent cells.

16.4 Model of cell locomotion in multicellular tissues

Based on the above considerations, a rough picture can be drawn of the mechanism of cell locomotion within a 3-dimensional tissue. For the sake of simplicity, we will consider an idealised slug with all its constituent cells heading in the same direction and moving by the myosin-dependent mode alone. Then the slug can be looked at as a tightly-packed mass of cells that are mechanically linked to each other with their lateral membranes via transmembrane cell adhesion molecules which in turn are connected to the actin cytoskeleton on their cytoplasmic side. In this architecture, the cytoskeleton of all cells form a large meshwork connected to the outermost layer of extracellular matrix covering the slug, and serves as the scaffold from which the cells can get traction [28]. In each cell, tail contraction of the cortical actin layer powered by myosin II would effectively generate forward flux of the cytosol by getting traction from this scaffold. The elongating actin filaments in the front of the cell would also obtain traction in the same manner.

In this way, all cells in the slug could contribute to its movement, in agreement with the earlier results that the motive force of a slug is proportional to its volume rather than the surface area [12, 13]. On the other hand, the major resistance forces for the movement of a slug cell are (1) the difference between the external forces acting on its front and back, (2) the force needed to break the adhesion with the neighbouring cells, and (3) the viscoelastic resistance of the cytosol and the cellular contents. The layer of extracellular matrix surrounding the entire tissue (surface sheath) is another sourse of resistance which acts on the anterior surface of the slug, thus affecting the movement of the internal cells indirectly. The latter two forces will be approximately proportional to the velocity of the cell.

The assemblage of the cortical actin layers have to bear these loads, but since the cells forming the outermost layer of the slug are the only interface between its inside and the outside, the maximal load is borne by those cells. Assuming the average motive force of the cells to be $6 \times 10^7 N \cdot m^{-3}$ [12, 13] and the radius of the cells to be 4 μm, the magnitude of cortical tension of the cells required to withstand the load is calculated to be approximately $10^{-2} N \cdot m^{-1}$ for a slug with a mean radius of 50 μm. Although this value is several-fold larger than the values reported for *Dictyostelium* cells at earlier stages of development [5, 9, 20, 24], it will be well within a realistic range for cortical tension of slug cells which have a considerably more developed cortical actin layer than solitary amoeboid cells.

The mechanism of cell sorting can also be discussed by considering the balance of force along the line where three cells meet. In the diagram shown in Fig. 16.2, Cell 3 extends a pseudopod into the cleft formed between Cell 1 and Cell 2. Whether Cell 3 moves forward relative to the other cells depends on the balance between the forces acting to break and keep the adhesion between them. At a stationary state, the difference between the two forces will be balanced with the force needed to unzip the adhesion at the same rate as the migration velocity of Cell 1 and Cell 2, but if the pressure of Cell 3 becomes larger, it will wedge into the cleft between Cell 1 and Cell 2. If Cell 3 is weaker than necessary to keep the constant geometry of the three cells, it will be delayed, and displaced by Cell 4 if that cell is stronger.

Transmission of the forces generated by the molecular motors via the tissue-wide meshwork of F-actin would be a complex process. However, by ignoring the elastic responses of the F-actin meshwork, the intuitive model described above can be formulated as a continuous model for a non-viscous, non-compressible fluid composed of self-propelled elements. (Non-viscous, because of the lack of slip between cells.) This is the basis of the models proposed by Umeda and Inouye as a unified framework for theoretical analysis of morphogenetic movements of multicellular tissues [11, 27, 28]. Some of the assumptions of the models have now better experimental evidence as discussed in the foregoing sections.

16.5 Coherent motion of cells in multicellular tissues

The multicellular stages of *Dictyostelium* development would not proceed at all without coherent motion of cells. Rotational movement is very common in mounds, and migration of a slug and fruiting body formation are nothing but the results of coherent motion of the cells, either parallel movement or spiral movement or a combination of them. These cell movements are usually attributed to the response of the cells to chemotactic signals emitted from a small population of cells and propagated across the tissue [7]. Although there is ample evidence for the control of cell movement in the multicellular stages by the signalling mechanism mediated by cyclic AMP [8, 21], this does not mean that is the only mechanism. If a mass of *Dictyostelium* cells behaves like a chunk of incompressible fluid composed of self-propelled elements, coherent motion would be expected to occur. Rotational cell movements have been reported with mutant cells lacking the ability to produce cyclic AMP signals, and this was reproduced by computer simulation in the system consisting of self-propelled deformable cells without global propagation of chemotactic signals [22].

Umeda and Inouye [29] studied the behaviour of a population of motile cells in 2-dimensional space using a mathematical model based on the balance between the motive force and the resistance forces described in the previous section. By analytical study of the model and computer simulation, they showed that rotational movement and parallel movement of the cells can emerge without a global signalling system, if there are (1) a local interaction between cells that has an effect to align the direction of their force vectors to the mean velocity vectors of the surrounding cells, and (2) a mechanism to hold the cells together. The second mechanism can be chemotaxis or cohesion of cells. The first mechanism also seems to exist; *Dictyostelium* cells have a property to follow the other cells with which they are in contact. This property, called "contact following", is independent of chemotaxis [18] and particularly strong at the multicellular stages of *Dictyostelium* development. For instance, when slugs are mechanically dissociated and placed between a thin film of buffer solution, two-dimensional arrays of cells are often observed to travel round by "gliding" motion. Similarly, disks of cells rotating about themselves are very common (the number of cells can be as few as two). Evidently, this type of collective movement does not involve long-range chemotactic signals, and the same type of mechanism is likely to be operating in undissociated slugs as well.

16.6 Conclusions

Cell movement in 3-dimensional tissues is a challenging subject, requiring combined experimental and theoretical approaches. *Dictyostelium* provides an excellent opportunity to study this interesting problem, with a rich accumulation of knowledge and experimental tools available as well as the structural simplicity which allows better comparison of the experimental data with theory and simulations. On the experimental side, identification of cell adhesion molecules and the mechanisms for regulating attachment/detachment of the cortical cytoskeleton with the membrane is crucial to the understanding of the process of cell migration in multicellular tissues. From the theoretical point of view, incorporation of the mechanism of cell movement in the models for morphogenesis in a more explicit way will be necessary to better understand the mechanism of cell sorting, whereas analytical approaches to the problem of coherent motion in 3-dimensional space is another interesting challenge.

Acknowledgements

I thank Dr. D. Knecht for GFP-ABD cells, Dr. T. Umeda for valuable comments on the manuscript, and Dr. K. Yoshida for discussion. Movies are available at the web site indicated below[1].

References

1. Armstrong, P.B. (1985). The control of cell motility during embryogenesis. *Cancer Metastasis Rev.* **4**, 59–79.
2. Borisy, G.G. and Svitkina, T.M. (2000). Actin machinery: pushing the envelope. *Curr. Opin. Cell Biol.* **12**, 104–112.
3. Clow, P. A. and McNally, J. G. (1999). In vivo observations of myosin II dynamics support a role in rear retraction. *Mol. Biol. Cell* **10**, 1309–1323.
4. Condeelis, J. (1993). Life at the leading edge: the formation of cell protrusions. *Ann. Rev. Cell Biol.* **9**, 411–444.
5. Dai, J. W., Ting-Beall, H. P., Hochmuth, R. M., Sheetz, M. P. and Titus, M. A. (1999). Myosin I contributes to the generation of resting cortical tension. *Biophys. J.* **77**, 1168–1176.

---

[1] http://cosmos.bot.kyoto-u.ac.jp/csm/mpb2002/

6. De Lozanne, A. and Spudich, J. A. (1987). Disruption of the *Dictyostelium* myosin heavy chain gene by homologous recombination. *Science* **236**, 1086–1091.

7. Dormann, D., Vasiev, B. and Weijer, C. J. (2000). The control of chemotactic cell movement during *Dictyostelium* morphogenesis. *Phil. Trans. R. Soc. Lond. B* **355**, 983–991.

8. Dormann, D. and Weijer, C. J. (2001). Propagating chemoattractant waves coordinate periodic cell movement in *Dictyostelium* slugs. *Development* **128**, 4535–4543.

9. Egelhoff, T. T., Naismith, T. V. and Brozovich, F. V. (1996). Myosin-based cortical tension in *Dictyostelium* resolved into heavy and light chain-regulated components. *J. Muscle Res. Cell Motil.* **17**, 269–274.

10. Hegerfeldt, Y., Tusch, M., Brocker, E. B. and Friedl, P. (2002). Collective cell movement in primary melanoma explants: plasticity of cell-cell interaction, beta1-integrin function, and migration strategies. *Cancer Res.* **62**, 2125–2130.

11. Inouye, K. and Takeuchi, I. (1979). Analytical studies on migrating, movement of the pseudoplasmodium of *Dictyostelium discoideum*. *Protoplasma* **99**, 289–304.

12. Inouye, K. and Takeuchi, I. (1980). Motive force of the migrating pseudoplasmodium of the cellular slime mould *Dictyostelium discoideum*. *J. Cell Sci.* **41**, 53–64.

13. Inouye, K. (1984). Measurement of the motive force of the migrating slug of *Dictyostelium discoideum* by a centrifuge method. *Protoplasma* **121**, 171–177.

14. Knecht, D. A. and Loomis, W. F. (1987). Antisense RNA inactivation of myosin heavy chain gene expression in *Dictyostelium discoideum*. *Science* **236**, 1081–1085.

15. Laevsky, G. and Knecht. D. A. (2001). Under-agarose folate chemotaxis of *Dictyostelium* amoebae in permissive and mechanically inhibited conditions. *Biotechniques* **31**, 1140–1149.

16. Lee, E., Pang, K. and Knecht, D. (2001). The regulation of actin polymerization and cross-linking in *Dictyostelium*. *Biochim. Biophys. Acta* **1525**, 217–227.

17. Moores, S. L., Sabry, J. H. and Spudich, J. A. (1996). Myosin dynamics in live *Dictyostelium* cells. *Proc. Natl. Acad. Sci. USA* **93**, 443–446.

18. Otsuka, H. (1994). Relationship between the movements of slug cells and chemotaxis in *Dictyostelium discoideum*. (in Japanese). Master's Thesis, Kyoto University.

19. Pang, K. M., Lee, E. and Knecht, D. A. (1998). Use of a fusion protein between GFP and an actin-binding domain to visualize transient filamentous-actin structures. *Curr. Biol.* **8**, 405–408.

20. Pasternak, C., Spudich, J. A. and Elson, E. L. (1989). Capping of surface receptors and concomitant cortical tension are generated by conventional myosin. *Nature* **341**, 549–551.

21. Patel, H., Guo, K. D., Parent, C., Gross, J., Devreotes, P. N. and Weijer, C. J. (2000). A temperature-sensitive adenylyl cyclase mutant of *Dictyostelium*. *EMBO J.* **19**, 2247–2256.

22. Rappel, W. J., Nicol, A., Sarkissian, A., Levine, H. and Loomis, W. F. (1999). Self-organized vortex state in two-dimensional *Dictyostelium* dynamics. *Phys. Rev. Lett.* **83**, 1247–1250.

23. Schindl, M., Wallraff, E., Deubzer, B., Witke, W., Gerisch, G. and Sackmann, E. (1995). Cell-substrate interactions and locomotion of *Dictyostelium* wild-type and mutants defective in three cytoskeletal proteins: A study using quantitative reflection interference contrast microscopy. *Biophys. J.* **68**, 1177–1190.

24. Schwarz, E. C., Neuhaus, E. M., Kistler, C., Henkel, A. W. and Soldati, T. (2000). *Dictyostelium* myosin IK is involved in the maintenance of cortical tension and affects motility and phagocytosis. *J. Cell Sci.* **113**, 621–633.

25. Shelden, E. and Knecht, D. A. (1995). Mutants lacking myosin II cannot resist forces generated during multicellular morphogenesis. *J. Cell Sci.* **108**, 1105–1115.

26. Trinkaus, J. P. (1988). Directional cell movement during early development of the teleost Blennius pholis: I. Formation of epithelial cell clusters and their pattern and mechanism of movement. *J. Exp. Zool.* **245**, 157–186.

27. Umeda, T. (1989). A mathematical model for cell sorting, migration and shape in the slug stage of *Dictyostelium discoideum*. *Bull. Math. Biol.* **51**, 485–500.

28. Umeda, T. and Inouye, K. (1999). Theoretical model for morphogenesis and cell sorting in *Dictyostelium discoideum*. *Physica D*, **126**, 189–200.

29. Umeda, T. and Inouye, K. (2002). Possible role of contact following in the generation of coherent motion of *Dictyostelium* cells. *J. theor. Biol.* **219**, 301–308.

30. Vicker, M. G. (2002). F-actin assembly in *Dictyostelium* cell locomotion and shape oscillations propagates as a self-organized reaction-diffusion wave. *FEBS Lett.* **510**, 5–9.

31. Wessels, D., Soll, D. R., Knecht, D., Loomis, W. F., De Lozanne, A. and Spudich, J. (1988). Cell motility and chemotaxis in *Dictyostelium* amebae lacking myosin heavy chain. *Dev. Biol.* **128**, 164–177.

32. Xu, X. X. S., Lee, E., Chen, T. L., Kuczmarski, E., Chisholm, R. L. and Knecht, D. A. (2001). During multicellular migration, myosin II serves a structural role independent of its motor function. *Dev. Biol.* **232**, 255–264.

33. Yoshida, K. and Inouye, K. (2001). Myosin II-dependent cylindrical protrusions induced by quinine in *Dictyostelium*: antagonizing effects of actin polymerization at the leading edge. *J. Cell Sci.* **114**, 2155–2165.

34. Yumura, S., Mori, H. and Fukui, Y. (1984). Localization of actin and myosin for the study of ameboid movement in *Dictyostelium* using improved immunofluorescence. *J. Cell Biol.* **99**, 894–899.

35. Yumura, S. (1996). Rapid redistribution of myosin II in living *Dictyostelium* amoebae, as revealed by fluorescent probes introduced by electroporation. *Protoplasma* **192**, 217–227.

36. Yumura, S. and Fukui, Y. (1998). Spatiotemporal dynamics of actin concentration during cytokinesis and locomotion in *Dictyostelium*. *J. Cell Sci.* **111**, 2097–2108.

Positioning of Cells at their Intrinsic Sites in Multicellular Organisms

Hisao Honda

Hyogo University, Kakogawa, Hyogo 675-0195 Japan
E-mail: hihonda@hyogo-dai.ac.jp

Summary. Certain types of cell are known to be under bilateral threshold control, e.g., cells receiving a signal of ligands do not respond until the ligand level comes close to a threshold. When the ligand level is around the threshold, the cells adhere to a surface that expresses the ligands, and again do not respond when the ligand level is over the threshold. The bilateral threshold control of membrane-bound ligands (ephrin) and their receptor (Eph) seems to govern positioning of cells at their intrinsic sites. Three examples are shown: (1) In the topographic projection of retinal ganglion axons to the midbrain, the axon terminals expressing Eph receptors crawl on the midbrain surface where the ligand density is graded. The axon terminals find their own sites on the midbrain where the ligand level is at their own threshold. (2) The bilateral threshold control does not only direct positioning of individual cells, but, under assumptions that the cells express ligands and receptors simultaneously in a single cell and the densities of the ligands and the receptors are reciprocal with each other, the bilateral threshold control organizes spontaneously a tissue of graded cell arrangement, which could provide positional information for morphogenesis and regeneration. (3) In a cell aggregate of two types of cells, cells co-expressing ligands and receptors (not necessarily reciprocally) can form curious cell patterns, a checkerboard pattern and a kagome (star) pattern that have been observed on the chick oviduct epithelium.

17.1 Introduction

In the present report I will demonstrate a mechanism to make a graded cell arrangement. Cells in Fig. 17.1 have specific proteins, membrane-bound ligands ephrin and their receptors Eph, and their expression levels differ from cell to cell. These molecules constitute a system of cell positioning. Although the cells are initially distributed at random (Fig. 17.1A), they exchange positions and finally make the graded cell arrangement (Fig. 17.1B).

Considering that a biological body has an axis along which something sets up a continuous gradient is useful for understanding morphogenesis and regeneration. For example, in an experiment on limb regeneration in amphibians or

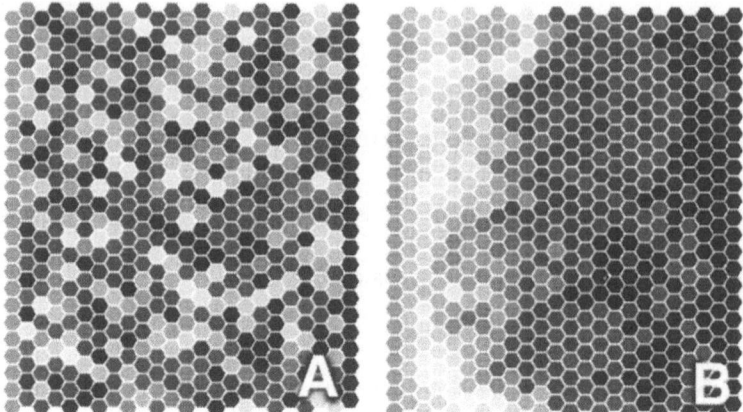

Fig. 17.1. Computer simulation of the formation of a graded pattern of cells. **A** An initial pattern. Cells expressing receptors with random R-value ($0.2 \sim 100$) were interspersed. The cells co-express ligands with reciprocal L-values ($L = S/R$). Boundary condition: Top and bottom are periodic, and left and right are fixed at $R = 0.2$ and 100, respectively. **B** Result of computer simulation under bilateral threshold control after 4.2×10^8 steps. Gray levels that have been converted from colors indicate R-value. See text (section 17.4) for details.

cockroachs (Fig. 17.2), a normal limb has a continuous gradient (Fig. 17.2A right). When the limb is partially removed, a discontinuity of the gradient arises (Fig. 17.2B). The cells at the junction begin to divide and the removed part is intercalated (Fig. 17.2C). What substance forms the gradient is a big puzzle.

Important observations come from cell sorting experiments [10]. For example, cells from the proximal part of a limb were mixed with cells from various parts of limb. When the two parts are close with each other, small cell patches are produced. But, when the two parts are far from each other, the cell patches produced are large. So although the graded substance along the axis was hypothetical, this experiment shows that something surely differs and works differently along this axis. Here, I would like to point out a close relationship between cell sorting and the body axis (Fig. 17.3). Cell sorting of two types of cells makes a cell aggregate of one cell type including a cell mass of another cell type (Fig. 17.3A). When other types of cells are added, a concentric circular pattern of cells is formed (Fig. 17.3B). This pattern could be topologically deformed to a rod (Fig. 17.3C) as indicated by [8]. The rod has an axis along which something is graded. The cell sorting has been explained by the differential cell adhesion hypothesis [9]. According to this hypothesis, the distal part of the limb is more adhesive than that of the proximal part. On the other hand, since ephrin ligands were observed at the proximal part

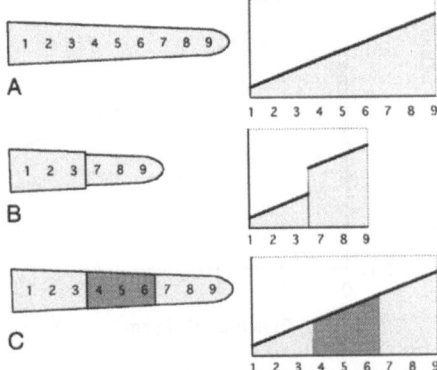

Fig. 17.2. Schematic presentation of the experiment of limb regeneration. **A** Left, Normal limb. Right, a graded pattern of something along the proximo-distal axis (1-9). **B** Part 4-6 was removed and a discontinuous gradient pattern appeared. **C** The removed part was regenerated and an intercalated gradient pattern was made.

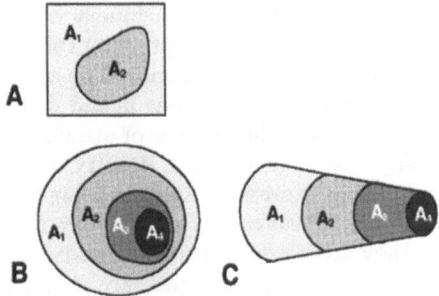

Fig. 17.3. A cell sorting pattern of two types of cells (**A**), a concentric circular pattern (**B**), and a graded pattern along the body axis (**C**).

of the limb and Eph receptors were observed at the distal part [3], I would like to point out another possibility for the explanation of cell sorting [6].

17.2 Ligand ephrin and their receptor Eph constitute a system of bilateral threshold control

Ligand ephrin is a membrane-bound ligand and the receptor for ephrin is called Eph. When receptor Eph collide with ephrins, they take on the activity of tyrosine-kinase, which catalyzes tyrosine phosphorylation. I would like to stress that ephrin and its receptor constitute a system of bilateral threshold control. Under bilateral threshold control, cells can perform positioning in a tissue or form a graded cell pattern. I will explain bilateral threshold control.

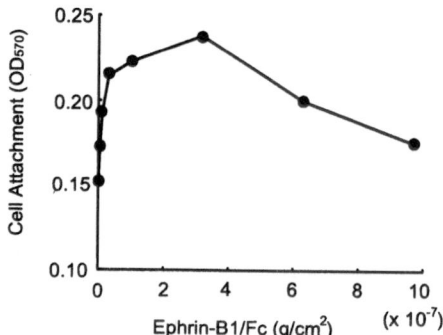

Fig. 17.4. Result of attachment assay of cells expressing Eph receptors to surfaces that express ephrin ligands at various levels [7]. Abscissa, density of ephrin ligands on a linear scale.

There is an experiment concerning attachment of endothelial cells to a substrate [7]. The cells expressing Eph receptors on their surfaces were put on a substrate that was coated with ephrin ligands. The amount of attached cells to the substrate was measured under various ligand densities (Fig. 17.4). As the ligand density was increased, the number of attached cells increased. But, when the ligand density was over a critical density, the number of attached cells decreased. Thus the cells expressing Eph receptors have an optimal ephrin density for adhesion to the ephrin-coated surface. We have to remember that Eph receptors, when they are bounded by the ligands, take on the activity of tyrosine phosphorylation. Eph receptors mediate signal transduction within the cells. It should be noted that the interaction between receptors and ligands does not cause mechanical binding between the cells and the substrate. The experimentalists elucidated that cell adhesion molecules, such as integrins, mediate the mechanical binding [7]. Thus we can reasonably consider that, when Eph receptors collide with the ligands, they produce signals and the signals induce the cell adhesion by integrin. The signal strength is assumed to be RL, where the receptor density is R, and the ligand density is L. The cell has a reference value S, and when the signal strength RL equals S, the cell begins to attach to the substrate strongly. I would like to call the reference value, S, the bilateral threshold. That is, when the signal strength is small ($RL < S$), the cell fluctuates. When the signal strength equals the reference value S ($RL = S$), the cell attaches to the substrate. When the signal strength is larger ($RL > S$), the cell again fluctuates. In other words, the cell of receptor density R attaches to the substrate whose ligand density is S/R where $|RL - S|$ is zero. We will call $|RL - S|$ repulsiveness.

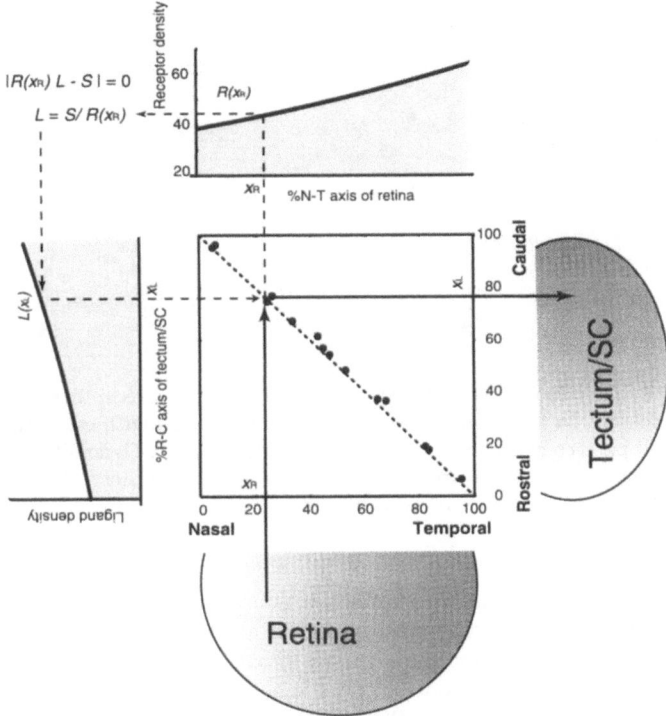

Fig. 17.5. A topographic map of retinal ganglion cells to the midbrain (tectum or superior colliculus). An axon from a site (x_R) in the retina has receptor density $R(x_R)$ as shown in the top panel. According to the servomechanism model $(|R(x_R)L - S_b| = 0)$, the axon terminal is mapped at a site where ligand density is $L = S/R(x_R)$. The value of x_L is obtained from $L = L(x_L)$ as shown in the left panel. A point (x_R, x_L) is plotted on the topographic map as shown in the central square. SC, superior colliculus. $L(x_L) = S^{1/2} \exp[a(x_L - 50)]$ and $R(x_R) = S^{1/2} \exp[a(x_R - 50)]$ where $a = 1/80$ and $S = 2500$.

17.3 Retinotectal projection of retinal ganglion cells

I will show evidence that ephrin and its receptor are molecules for cell positioning. There was a puzzling phenomenon that an optical pattern in the retina is projected on the midbrain, the tectum (fishes and birds) or the superior colliculus (SC, mammals). The problem is how the retinal neuron finds its intrinsic site in the midbrain. The breakthrough is the finding that the retina has an increasing nasal-to-temporal gradient of receptor Eph and the SC has an increasing rostral-to-caudal gradient of ephrin [1, 2]. A retinal neuron whose receptor density is low projects to a site of high ligand density in the SC, and *vice versa* (Fig. 17.5). A retinal neuron extends an axon to the

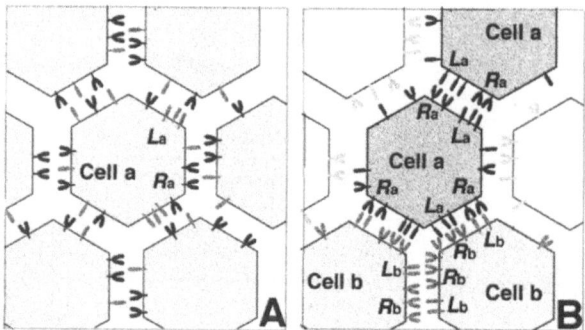

Fig. 17.6. A Assumption of co-expression of ligands and receptors in a single cell. Ligand density La and receptor density Ra are reciprocal with each other ($R_a L_a = S$). **B** Cell pattern consisting of two types of cells a and b. Generally ligand density L and receptor density R are not reciprocal with each other. There are cases in which $R_a L_a = S, R_b L_b = S, R_a L_b = S$, and/or $R_b L_a = S$.

SC and the axon terminal contacts the SC. Receptors of the axon terminal collide with ligand ephrins and the bilateral threshold control takes place. An axon terminal with receptor density R attaches to the site of ligand density of S/R, because the repulsiveness is zero ($RL - S = 0$). A retinal axon with higher receptor density R_H projects to the site of low ligand density $L = S/R_H$, and a retinal axon with lower receptor density R_L projects to the site of high ligand density $L = S/R_L$. The axons behave so that the repulsiveness becomes zero. We called this model *the servomechanism model*. In the retinotectal projection, the ephrin and its receptor work as molecules for cell positioning.

17.4 Autonomous formation of a graded cell pattern

Now, I will show that ligand ephrins and their receptor Eph's can construct a system to produce a graded cell pattern autonomously [6]. We assumed that the two types of molecules, ephrin and its receptor, are expressed simultaneously in a single cell (Fig. 17.6A). Experimentalists already showed that the retinal neuron expresses small amounts of ephrins in addition to Eph receptors [2]. So, our assumption is not unnatural. Secondly we assumed the cell expresses these two types of molecules reciprocally, that is, $RL = S$. In these assumptions the cells attached to each other under the bilateral threshold control. Generally we can expect similar cells gather with each other.

Cells with various ligand densities were randomly distributed (Fig. 17.1A). As the expression of ligand and receptor was reciprocal, the receptor densities of the cells were also distributed at random. Left and right boundaries were

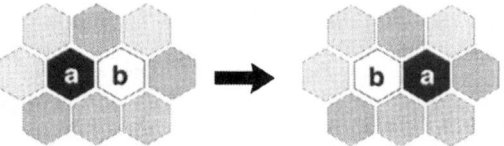

Fig. 17.7. Cell exchange in a neighboring cell pair in computer simulations. A pair of neighboring cells a and b and their neighbors are shown. The total repulsiveness $\sum_i |R_a L_i - S| + \sum_i |R_i L_a - S| + \sum_j |R_b L_j - S| + \sum_j |R_j L_b - S|$ is calculated before and after cell exchange. When the total repulsiveness decreases, the cell exchange is performed with high probability according to [6].

fixed at low and high receptor densities, respectively ($R = 0.2$ and 100). A pair of neighboring cells a and b was picked (Fig. 17.7) and the total repulsiveness around cell a and cell b was calculated (see the legend of Fig. 17.7). After exchange of the cells a and b, the total repulsiveness was again calculated. If the total repulsiveness increased, the cell exchange was reversed. When we continued the process, we obtained a graded cell pattern as shown in Fig. 17.1B. The graded cell arrangement of the simulation has the same properties of cell sorting as previously mentioned [10]. That is, when cells from two separated parts were mixed, the cells showed a pattern with large patches (Fig. 17.8C), whereas cells were intermixed when they were from neighboring parts (Fig. 17.8A).

I have to comment on actual limbs. The distal part of limbs has actively dividing cells. The limb end grows by leaving cells behind. Thus, the cells at the distal part have divided many times. It is reasonable that properties of cells are graded systematically along the axis. We made the initial condition of completely random cell pattern for the simulation shown in Fig. 17.1A, but actual cell patterns may be graded somehow. The ephrin and its receptor may work for maintenance of the graded cell pattern, rather than for producing the gradient. In addition, the ephrin or Eph receptor is not necessarily a morphogen itself. It may be more plausible that cells expressing the ephrins and Eph receptors are carriers of morphogens.

17.5 Cell patterns consisting of two types of cells

Finally I would like to indicate that the molecules for cell positioning can make interesting cellular patterns. A case is known in which two types of cells make a checkerboard pattern, that is, the cells are alternately arranged or in a kagome (star) pattern in which one cell type encloses the other cell type. These cell patterns are actually observed on the epithelial surface of the chick oviduct [5, 11, 12].

We assumed that a cell co-expresses ephrin and its receptor Eph in a single cell, but the expressions are not always reciprocal here [6]. When there are two

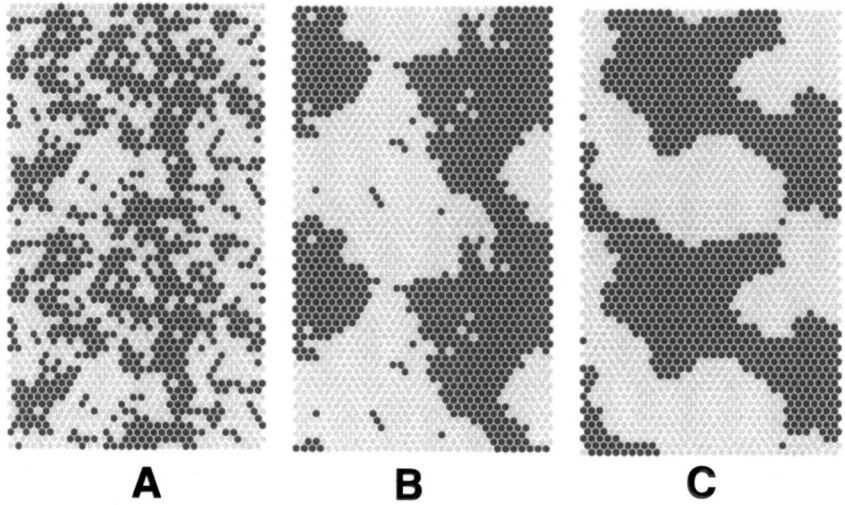

Fig. 17.8. Results of computer simulations of cell sorting. Two types of cells (cell number ratio about 1:1) were initially distributed at random in the simulation field, and simulated under bilateral threshold control for 10^7 steps. A complete periodic boundary condition was used. Each figure consists of two rectangles of the boundary condition. Cells co-express reciprocally receptors and ligands $[R, L]$ as $[50, 50]$ and $[60, 41.6]$ (A), $[45, 55.6]$ and $[65, 38.5]$ (B), $[40, 62.5]$ and $[70, 35.7]$ (C). (Italics and bold indicate reciprocal relations.)

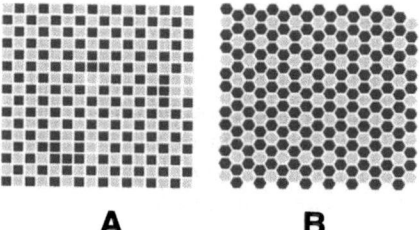

Fig. 17.9. Results of computer simulations of cell pattern formation. Patterns consist of two types of cells a and b whose $R-$ and $L-$values are shown in $[R_a, L_a]$ and $[R_b, L_b]$. Patterns are a part of the original figures. Italics and bold indicate reciprocal relations. $S = 2500$. **A** Checkerboard pattern obtained when $R_b L_a = S$ and $R_a L_b = S$, $[33, 42]$ and $[60, 76]$. **B** Kagome (star) pattern obtained when $R_b L_a = S$ and $R_a L_b, R_a L_a$ and $R_b L_b$ are not equal to S, $[33, 36]$ and $[110, 76]$. Cell number ratio is about 1 : 2 [6]. From Fig. 17.5A h and g in [6] with permission.

types of cells a and b, we have several cases, that is, the expression of ligands and receptors is reciprocal in cell a ($R_aL_a = S$), or in cell b ($R_bL_b = S$), or between cell a and cell b ($R_aL_b = S$ or $R_bL_a = S$). We examined all cases. Here we will mention two patterns, the checkerboard pattern (Fig. 17.9A) and the kagome (star) patterns (Fig. 17.9B). The checkerboard pattern was constructed when the two types of molecules in cell a and cell b are reciprocal, that is, cell a and cell b adhere to each other. The kagome pattern was constructed, for example, when $R_bL_a = S$ and other combinations (R_aL_b, R_aL_a and R_bL_b) are not equal to zero.

In conclusion, the membrane-bound ligand ephrin and its receptor Eph can construct a system of bilateral threshold control and perform cell positioning. Such types of molecules could have remarkable abilities for morphogenesis in multicelluar organisms.

References

1. Cheng, H. J., Nakamoto, M., Bergemann, A. D. and Flanagan, J. G. (1995). Complementary gradients in expression and binding of ELF-1 and Mek4 in development of the topographic retinotectal projection map. *Cell* **82**: 371-81

2. Drescher, U., Kremoser, C., Handwerker, C., Löschinger, J., Noda, M. and Bonhoeffer, F. (1995). In vitro guidance of retinal ganglion cell axons by RAGS, a 25 kDa tectal protein related to ligands for Eph receptor tyrosine kinases. *Cell* **82**, 359-370

3. Gale, N. W., Holland, S. J., Valenzuela, D. M., Flenniken, A., Pan, L., Ryan, T. E., Henkemeyer, M., Strebhardt, K., Hirai, H., Wilkinson, D. G., Pawson, T., Davis, S. and Yancopoulos, G. D. (1996). Eph receptors and ligands comprise two major specificity subclasses and are reciprocally compartmentalized during embryogenesis. *Neuron* **17**: 9-19

4. Honda, H. (1998). Topographic mapping in the retinotectal projection by means of complementary ligand and receptor gradients: a computer simulation study. *J. Theor. Biol.* **192**: 235-246

5. Honda, H., Yamanaka, H. and Eguchi, G. (1986). Transformation of a polygonal cellular pattern during sexual maturation of the avian oviduct epithelium. *J. Embr. Expl. Morph.* **98**: 1-19

6. Honda, H., Mochizuki, A. (2002). Formation and maintenance of distinctive cell patterns by co-expression of membrane-bound ligands and their receptors. *Devel. Dynamics* **223**: 180-192

7. Huynh-Do, U., Stein, E., Lane, A. A., Liu, H., Cerretti, D. P. and Daniel, T. O. (1999). Surface densities of ephrin-B1 determine EphB1-coupled activation of cell attachment through a_vb_3 and a_5b_1 integrins. *EMBO J.* **18**: 2165-2173

8. Mittenthal, J. E. and Mazo, R. M. (1983). A model for shape generation by strain and cell-cell adhesion in the epithelium of an arthropod leg segment. *J. Theor. Biol.* **100**: 443-483

9. Steinberg, M. S. (1962). Mechanism of tissue recognition by dissociated cells II: Time-course of events. *Science* **137**: 762-763

10. Wada, N., Kimura, I., Tanaka, H., Ide, H. and Nohno, T. (1998). Glycosylphosphatidylinositol-anchored cell surface proteins regulate position-specific cell affinity in the limb bud. *Dev. Biol.* **202**: 244-52
11. Yamanaka, H. I. (1990). Pattern formation in the epithelium of the oviduct of Japanese quail. *Intern J. Dev. Biol.* **34**: 385-390
12. Yamanaka, H. I. and Honda, H. (1990). A checkerboard pattern manifested by the oviduct epithelium of the Japanese quail. *Intern J. Dev. Biol.* **34**: 377-383

Part V

Spatial Pattern and Structure Formation in Ecological Systems

Biological Invasion into Periodically Fragmented Environments: A Diffusion-Reaction Model

Nanako Shigesada [1], Noriko Kinezaki[1], Kohkichi Kawasaki[2], and Fugo Takasu[1]

[1] Department of Information and Computer Sciences, Nara Women's University, Kita-Uoya Nishimachi, Nara 630-8506, Japan
telephone & fax:+81-742-20-3438
e mail:sigesada@ics.nara-wu.ac.jp
[2] Department of Knowledge Engineering and Computer Sciences, Doshisha University, Kyotanabe 610-0321, Japan

18.1 Introduction

Range expansions of invading species in homogeneous environments have been extensively studied since the pioneering work of Fisher 1937 and Skellam 1951 [1, 4, 5, 7, 8, 10]. Here we focus on range expansion of a species in a two-dimensional heterogeneous environment that is generated by segmenting an original favorable habitat into a regularly striped or criss-cross pattern as shown in Fig. 18.1:

(a) Striped fragmentation - an environment is segmented into belts in such a way that favorable and unfavorable habitats with widths l_1 and l_2 respectively, are arranged alternately [6].

(b) Criss-cross fragmentation - an environment is segmented in both horizontal and vertical axes in such a way that square-shaped favorable habitats with a side l_1 are regularly distributed to leave the criss-cross unfavorable belt with width l_2 in the background.

To deal with range expansion in these two fragmented environments, we employ an extended Fisher model, in which the intrinsic growth rate and diffusion coefficient vary depending on habitat. Let $n(x, y, t)$ denote the population density at time t and spatial coordinate (x, y). Then the extended Fisher equation for a fragmented environment is generally expressed as:

$$n_t = (D(x,y)n_x)_x + (D(x,y)n_y)_y + (\epsilon(x,y) - n)n \qquad (18.1)$$
$$(t > 0, -\infty < x, y < \infty)$$

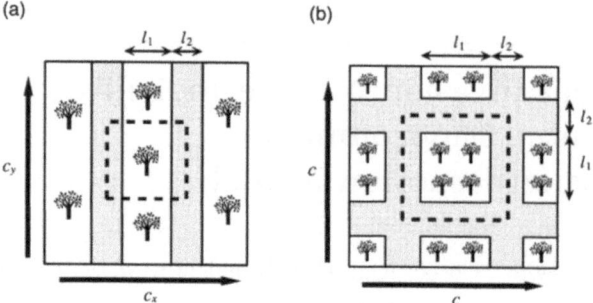

Fig. 18.1. (a) The striped environment. (b) The criss-cross environment. c_x, c_y and c are the rates of spread in the directions of the arrows. The area in the dotted square in each environment indicates a unit structure.

where $\epsilon(x,y)$ and $D(x,y)$ are the intrinsic growth rate and the diffusion coefficient, respectively. We set both of them equal to unity in the favorable habitat without loss of generality. Thus

$$\epsilon(x,y) = 1, \qquad D(x,y) = 1 \qquad \text{(in the favorable habitat)},$$
$$\epsilon(x,y) = e(\leq 1), \;\; D(x,y) = d(> 0) \;\; \text{(in the unfavorable habitat)}.$$

Obviously, in the case of $e = 1$ and $d = 1$, (18.1) is reduced to the Fisher equation for a homogeneous environment. Thus our model involves four parameters, e, d, l_1 and l_2.

In the following, we analyze (18.1) for the environments (a) and (b) to examine how the spread of organisms is affected by the four parameter values that specify the qualities of habitats, and the pattern, size and scale of habitat fragmentation.

18.2 Striped fragmentation

Figure 18.2 shows a snapshot of a numerical solution of (1.1) for the striped environment, obtained when a few propagules are initially released at the origin. The population grows faster in the favorable habitat than in the unfavorable habitat, resulting in range expansion with a wavy range front. As seen in Fig. 18.2b, the global shape of the contour map is oval-like. The rate of spread in each radial direction tends to be periodic in accordance with the spatial period so that its average speed becomes constant, which is analytically determinable [3]. Overall, the range expands self-similarly keeping its shape.

Fig. 18.3 illustrates snapshots of envelopes of the contour maps for varying sets of e and l_2 with fixed $l_1 = d = 1$. Note that these snapshots indicate the

(a)   (b)

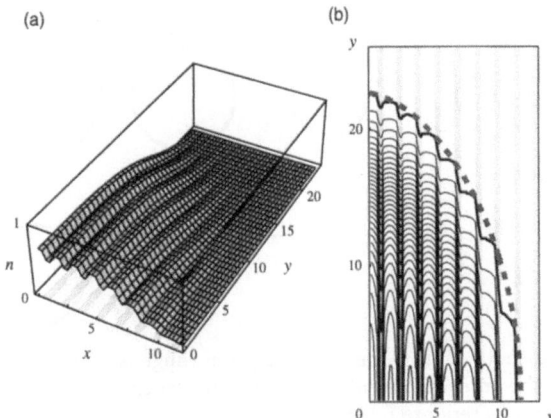

Fig. 18.2. Snapshot of numerical solution of (18.1) when a few propagules are initially released at the center of a fragmented environment. (a) The spatial pattern of population density, $n(x, y, t)$. (b) The contour map of the population density. The dashed line indicates the envelope. Parameters are chosen as $e = -0.5$, $d = 0.1$, $l_1 = 1.0$ and $l_2 = 0.5$.

relative size and shape of range at any arbitrary time after the lapse of the transient phase. Similarly, snapshots of envelopes are shown for varying sets of d and l_2 in Fig. 18.4, and for varying sets of l_1 and l_2/l_1 in Fig. 18.5. Generally speaking, the envelope shows various patterns, either nearly circular, oval-like, spindle-like or vanishing in the extreme case, depending on parameter values. All these patterns are elongated in the direction of the y axis. In other words, the speed is fastest along the stripes and slowest across the stripes.

Let us see how the pattern of the envelope is controlled by each of the four basic parameters, e, d, l_1 and l_2. As e decreases or l_2 increases, the relative size of the envelope monotonically shrinks, when other parameters are fixed at any values (Fig. 18.3). On the other hand, d exerts widely variable effects on the shape and size of the envelope depending on the values of l_2 as seen in Fig. 18.4. When $l_2 = 2.5$, the envelope changes from a spindle-like shape to more rounded shapes with concomitant decreases in size ultimately going extinct, as d increases. In contrast, when $l_2 = 1$, the envelope changes from a spindle-like pattern to a more rounded and enlarged pattern with increasing d.

More interestingly, when l_1 is increased together with l_2 with their ratio l_2/l_1 fixed (i.e., the scale of fragmentation is enlarged without changing the relative spatial pattern), the size expands monotonically (see Fig. 18.5). This scale effect is particularly enhanced when the fraction of unfavorable habitat is large. This consequence has a potential relevance to environmental management. If one wants to control a spreading pest, a finely fragmented

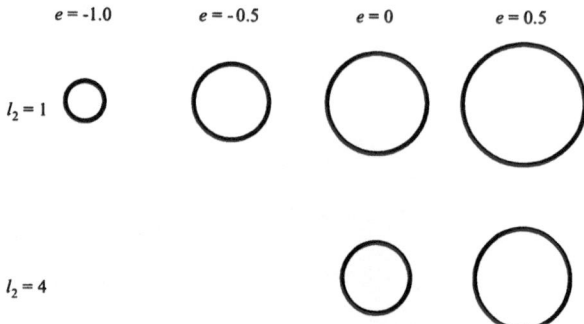

Fig. 18.3. Snapshots of envelopes of population range as a function of the intrinsic growth rate on the unfavorable habitat, e, and the width of the unfavorable habitats, l_2. The other parameters are fixed as $l_1 = 1$ and $d = 1$.

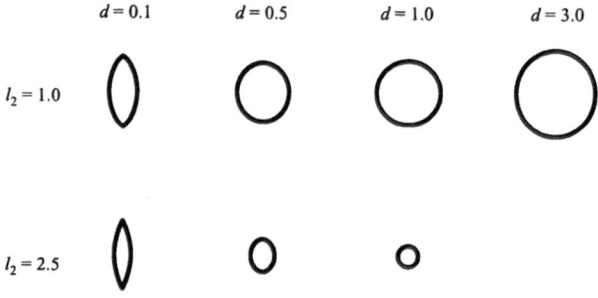

Fig. 18.4. Snapshots of envelopes of population range as a function of the diffusion coefficient on the unfavorable habitats, d, and the width of the unfavorable habitats, l_2. Other parameters are fixed as $e = -0.5$ and $l_1 = 1$.

environment should be more effective. Conversely, a largely fragmented environment is more favorable for conservation of an endangered species.

18.3 Criss-cross fragmentation

For a criss-cross environment, we numerically solve (18.1) with an initial distribution localized at the origin, and compare the results with those from the striped environment. To this end, we first specify the scale of fragmentation. Let us define unit structures for striped and criss-cross environments as enclosed by the dotted square frames in Fig. 18.1. In each unit structure, the areas of favorable and unfavorable habitats are denoted by G and N, respectively. We first examine how the speed changes with G, while the total unit area $G + N$ is fixed for both fragmented environments.

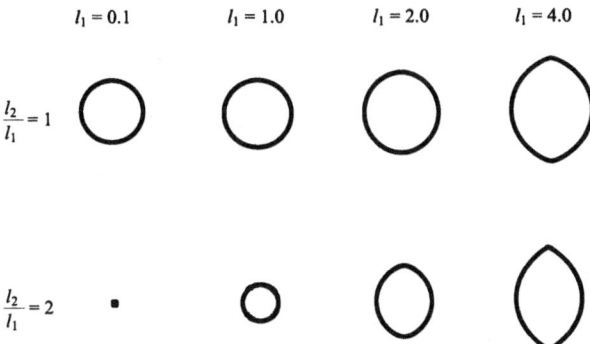

Fig. 18.5. Snapshots of envelopes of population range as a function of the width of the favorable habitat l_1 and the ratio of the widths of the unfavorable and favorable habitats, l_2/l_1. The other parameters are fixed as $d = 1$ and $e = -0.5$.

Fig. 18.6 shows the speeds against the favorable area G, when the total unit area $G+N$ is set to be 64 and 100, respectively. The solid curves represent the speed c along the x axis in the criss-cross environment, and the dotted and dashed curves are the speeds along the stripes and across the stripes (say, c_y and c_x, respectively) in the striped environment. In each curve, there is a minimum value of G, G_{min}, below which the speed is zero, so that invasion fails. Above G_{min}, the speed increases monotonically with G, reaching the maximum value when G becomes equal to the total unit area. Overall, the speed in the criss-cross environment, c_y, is similar to, but slightly higher than, the speed across the stripes, c_x. In contrast, the speed c_y along the stripes is higher than the others except where c_y is close to zero.

The values of G_{min} are different between the criss-cross and striped environments. When $G+N = 64$, G_{min} is 12.8 in the criss-cross environment, while G_{min} is 18.4 in the striped environment. When the total unit area, $G + N$, is increased to 100, G_{min} remains at the same value, 12.8, for the criss-cross environment, while G_{min} increases to 23 in the striped environment. Because G_{min} is smaller in the criss-cross environment, it seems as though invasion is easier in the criss-cross environment than in the striped environment.

To confirm whether this trend is generally true, we calculate G_{min} as a function of total unit area, $G+N$, as shown in Fig. 18.7a. G_{min} for the striped environment (dashed line) increases monotonically with the total area, while for the criss-cross environment (solid line) it increases sharply at first but quickly reaches a plateau. Consequently, the two curves cross each other at an intermediate value of $G + N$. Thus we may conclude that the criss-cross environment can be more easily invaded than the striped environment, only when the total unit area is relatively large.

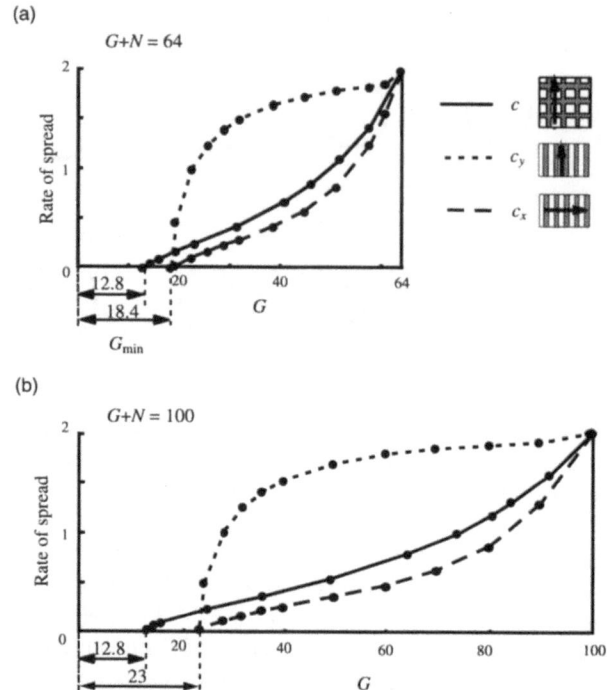

Fig. 18.6. The rates of spread as a function of the area of the favorable habitat in the unit structure, G. The solid curve represents the speed c in the criss-cross environment. Dotted and dashed curves are the speed along the stripes, c_y, and the speed across the stripes, c_x, respectively. The size of the unit structure is fixed as 64 in (a) and 100 in (b). The other parameters are fixed as $e = -10$ and $d = 0.5$.

To explain why the speeds in the two environments have such different dependencies on the scale of fragmentation, we diagramatically illustrate the unit structures of the areas, 64 and 100, in Fig. 18.7b, in which the minimum areas for survival, G_{min}, are specified in white. In the striped environment, G_{min} changes from 18.4 to 23 when the total unit area is increased from 64 to 100, whereas the width of the favorable stripe remains the same at 2.3. This means that, in the striped environment, invasion succeeds if the width of the favorable stripe is at least 2.3, no matter how large is the unfavorable area. Accordingly G_{min}, which is given by the product of 2.3 and the side length of the unit structure, increases monotonically with the total unit area. On the other hand, in the criss-cross environment, invasion succeeds if the favorable area exceeds 12.8, no matter how large is the unfavorable area. Thus G_{min} tends to plateau as $G + N$ increases.

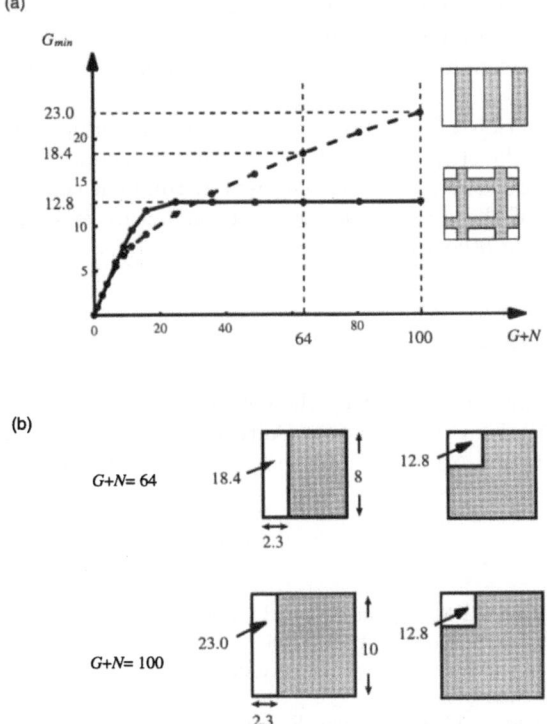

Fig. 18.7. (a) The minimum area G_{min} of the favorable habitat in a unit structure for species survival as a function of the area of the total unit structure, $G + N$. The dashed and solid curves represent G_{min} for the striped environment and the criss-cross environment, respectively. (b) The unit structures with a minimum favorable area (white) in the striped and criss-cross environments, when the total area is 64 or 100. The other parameters are fixed as $e = -10$ and $d = 0.5$.

Translating this result into the context of environmental conservation, we may say that the criss-cross fragmentation is more favorable for species survival than the striped fragmentation, if the area of the unit structure exceeds a certain value ($G + N > 30$ in the present case) and vice versa if otherwise.

18.4 Conclusions

To address the effect of habitat fragmentation on the speed and spatial pattern of species range expansion, we presented an extended Fisher equation. The model is applied to two types of fragmentation, striped and criss-cross environments. The results are summarized as follows:

1. In the striped environment, the speed transversally crossing the stripes, c_x, is smaller than the speed advancing parallel to the stripe, c_y.
2. In the criss-cross environment, the speed c is intermediate between c_x and c_y.
3. When the fraction of the unfavorable patch in a unit structure is the same for both the striped and criss-cross environments, the persistence of species is more feasible in the criss-cross environment than in the striped environment, if the area of the total unit structure is sufficiently large, and vice versa if otherwise.

Acknowledgements

This work was supported in part by the Grant-in-Aid for Scientific Research from the Japan Ministry of Education, Culture, Sports, Science, and Technology, No. 08640804 and No. 09NP1501 to N.S. and No. 13640141 to K.K.

References

1. Andow, D., Kareiva, PM., Levin, SA. and Okubo, A. (1990). Spread of invading organisms. *Landscape Ecology* 4: 177-188.
2. Fisher, R.A. (1937). The wave of advance of advantageous genes. *Ann. Eugen.* (Lond.) **7**: 255-369.
3. Kinezaki, N., Kawasaki, K., Takasu, F. and Shigesada, N. Modelling biological invasions into fragmented environments. *Theor Popul Biology.* (in press).
4. Kot, M., Lewis, MA. and Driessche, P. (1996). Dispersal data and the spread of invading organisms. *Ecology* **77**: 2027-2042.
5. Okubo, A. and Levin, SA. (2002). *Diffusion and Ecological Problems: Modern Perspectives.* Second Edition. Springer-Verlag.
6. Shigesada, N., Kawasaki, K. and Teramoto, E. (1986). Traveling periodic waves in heterogeneous environments. *Theor. Popul. Biology* **30**: 143-160.
7. Shigesada, N. and Kawasaki, K. (1997). *Biological Invasions: Theory and Practice.* Oxford Series in Ecology and Evolution. Oxford University Press. pp.205.
8. Shigesada, N. and Kawasaki, K. (2002). Invasion and species range expansion: effects of long-distance dispersal. In: *Dipsersal Ecology* (eds Bullock J, Kenward R, Hails R). Blackwell Science. pp. 350-373.
9. Skellam, JG. (1951). Random dispersal in theoretical populations. *Biometrika* **38**: 196-218.
10. Weinberger, HF. (1982). Long-time behavior of a class of biological models. *SIAM J. Math. Anal.* **13**: 353-396.

Spatial Pattern Formation in Plant Communities

Tomáš Herben[1] and Toshihiko Hara[2]

[1] Institute of Botany, Academy of Sciences of the Czech Republic, CZ-252 43 Pruhonice, and Department of Botany, Faculty of Science, Charles University, Praha, Czech Republic
E-mail: herben@site.cas.cz

[2] Institute of Low Temperature Science, Hokkaido University, Sapporo 060-0819, Japan
E-mail: t-hara@lowtem.hokudai.ac.jp

19.1 Introduction

Horizontal spatial pattern is one of the most conspicuous features of plant communities. Most air photographs of any habitat show unequal arrangement of individuals in horizontal space, aggregation of individuals belonging to one plant species, and many different types of spatial correlation if many species are involved. This horizontal spatial heterogeneity was noticed by early botanists and has spawned a large body of literature on its identification and interpretation (for a review, see [11]). Spatial patterning is one of the major research subjects in plant ecology: understanding how this ubiquitous phenomenon comes into being is likely to be one of the essential elements in understanding how plant communities are assembled and how they work. However, spatial patterns are often much noisier than many other biologically interesting patterns, highlighting the role of stochastic events that can overwhelm the underlying regularities - or questioning the existence of such regularity at all. Spatial pattern has also been invoked as having important dynamical consequences for plant communities [32, 35]. Widespread as the patterns in plant communities may be, there is still no complete consensus on the processes that generate and maintain them, and on the dynamical consequences they may have. In this paper, we will briefly review current research on this subject, and try to highlight current developments in the area.

19.2 Spatial and spatiotemporal pattern in plant communities

Plant communities are three-dimensional entities; yet most plant ecologists talking about spatial pattern understand the term as referring to the two-dimensional projection of plant bodies onto the earth's surface. This is made possible by the fact that the vertical dimension is the mere height of the (already horizontally arranged) plant bodies themselves. The vertical dimension plays an essential role in shaping horizontal spatial pattern since plants often compete by vertical growth, and the result of this competition is often death of some individuals that changes the horizontal spatial pattern as well; still the height they attain is largely determined by the biomechanical constraints of the supporting organs. In contrast, the horizontal dimensions have no such constraints and can thus exhibit a wider potential range of phenomena; as a result, they are often studied independently of the vertical dimension.

Thus the common understanding of spatial pattern refers to horizontal spatial arrangement of individuals of species present in the community, both within species, and between species; these form the widespread patchiness of a plant community. Although this kind of horizontal spatial pattern in plant communities is extremely varied, still there are several rather consistent features. Within a single species, the most common pattern is for individuals to be aggregated, i.e., closer together than expected randomly ([11]; although different patterns are often found in tropical forests, [5]). In contrast to intraspecific patterns, patterns of two or more species are much more varied, ranging from segregation of two species in space through absence of correlation up to a positive correlation. These are often deemed to be due to different types of positive functional effects of one species on another. Several approaches have been used to identify spatial pattern in plant communities. Nearest-neighbour distance analysis [6] is a standard tool to identify whether the spatial arrangement of individuals of one species is random (it almost never is) and to determine whether two species are aggregated or segregated. Different kinds of (auto)correlation techniques are also often used [40]. Finally, there is a long tradition of using variance/mean ratio analysis at different scales [11].

Most patterns found to date are, not surprisingly, scale-dependent (e.g. [43]). There seems to be a range of "correct" scales at which the study of horizontal spatial pattern is the most interesting: if the scale is too short, patchiness becomes trivial because it reflects variations in size of single individuals (aggregations of trees tend to be larger than those of small grasses); at very large scales patterns are obviously due to differences in the external environment that are independent of processes within plant communities. Quite expectedly, individual plant species differ markedly in intensity and range of their aggregation. While much of the variation in aggregation range can be ascribed to trivial differences in individual size, differences in aggregation are also commonly observed when species of similar sizes are being compared [16].

Another common feature of spatial patterns in plant communities is their dynamic nature. In most communities, the general features of spatial patterns tend to be quite persistent. However, this does not mean that patterns remain "frozen"; in most cases the overall parameters of the spatial pattern persist, but individual aggregations move or disappear, and new ones establish (Fig. 19.1). Spatio-temporal autocorrelation analysis almost always identifies decay of species autocorrelation in time. In some cases, this is accompanied by a positive spatio-temporal autocorrelation over non-zero spatial and temporal lags, indicating movement of species through physical space. While frozen patterns have been identified in plant communities ([4], see also [18]), they seem to be the exception rather than the rule.

An obvious question from the dynamical point of view is how the spatial patterns are initiated when plant cover begins to develop in an open area. In vegetation succession starting with open space, plant distribution is initially affected by environmental heterogeneities and strong stochastic events due to unequal propagule distribution. Therefore spatial pattern almost never starts developing from an initially homogeneous stand.

In summary, it should perhaps be made clear that while spatial patterns in plant communities are usually easily identifiable, as a rule there is always a lot of noise, both within a community and between communities. Patterns found in two otherwise similar communities almost always differ [11]; in one community, there is always a mixture of patches of varying sizes and statistical analysis almost never identifies one dominant "wavelength". They have much more of a stochastic nature than many other biologically interesting patterns. This makes understanding the underlying processes much more difficult. In a few cases, regular periodic structures at the level larger than a plant individual have been described, such as wave regeneration of forests, or tiger-bush in semidesert vegetation; while these patterns attract the interest of theoreticians [25, 34, 45], they represent more an exception than the rule. The overwhelming majority of patterns is much less regular: but the same overwhelming majority of plant communities shows some spatial patterns that call for an explanation.

19.3 Dynamical processes involved in spatial coupling

Since plant communities are not easy to experiment with, the role of individual generating processes for the spatial pattern found there has to be inferred by indirect means. The crucial question here is to identify whether a particular process (such as localised dispersal, for example) is operating and contributes to the spatial patterns found. Many different theoretical models, differing in the sets of assumptions they take and in formalisms they use (such as whether time and space are treated as continuous vs. discrete) can generate spatial patterns. Unfortunately qualitative correspondence of spatial patterns generated by a model and those found in the field cannot be taken as a demonstration that the particular generating process was involved in formation of that spa-

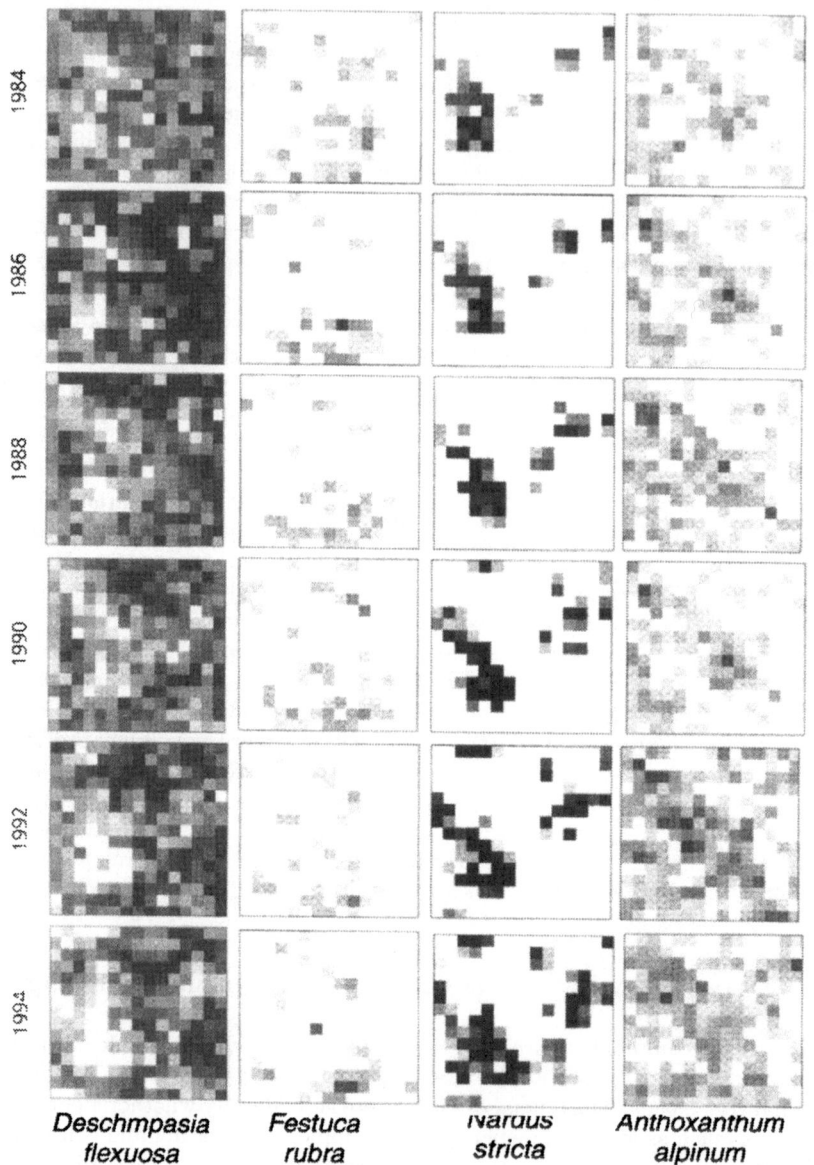

Fig. 19.1. A spatiotemporal process in a montane grassland community at the Krkonoše, Czech Republic. The plot (50x50 cm) is divided into 15 x 15 cells and is dominated by four grass species. The depth of shading is proportional to abundance of the species; white indicates absence of the species. From [18]. Copyright Cambridge University Press 2000. Reproduced with permission.

tial pattern. There are two reasons for this. Firstly, one particular generating mechanism (such as a reaction-diffusion system with an activator and an inhibitor [7]) may generate widely varying spatial patterns depending on the parameter values taken. Secondly, one particular type of spatial pattern may be generated by several, often substantially differing, theoretical models [26]. Therefore the mechanisms that account for the observed spatial patterns have to be identified at a lower level, by examination of low-level processes that operate locally and thus can produce spatial patterns. In the following section, I will briefly review major locally acting processes in plant communities.

Two major internal processes operate in plant communities that could generate the spatial patterns seen: natality (associated with dispersal) and interactions between individuals. Both these processes operate on a spatial scale that is similar to the scale over which spatial pattern of plant individuals is found, and are thus likely to contribute to the formation of these patterns.

19.4 Natality and dispersal

While plant individuals themselves usually do not move, new individuals typically establish at a distance from the mother individual, a process common to all sedentary organisms. Dispersal distance that is associated with establishment is highly variable among species, and may range from a few millimetres up to distances of hundreds kilometres, although such extremes are rare. If new individuals establish by means of seeds or other propagules that detach from the mother individual before establishment, dispersal distances are typically larger and usually follow a characteristic exponential or Gaussian decay curve. In contrast, new individuals establishing through vegetative (clonal) growth (i.e. the connection to the mother individual is maintained for a variable period after the establishment) tend to form at a small and rather constant distance to the mother individual, depending on the morphology of the connection between mother and daughter. This distance is often highly species-specific [22]. Morphology also restricts directions/angles at which new plants form; while some species have generalised morphologies with few constraints, some species possess very specialised morphologies that determine exactly where the new plantlet will establish [22]. Such morphologies have been successfully modelled, but very few attempts have been made to link architectural limitations with spatial patterns of plant communities (but see [1]). Clonal growth is a widespread feature of plants, particularly in habitats where possibilities of vertical growth are restricted by the ecological regime of the habitat, such as low productivity or predictable frequent disturbance or seasonality [39]. Clonal growth in plants has also been proposed as one of the major intrinsic processes underpinning spatial pattern in plant communities [21].

19.5 Interactions between individuals

Because plants are immobile, they interact only with their immediate neighbours. Interaction here refers to any kind of effect, both positive and negative, that one plant individual may exert on another individual; however, the majority of interactions are due to resource competition for nutrients (by roots) and for light (by above ground organs). Positive interactions (such as sheltering) also may come into play and underlie spatial patterns in some environments [13].

While the local nature of plant interactions is obvious, it is much less obvious over what distance such interactions do take place. In resource competition, the amount of resource acquired increases (often faster than in a linear fashion) with the size or surface of the resource-acquiring organ (roots for nutrients, leaves for light). Hence competition favours bigger individuals; the upper limit of their size is ultimately determined by biomechanical and physiological constraints put on the plant. As a consequence, in many plant communities there may be individuals of very different sizes; interaction *range* then depends on the size of individuals that interact [30, 38].

The outcome of the interaction is determined by several factors: (i) size and distance of neighbouring plants, and (ii) their species identity. Much of the research in the past decades has shown that the outcome of interactions is primarily determined by size and number of neighbours [10, 30, 38] whereas neighbour identity (which species they belong to) matters much less. This essentially reflects the fact that all plants require only a few resources, namely light and nutrients, and there is therefore little opportunity for intricate niche specialisation among species. Interactions among plant species are therefore not strongly species-specific; while per-unit-biomass effects of individual species on a target species may differ [9], they are often overwhelmed by size differences among individuals, both intra- and interspecific. At a population level, this often means that a species with higher maximum size is likely to win in the long term [20], no matter what the per-unit-biomass effects may be. This may be modified depending on the degree of competition asymmetry (i.e. the disproportionality in acquiring resources as a function of difference in size [14]), but the general pattern remains.

In some communities however, the maximum size is constrained by external factors (such as predictable periodical disturbance or very low productivity of the environments). If size of individuals is constrained in such a way, success of an individual in competition cannot be measured by its size; instead, it is a function of the number of offspring individuals that can establish locally and occupy the available space. This establishes a close link between processes of interaction and natality/dispersal in plants with restricted size variation. Such habitats are often occupied by vegetatively spreading (clonal) plants; these are particularly successful in placing their offspring in the horizontal direction while not investing much on vertical growth. This process has been termed horizontal competition; since individuals of different species here differ much less in size, species identity (difference in per-unit-biomass effects) are likely to become more important. The differences among species in the ability of spatial expansion are indeed larger than that of shoot competition [17]. In such a case the outcome of interaction can be captured by an "interaction matrix" telling which species prevails if two species meet in space. Long-term dynamics and spatial structure of such communities can be easily derived from the structure of the interaction matrix, namely presence and number of circular loops.

19.6 Models of spatial pattern formation in plant communities

In the past two decades, theoreticians were studying effects of these two locally-acting processes in plant communities; many models of plant communities that involved spatially-explicit processes have been published [8]. These models used different assumptions (often depending on particular biological features of the plants studied) and different formalisms, ranging from simple cellular automata to elaborate individual-based models [8]. However, most of these models have focused on finding the conditions under which different species can coexist [32], rather than how spatial pattern might be formed (but see [15, 19, 25, 28, 41, 42]). Indeed, coexistence is of crucial relevance in plants since their permanent species coexistence is limited by the low number of independent resources for which plants compete. If species coexistence cannot be attained in a model, spatial pattern will be only transient and in the end the system will become perfectly homogeneous except for size structure variation of a single dominant population. Given the spatially-constrained nature of plant interactions, formation of a spatial pattern in a model is often viewed as a means for several species of plants to coexist.

Still there have been major achievements in understanding how spatial pattern comes into being and the link between spatial pattern and species coexistence. A number of theoretical and empirical studies have established that one of the major factors accounting for the coexistence is the trade-off between colonisation ability and competitive ability in plants [12] and differentiation of plant species along this axis. This has a direct bearing on spatial pattern formation, because wave-like travelling structures often form when interactions of several species along this axis are modelled [27, 44]. The spatial pattern could then constrain the extent of interspecific interactions and lead to permanent species coexistence; dynamics of the spatially extended system is not necessarily the same as of its mean-field model approximation, particularly if the response of a plant to the mean neighbourhood is not the mean of responses to all neighbourhoods [24]. This pattern-coexistence relationship has recently been explored theoretically by a number of studies (e.g. [3, 23, 32]). Further, several studies have shown how different dispersal mechanisms (namely vegetative growth vs. seed dispersal, i.e. mechanisms differing in the range over which they operate) account for very different spatial patterns [15], again stressing the link between coexistence and spatial patterns. A third group of studies showed how small scale processes could give rise to behaviours synchronised at a large scale [19, 25].

Most of these models, however, were used heuristically to demonstrate that a particular mechanism is *able* to produce species coexistence and spatial pattern qualitatively similar to that observed in the field. This is a necessary, but not a sufficient, demonstration that it indeed does operate and thus can be held responsible for the spatial patterns found. The latter would require, in addition, a good parameterisation of the model and a quantitative comparison of the predicted spatial and, if possible, spatio-temporal patterns. Surprisingly enough, such models are infrequent in the ecological literature; this is perhaps because it is a marginal area which is too empirical for theoreticians, but too theoretical for field ecologists (but see [24, 32, 41, 44]).

19.7 Field evidence of dynamical effects of spatial pattern

An essential feature of spatially extended heterogeneous systems is the bi-directional relationship between spatial pattern and dynamics: not only is the pattern formed as a consequence of a certain generating mechanism, but it also constrains the ways species in the system can interact. However, it still remains to be determined to what extent this theoretical result applies in the field.

Silvertown et al. [36] used a field-derived transition matrix of five clonal species and a simple cellular automaton model to simulate their dynamics starting from several initial configurations that differed in spatial arrangement of species while their overall frequencies were constant. Different spatial arrangements resulted in qualitatively different outcomes, both over short and long time scales. This model prediction of a pattern-on-process effect has been tested directly by several recent experimental studies [33, 37]. They manipulated spatial patterns of an experimental multispecies community via changing initial spatial aggregation (by sowing/seeding). This essentially amounts to establishing communities with identical species proportions but varying average local neighbourhood composition. When dynamics of these communities were compared, initial spatial arrangement often had significant effects on the outcome of the experiment.

19.8 Spatial pattern formation in heterogeneneous environments

In addition to the processes above that act even in completely homogeneous physical environments, spatial patterns of plant communities are likely to be

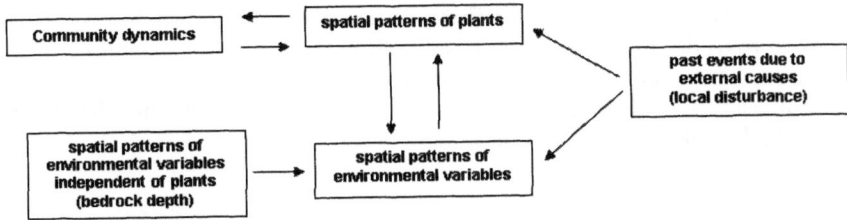

Fig. 19.2. Different pathways through which external variation in space or space-time may determine spatial patterns.

influenced by heterogeneity in environmental conditions. While environmental heterogeneity is not a parameter commonly included in theoretical analyses of spatial pattern-coexistence relationships, it is a most widespread feature of real communities that likely has major impact on spatial pattern in plants. Virtually every study that has attempted to quantify environmental heterogeneity has found some significant variation of ecologically relevant variables (for a systematic approach to the issue see e.g. [2]); this heterogeneity may either be stable or may vary in time. Environmental heterogeneity may be due to (i) stable patterns independent of vegetation (such as bedrock depth or altitude), (ii) historical events (such as time since last disturbance), or (iii) the effects of vegetation itself through some feedback effect (such as soil development during succession on volcanic substrata). While these distinctions are conceptually simple, without carefully designed studies it is very difficult to separate the kinds of heterogeneity in the field (for a modelling approach to a similar problem, see [45]). The main methodological difficulties involved are twofold: (i) it is rarely known what parameters and what ranges of their values are relevant for plant distribution in the field, (ii) without manipulative experiments it is difficult to establish the cause-and-effect structure in plant-environment correlations. In particular, correlation of plant distribution with environmental variation may be due both to effects of plants themselves (which does not bring any external forcing into the system) and to unidirectional effects of external variables. Nevertheless, disentangling the effects of extrinsic environmental heterogeneity and internal processes of local dispersal and local interaction is a prerequisite for deeper understanding of spatial pattern formation in plant communities (Fig. 19.2) and it is surprising how little systematic attention has been paid to it.

Out of the three types of heterogeneity, patchy disturbance is most likely to produce small-scale species correlations that are so typical of plant communities and thus to mimic effects of internal dynamics within the community. Stable patterns independent of vegetation (such as bedrock depth), are perhaps less relevant for the discussion here mainly because they underlie patterns that remain "frozen" in time; however because of the methodological difficulties involved in separation of the three sources of environmental heterogeneity they have to be taken into account as well. Further, several theoretical studies have shown that this heterogeneity may interact in a non-trivial way with the internal dynamical processes that generate spatial patterns themselves ([1, 29], J. Molofsky pers. comm.).

19.9 Conclusions

Spatial patterns in plant communities are strong and persistent. Current knowledge of functioning of plant communities supports the view that locally-acting processes that are known to generate spatial pattern in theory do operate in plant communities. This does not necessarily mean that the pattern we find is indeed generated by them. Surprisingly little work has been done to identify whether the field-parameterised versions of models involving these mechanisms lead to predictions that are qualitatively and quantitatively correct. In contrast, recent research did show that another key prediction, effect of pattern on dynamics, can be experimentally demonstrated.

The variety of spatial patterns found and the high degree of noise in these patterns seems to indicate that there is not a general mechanism accountable for these patterns; dominant forces are likely to vary to some extent from one habitat type to another. In addition, stochastic events due to low numbers (both in space and time) have large effects on the presence and subsequent spatial distribution of species; long-range dispersal events are particularly prone to generate this kind of effect. Further, there is a large (and not always fully known and appreciated) amount of spatial pattern in environmental parameters that underlie many plant spatial patterns, particularly at larger scales. Models have also shown that these may also interact in a non-trivial fashion with patterns generated by within-community processes.

Acknowledgements

We thank Toshio Sekimura for organizing the conference, "Morphogenesis and Pattern Formation in Biology" in Nagoya in 2002 and for the possibility to present a contribution there on which this paper is based; the stimulative atmosphere of people from many different specialities was essential. We also thank Deborah Goldberg and Jun-ichirou Suzuki for their comments on earlier drafts of this paper which were essential for the shaping of this contribution. The first author was partly supported by GAČR grants 206/02/0953 and 206/02/0590.

References

1. Bell, A. D. (1984). Dynamic morphology: a contribution to plant population ecology. In: Dirzo R. and Sarukhán J. (eds.): *Perspectives in Plant Population Ecology*, Sinauer, Sunderland, pp. 48-65.
2. Bell, G., Lechowicz, M. J. and Waterway, M. J. (2000). Environmental hetero-geneity and species diversity of forest sedges *J. Ecol.* **88**: 67-87
3. Bolker, B. and Pacala, S. W. (1999). Spatial moment equations for plant com-petition: understanding spatial strategies and the advantages of short dispersal. *Am. Nat.* **153**: 575-602.
4. Casparie, W. A. (1972). Bog development in Southeastern Drente (The Nether-lands). *Vegetatio* **25**, 1-271.
5. Condit, R. et al. (2000). Spatial patterns in the distribution of tropical tree species. *Science* **288**: 1414-1418.
6. Cressie, N. A. C. (1991). *Statistics for Spatial Data.* J. Wiley, New York.
7. Cronhjort, M. B. (2000). The interplay between reaction and diffusion. In: Dieck-mann U., Law R. and Metz J.H.J. (eds.) *The Geometry of Ecological Interac-tions: simplifying spatial complexity*, Cambridge University Press, pp.151-170.
8. Czárán, T. (1998). *Spatiotemporal Models of Population and Community Dy-namics.* - Chapman & Hall, London.
9. Goldberg, D. E. and Landa, K. (1991). Competitive effect and response: hier-archies and correlated traits in the early stages of competition. - *J. Ecol.* **79**: 1013-1030
10. Goldberg, D. E. and Werner, P. A. (1983). Equivalence of competitors in plant communities: a null hypothesis and a field experimental approach. *Amer. J. Bot.* **70**: 1098-1104.
11. Greig-Smith, P. (1979). Pattern in vegetation. - *J. Ecol.* **67**: 755-779.
12. Grubb, P. J. (1977). The maintenance of species richness in plant communities: the importance of the regeneration niche. *Biol. Rev.* **52**: 107-145.
13. Hacker, S. D. and Gaines, S. D. (1997). Some implications of direct positive interactions for community species diversity *Ecology* **78**: 1990-2003.
14. Hara, T. and Wyszomirski, T. (1994). Competitive asymmetry reduces spatial effects on size structure dynamics in plant populations. - *Ann. Bot.* **73**: 285-297.
15. Harada, Y. and Iwasa, Y. (1994). Lattice population dynamics for plants with dispersing seeds and vegetative propagation. *Researches on Population Ecology* **36**: 237-249.
16. Herben, T., During, H. J. and Krahulec, F. (1995). Spatiotemporal dynamics in mountain grasslands: species autocorrelations in space and time. *Folia Geobot. Phytotax.* **30**, 185-196.
17. Herben, T. and Hara, T. (1997). Competition and spatial dynamics of clonal plants. In: de Kroon H. and van Groenendael J. (eds.) *The Ecology and Evolution of Clonal Plants*, Backhuys Publ., Leiden, pp. 331-357.
18. Herben, T, During, H. J. and Law, R. (2000). Spatio-temporal patterns in grass-land communities. In: Dieckmann U., Law R. and Metz J.H.J. (eds.) *The Ge-ometry of Ecological Interactions: Simplifying Spatial Complexity*, Cambridge University Press, pp.48-64.
19. Iwasa, Y. and Kubo, T. (1995). Forest gap dynamics with partially synchronized disturbances and patch age distribution. *Ecol. Model.* **77**: 257-271

20. Keddy, P.A, Twolan-Strutt, L. and Wisheu, I. C. (1994). Competitive effect and competitive response in rankings in 20 plants: are they consistent across environments? *J. Ecol.* **82**: 635-643.

21. Kershaw, K. A. and Looney, H. H. (1985). *Quantitative and Dynamic Plant Ecology.* 3rd ed., Edward Arnold, London.

22. Klimeš, L. et al. (1997). Clonal plant architecture: a comparative analysis of rom and function. In: de Kroon H. and van Groenendael J. (eds.) *The Ecology and Evolution of Clonal Plants*, Backhuys, Leiden, pp. 1-30.

23. Law, R. and Dieckmann, U. (2000) Moment approximations of individual-based models. In: Dieckmann, U., Law, R., and Metz, J. A. J. (eds.) *The Geometry of Ecological Interactions: Simplifying Spatial Complexity.* Cambridge University Press, Cambridge, pp. 252-270.

24. Law, R., Purves, D. W., Murrell, D. J. and Dieckmann, U. (2001). Causes and effects of small scale spatial structure in plant populations. In: Silvertown J and Antonovics J (eds.) *Integrating Ecology and Evolution in a Spatial Context*, Blackwell, Oxford, pp. 21-44.

25. Lejeune, O. and Tlidi, M. (1999). A model for the explanation of vegetation stripes (tiger bush). *J. Veg. Sci.* **10**: 201-208.

26. Lepš, J., Goldberg, D. E., Herben, T. and Palmer, M. (1999). Mechanistic explanations of community structure: Introduction. *J. Veg. Sci.* **10**: 147-150.

27. Molofsky. J,, Durrett, R., Dushoff, J., Griffeath, D. and Levin, S. (1999). Local frequency dependence and global coexistence. *Theor. Popul. Biol.* **55**: 270-282

28. Molofsky, J., Bever, J. D., Antonovics, J. and Newman, T. J. (2002). Negative frequency dependence and the importance of spatial scale. *Ecology* **83**: 21-27.

29. Muko, S., Iwasa, Y. (2000). Species coexistence by permanent spatial heterogeneity in a lottery model. *Theor. Popul. Biol.* **57**: 273-284.

30. Pacala, S. W. and Silander, J. A. (1985). Neighborhood models in plant population dynamics. I. Single species models of annuals. *Am. Nat.* **125**: 385-411.

31. Pacala, S. W., Canham, C. D., Saponara, J., Silander, J. A., Kobe, R. K. and Ribbens, E. (1996). Forest models defined by field measurements: estimation, error analysis and dynamics. - *Ecol. Monogr.* **66**: 1-43.

32. Pacala, S. W. and Levin, S. A. (1996). Biologically generated spatial pattern aned the coexistence of competing species. In: Tilman D. and Kareiva P. (eds.) *Spatial Ecology, Monographs in Population Biology*, Princeton University press, pp. 304-232.

33. Rejmanek, M. (2002). Intraspecific aggregation and species coexistence. *Trends in Ecology and Evolution* **17**: 209-210.

34. Sato, K. and Iwasa, Y. (1993). Modeling of wave regeneration (shimagare) in subalpine Abies forests: Population dynamics with spatial structure. *Ecology* **74**: 1538-1550.

35. Silvertown, J. and Wilson, J. B. (2000). Spatial interactions among grassland plant populations. In: Dieckmann U., Law R., Metz J.A.J. (eds.) *The Geometry of Ecological Interactions: Simplifying Spatial Complexity.* Cambridge University Press, Cambridge, pp.28-47.

36. Silvertown, J., Holtier, S., Johnson, J. and Dale, P. (1992). Cellular automaton models of interspecific competition for space: the effect of pattern on process. *J. Ecol.* **80**: 527-534.

37. Stoll, P. and Prati, D. (2001). Intraspecific aggregation alters competitive interactions in experimental plant communities. *Ecology* **82**: 319-327.

38. Stoll, P. and Weiner, J. (2000). A neighbourhood view of interactions among individual plants. In: Dieckmann U., Law R. and Metz J.H.J. (eds.) *The Geometry of Ecological Interactions: Simplifying Spatial Complexity*, Cambridge University Press, pp.11-27.
39. Suzuki, J. and Hutchings, M. J. (1997). Interactions between shoots in clonal plants and the effects of stored resources on the structure of shoot populations. - In: de Kroon, H. and van Groenendael, J. (eds.): *The Ecology and Evolution of Clonal Plants*, pp. 311-330. Backhuys Publishers, Leiden.
40. Upton, G. J. G. and Fingleton, B. (1985). *Spatial Data Analysis by Example. Vol. I. Point Pattern and Quantitative Data.* Wiley & Sons, Chichester.
41. Wiegand, K., Schmidt, H., Jeltsch, F., and Ward, D. (2000). Linking a spatially-explicit model of acacias to GIS and remotely-sensed data. *Folia Geobot* **35**: 211-230
42. Wiegand, K., Henle, K., and Sarre S. D. (2002) Extinction and spatial structure in simulation models. *Conserv. Biol.* **16**: 117-128
43. Wilson, J. B. (1995). Testing for community structure: A Bayesian approach. *Folia Geobotanica & Phytotaxonomica* **30**: 461-469
44. Wissel, C. (2000). Grid-based models as tools for ecological research. In: Dieckmann U., Law R. and Metz J.H.J. (eds.) *The Geometry of Ecological Interactions: Simplifying Spatial Complexity*, Cambridge University Press, pp.94-114.
45. Yokozawa, M., Kubota, Y. and Hara, T. (1999). Effects of competition mode on the spatial pattern dynamics of wave regeneration in subalpine tree stands. *Ecol. Model.* **118**: 73-86.

The Mode of Competition and Spatial Pattern Formation in Plant Communities

Masayuki Yokozawa

National Institute for Agro-Environmental Sciences, Tsukuba 305-8604, Japan
myokoz@niaes.affrc.go.jp

20.1 Introduction

The spatial patterns emerging in plant communities are outcomes of differences in growth rates of individuals. The interplay of endogenous and exogenous factors affecting plant growth generates the differences. The endogenous factors in plant populations imply processes of two kinds: i) neighborhood/spatial effects and ii) the mode of competition between individuals [21, 27]. The competition among plants is usually asymmetric, i.e. larger plants capture more resources, because they can pre-empt resources from their smaller neighbors. The higher the degree of competitive asymmetry, the greater the reduction in individual relative growth rate (absolute growth rate per unit biomass) for smaller individuals. Then, variation in growth rates, caused by differences in local neighborhood conditions, leads to individual size differences and spatial patterns [1]. Exogenous factors can cause disturbances in the environment of individual plants, for example, typhoon, abrupt climate change [12, 15, 16, 17, 18, 22, 23]. Falling down of large trees can generate canopy gaps, where there is lower or no canopy. Canopy gaps allow new plants to grow. The gap dynamics and its spatial distribution, therefore, are related to the emergence and maintenance of species diversity in forest ecosystems. As another example of spatial patterns in plant community, the wave-shaped pattern ('Shimagare' in Japanese) is well-known. There, we can see stripes of dieback and zones of growing trees appear alternately almost perpendicular to the mountain slope, resulting in a wave-shaped spatial pattern [5, 10]. Many ecological and physiological studies have been made on the wave regeneration and it has been pointed out that the prevailing upward unidirectional wind along the slope causes such wave-shaped spatial pattern. However, the wave-shaped spatial pattern is restricted almost to pure *Abies* stands, conifer trees, in sub-alpine regions. In sub-alpine regions, *Betula*, broad-leaved trees, also form pure stands, but the wave-shaped spatial pattern has never been observed.

The objective of this study is to clarify how the interplay of endogenous and exogenous factors affects the spatial pattern formation of plant communities. We studied spatial pattern formation in plant communities using a two-dimensional lattice model and incorporating competition and disturbance regimes.

20.2 Model

20.2.1 Competition among individuals

Firstly, we model the competition process between individual plants on a two-dimensional lattice [32, 33]. Let us denote the biomass of an individual at site $\mathbf{x}_i$ as $S_n(\mathbf{x}_i)$ $(i = 1, 2, \ldots, N)$. The growth in biomass at the discrete time step, n, is governed by the following equation:

$$\Delta S_n(\mathbf{x}_i) = S_{n+1}(\mathbf{x}_i) - S_n(\mathbf{x}_i)$$

$$= S_n(\mathbf{x}_i) \left[c_0 - c_1 S_n(\mathbf{x}_i) - \sum_{\mathbf{x}_j \in \Lambda(\mathbf{x}_i)} W\left(S_n(\mathbf{x}_j), S_n(\mathbf{x}_i)\right) \right] \quad (20.1)$$

where $\Lambda(\mathbf{x}_i)$ is the surrounding eight neighbors of an individual at $\mathbf{x}_i$, $c_0(> 0)$ is the potential maximum relative growth rate and $c_0/c_1(> 0)$, the potential maximum biomass of all the individuals under non-competing conditions. $W(S_n(\mathbf{x}_j), S_n(\mathbf{x}_i))$ represents the competition function between the i-th and j-th individuals.

In order to define the competition function, we use the 'zone of influence' concept. Since a plant grows by acquiring resources from a zone of influence depending on its size, let us consider the zone of influence illustrated in Fig. 20.1(a). We assume the partition of resources between individuals is given by the following equation:

$$\frac{a(\mathbf{x}_i, \mathbf{x}_j)}{r(\mathbf{x}_i, \mathbf{x}_j)} = \frac{S_n(\mathbf{x}_i)^\beta}{S_n(\mathbf{x}_i)^\beta + S_n(\mathbf{x}_j)^\beta} \quad (20.2)$$

where $r(\mathbf{x}_i, \mathbf{x}_j)$ is the available resource in the overlapping zone of influence between the i-th and j-th individuals, $a(\mathbf{x}_i, \mathbf{x}_j)$ is the amount of resource the i-th individual can capture from the overlapping zone, and $\beta(\geq 1)$, represents the degree of asymmetry in competition, i.e. the disproportionality of resource partition. It has been shown both theoretically and by field data analysis that the degrees of asymmetry in competition are different among plant species: coniferous trees and grasses tend to show lower asymmetric competition; broad-leaved trees and forbs tend to show higher asymmetric competition [7, 11, 14, 24, 31]. Such differences can be expressed by the values of β in the model.

(a)

(b)

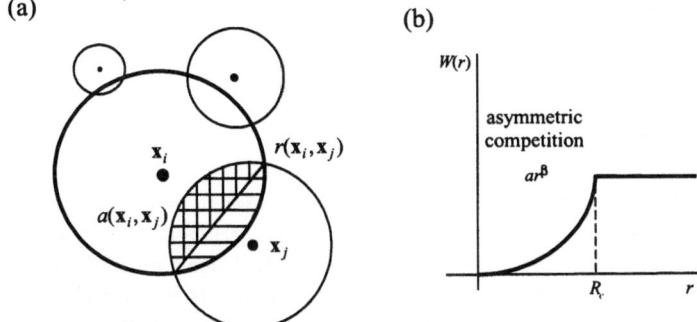

Fig. 20.1. (a) Zones of influence of individuals, from which each plant grows by acquiring resources. Plants begin to compete when their zones of influence overlap. $r(\mathbf{x}_i, \mathbf{x}_j)$ denotes the amount of resource in the overlapping zone of influence. (b) Competition function form, Eq. (20.3); $W(r)$ [$r = S_n(\mathbf{x}_j)/S_n(\mathbf{x}_i)$]. The interval, $0 < r < R_c$ corresponds to the asymmetric competition regime; the interval, $R_c < r$, the size-independent competition regime.

This relationship implies that only individual biomass ratios determine the interaction with competitors, not the absolute biomass. Since there is an upper limit to the extent of resources captured by an individual due to the finiteness of body size (canopy, stem, and root system), the competition interference is restricted. Thus, the competitive effect should become saturated as the difference in the individuals' size increases. Consequently, the competition function can be defined as follows:

$$
W\left(S_n(\mathbf{x}_j), S_n(\mathbf{x}_i)\right) = \begin{cases} a(\mathbf{x}_i, \mathbf{x}_j)\left(\dfrac{S_n(\mathbf{x}_j)}{S_n(\mathbf{x}_i)}\right)^{\beta} & \text{if } 0 \le \dfrac{S_n(\mathbf{x}_j)}{S_n(\mathbf{x}_i)} \le R_c \\ a(\mathbf{x}_i, \mathbf{x}_j)R_c^{\beta} & \text{if } R_c < \dfrac{S_n(\mathbf{x}_j)}{S_n(\mathbf{x}_i)} \end{cases} \quad (20.3)
$$

where $R_c(> 1)$ denotes the upper limit of the biomass ratio for asymmetric competition. The functional form of $W(\cdot)$ in Eq. (20.3) is shown in Fig. 20.1(b).

In the dynamics of competition among individuals, we further consider death and birth processes as follows: (i) death process; if $S_n(\mathbf{x}_i) \ge s_d$ (constant), the individual at $\mathbf{x}_i$ dies with probability p_d due to senescence. In addition, we assume that an individual dies instantaneously irrespective of its size if the growth rate, $\Delta S_n(\mathbf{x}_i)$, is negative due to competition with the surrounding neighbors. (ii) Birth process; a new individual appears with probability p_b at an empty lattice point after the death of an individual. The size of the new born individual was identically set to s_0.

20.2.2 Disturbance regimes

Next, as exogenous factors against a plant community (i.e. disturbance), we introduce two rules into the model.

Episodic disturbance

When an individual of size $S_n(\mathbf{x}_i)$ dies due to episodic external forcing, e.g. typhoon, individuals belonging to the surrounding neighbors, $\Lambda(\mathbf{x}_i)$, of the dead individual and being smaller than the central dead individual also die. This mimics simultaneous fall of surrounding smaller trees caused by the fall of the central dead tree, i.e. gap formation due to natural disturbances. We assume that only individuals larger than s_d can make a gap.

Unidirectional disturbance

Let us consider the prevailing wind as blowing in one direction constantly irrespective of time. We here assumed that the prevailing wind comes from the top left of the square lattice (i.e. northwest) and that an individual dies with probability one if it is taller than the average of the three nearest neighbors in the windward direction by more than a critical height difference dh as follows:

$$S_{n+1}(\mathbf{x}_i) = 0 \text{ if } S_n(\mathbf{x}_i) - \frac{S_n(\mathbf{x}_{i1}) + S_n(\mathbf{x}_{i2}) + S_n(\mathbf{x}_{i3})}{3} > dh \qquad (20.4)$$

where $\mathbf{x}_{i1}, \mathbf{x}_{i2}, \mathbf{x}_{i3} \in \Lambda(\mathbf{x}_i)$ are in the windward direction, i.e. west, north and northwest. This mimics the death process of trees due to wind effects (e.g. winter desiccation, summer cooling), which is the same process as assumed in the study of Sato and Iwasa [20].

20.3 Results and discussion

Simulations were carried out on a 100×100 ($N = 10000$) square lattice with periodic (wrapped-around) boundary conditions, i.e. opposite edges of the lattice were joined to form a torus. In the simulations, we studied the following two types of competing populations: a lower asymmetrically competing population with $\beta = 1$ and a higher asymmetrically competing population with $\beta = 5$. As the initial condition, a random spatial configuration of individuals was used with a random initial size distribution given by $S_0(\mathbf{x}_i) = s_0(1 + \xi(\mathbf{x}_i))$, where ξ denotes a uniform random number ranging from -0.5 to +0.5 and the initial mean size s_0 (= the size of the new born individual) was set to 5. The parameters c_0 and c_1 were set to 0.1 and 50 respectively for all the simulations.

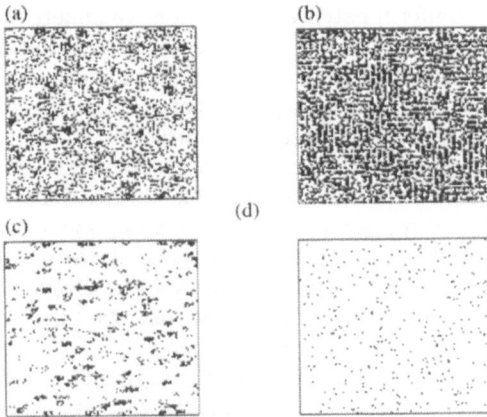

Fig. 20.2. Snapshot spatial configurations of individual size. Higher asymmetric competing population with gap formation (a) and without gap formation (b). Lower asymmetric competing population with gap formation (c) and without gap formation (d). Darker points represent canopy gap sites.

20.3.1 Episodic disturbance

Snapshot spatial configurations of individual size are shown in Fig. 20.2. The size of local clumps (or patches) formed by living individuals was larger under lower asymmetric competition (Fig. 20.2(c), (d)) than under higher competition (Fig. 20.2(a), (b)). This can be quantitatively confirmed by the patchiness index and semivariogram [32]. Individuals larger than the surrounding ones at the initial time were able to survive and depressed the growth of smaller ones. Therefore, surviving individuals were distributed randomly in space. The difference in the spatial configuration pattern of individual size became small between the higher asymmetrically competing population and the lower one when the gap formation process was included (Fig. 20.2(a), (c)).

These results may be most relevant to forests: climax forests consist of a number of local patches with different size distributions or developmental stages (i.e. mosaic structure). It has been demonstrated that the mosaic structure is different among forests. For example, the mosaic structure of temperate hardwood forests is patterned by fine-scale patches (0.1-1.0 ha) [15, 16, 17, 18, 23]. In contrast, northern coniferous forests consist of large and relatively uniform patches (0.1-100000 ha) [12, 22]. It has been hypothesized that the difference in mosaic structure among forests is generated by the difference in the magnitude (fine-scale vs. large-scale gaps) and frequency (long vs. short intervals) of natural disturbances. In addition to this hypothesis, the present results show that the mosaic structure of forests is also influenced by the mode of competition between neighboring trees. The spatial patterns in the present model show finer mosaic patches under higher asymmetric competition and more uniform patches under lower asymmetric competition. Our results therefore suggest that the relatively homogeneous spatial pattern of coniferous forests is likely to be due to lower asymmetric competition between coniferous trees [12, 22]. Fine-scale patches of temperate hardwood forests are likely to be due to higher asymmetric competition.

Spatial heterogeneity such as mosaic structure results in habitat diversity in terms of the regeneration of juveniles and facilitates the coexistence between shade-tolerant climax species and light-demanding pioneer species in forests [29]. Therefore, it has been suggested that natural disturbances, which bring about spatial heterogeneity, contribute to species diversity in forests [2, 3, 9, 13]. The present simulations also show that the effect of gap formation on spatial pattern dynamics is larger under lower asymmetric competition than under higher competition. Under higher asymmetric competition, spatial pattern dynamics are similar in both the gap formation and non-gap cases. The difference in the effect of competition mode on spatial pattern dynamics is smaller in the gap formation case than in the non-gap case. In other words, the gap formation process (i.e. disturbances) under lower asymmetric competition has increased the degree of asymmetry in competition in terms of spatial pattern. Therefore, spatial heterogeneity (structural diversity), which is to lead to high species diversity, is more enhanced by gap formation in plant communities under lower asymmetric competition than in those under higher asymmetric competition.

20.3.2 Unidirectional disturbance

Spatial patterns of individual size under unidirectional disturbance (Eq. (20.4)) are shown in Fig. 20.3. For the case without competition, the wave-shaped spatial patterns (Shimagare) emerged at a critical height difference. After simulations with changing the critical height difference, the wave fronts of spatial pattern were almost linear at $dh = 10$ (Fig. 20.3(a)). The shape pattern became more blurred with an increase in dh [33]. The spatial patterns of individual size for the case with competition are shown in Fig. 20.3(b) and (c). The wave-shaped spatial pattern was much clearer under lower asymmetric competition than under higher competition. In lower asymmetrically competing populations, the wave fronts were linear and perpendicular to the wind direction. For the other cases with the different critical height difference, the wavefronts of spatial patterns were not linear but more or less wound. In higher asymmetrically competing populations, all the spatial patterns wound. When the intensity of competition, $a(\mathbf{x}_i, \mathbf{x}_j)$, was high, wave-shaped spatial pattern did not emerge.

From Fig. 20.3(d), the domain of wave-shaped spatial pattern is restricted to the weaker intensity of competition and the lower degree of asymmetry in competition. The transient boundary varied with the critical height difference dh. We can assume that the intensity of disturbance increases with a decrease in dh. As dh was decreased, the domain of wave-shaped spatial pattern got

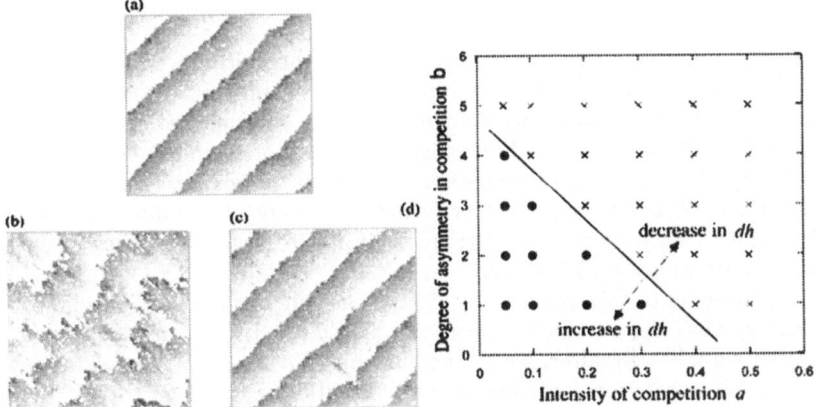

Fig. 20.3. Snapshot spatial configurations of individual size without competition (a), with higher asymmetric competition (b) and lower asymmetric competition (c). The prevailing wind comes from the top left. If an individual is larger than the average of the three windward nearest neighbours by more than a critical value, dh, the individual dies. Brighter points represent smaller individuals. (d) Phase diagram of spatial pattern for the intensity of competition and the degree of asymmetry in competition at the critical height difference $dh = 10$. The $\bullet$ symbols indicate the cases when wave-shaped pattern emerged and $\times$ the cases when wave-shaped pattern did not emerge. The transient boundary moved with the critical height difference as shown by the arrows.

larger. This implies that the wave-shaped spatial pattern is more likely to emerge under more intense disturbance. It is also shown that the wave-shaped spatial pattern is more likely to emerge under lower asymmetric competition than under higher competition, irrespective of the intensity of competition and disturbance. Both *Abies* and *Betula* are dominant trees in subalpine regions. However, wave-shaped spatial pattern has been observed only in *Abies* stands but not in *Betula* stands. By taking into account the species-specific degree of asymmetry in competition (lower asymmetric competition for *Abies* and higher asymmetric competition for *Betula* [7, 11, 14, 24, 31]), we can explain this phenomenon based on the results.

Under higher asymmetric and intense competition, the spatial pattern with unidirectional disturbance is similar to that with episodic disturbance. Therefore, the spatial pattern under higher asymmetric competition is little affected by the disturbance regime (unidirectional prevailing winds as deterministic disturbance vs. gap formation caused by stochastic disturbance). On the contrary, the spatial pattern under lower asymmetric competition is greatly affected by the disturbance regime: wave-shaped regular spatial pattern under unidirectional prevailing winds as deterministic disturbance and random spatial pattern under gap formation caused by stochastic disturbance.

20.3.3 Vertical structure

Higher asymmetric competition can amplify the initially small differences in seedling size and result in large differences in the fate of individual plant species. Populations with small size, late germination or slow growth would be excluded faster with increasing intensity of asymmetric competition [4, 6, 28, 34]. Thus higher asymmetric competition between individuals can generate a more multi-layered hierarchical structure of size distribution (size hierarchies) than lower competition, resulting in large differences in fitness among individuals [8, 19, 25, 26].

To investigate this analytically [30], we use the mean-field approximation assuming that the parameters, $a(\mathbf{x}_i, \mathbf{x}_j)$, in Eq.(20.3) are constant among all the individuals (hereafter, we define the constant as α). Then, a stationary homogeneous solution (uniform biomass distribution) is given by:

$$S(\mathbf{x}_i) = S_s \equiv \frac{c_0 - (N - 1)\alpha}{c_1} \tag{20.5}$$

for all the individuals. The linearized equation for small perturbation around the above uniform distribution, $\delta S(\mathbf{x}_i) = S(\mathbf{x}_i) - S_s$, is given by

$$\delta\dot{S}(\mathbf{x}_i) = \{c_0 - 2c_1 S_s - (N-1)\alpha(1-\beta)\}\,\delta S(\mathbf{x}_i) - \alpha\beta\sum_{j\neq i}\delta S(\mathbf{x}_j)$$

$$= \{-c_0 + (N-1)\alpha(1+\beta)\} - \alpha\beta\sum_{j\neq i}\delta S(\mathbf{x}_j) \tag{20.6}$$

where the higher order terms above $\delta S(\mathbf{x}_i)^2$ were neglected. Since all the off-diagonal entries are $-\alpha\beta$, the dominant eigenvalue is given by $-c_0 + (N-1)\alpha(1+\beta) + \alpha\beta$. Therefore, the uniform biomass distribution is stable under the condition:

$$N < 1 + \frac{c_0 - \alpha\beta}{\alpha(1+\beta)}. \tag{20.7}$$

From Eq. (20.7), we can see that as N and/or β increases [decreases], the multi-layered [mono-layered] size structure becomes stable, i.e. the multi-layered size structure is more likely to be stable under higher asymmetric competition (larger β) than under lower competition (smaller β).

20.4 Conclusion

Spatial pattern formation is not determined only by the disturbance regime (exogenous factor) but by the interaction between competition (endogenous biological factor) and disturbance regime. This study shows that the lower asymmetrically competing populations are likely to form a spatial pattern depending on disturbance regime while the higher asymmetric competing populations are likely to form a vertical structure. It is important to consider the interaction between these two factors for the study of spatial pattern formation in relation to the regeneration and structural diversity of plant communities.

References

1. Bonan, G.B. (1993). Analysis of neighbourhood competition among annual plants: implications of a plant growth model. *Ecol. Model.*, **65**, 123–136.
2. Cornell, H.V. and Lawton, J.H. (1992). Species interactions, local and regional processes, and limits to the richness of ecological communities: a theoretical perspective. J. Anim. Ecol., **61**, 1–12.
3. Denslow, J.S. (1985). Disturbance-mediated coexistence of species. In: Pickett, S.T.A., White, P.S. (eds) *The Ecology of Natural Disturbance and Patch Dynamics.* Academic Press, New York, London, pp. 307–323.
4. Firbanks, L.G. and Watkinson, A.R. (1985). A model of interference within plant monocultures. *J. Theor. Biol.*, **116**, 291–311.
5. Foster, J.R. (1988). The potential role of rime ice defoliation in tree mortality of wave-generated balsam fir forests. *J. Ecol.*, **76**, 172–180.
6. Fowler, N.L. (1988). What is a safe site?: neighbor, litter, germination date and patch effects. *Ecology*, **69**, 947–961.
7. Hara, T., Kimura, M. and Kikuzawa, K. (1991). Growth patterns of tree height and stem diameter in populations of *Abies veitchii*, *A. mariesii* and *Betula ermanii*. *J. Ecol.*, **79**, 1085–1098.
8. Hartgerink, A.P. and Bazzaz, F.A. (1984). Seedling-scale environmental heterogeneity influences individual fitness and population structure. *Ecology*, **65**, 198–206.
9. Hiura, T. (1995). Gap formation and species diversity in Japanese beech forests: a test of the intermediate disturbance hypothesis on a geographic scale. *Oecologia*, **104**, 265–271.
10. Iwaki, H. and Totsuka, T. (1959). Ecological and physiological studies on the vegetation of Mt. Shimagare. II. On the crescent-shaped 'dead trees strips' in the Yatsugatake and the Chichibu mountains. *Bot. Mag. Tokyo*, **72**, 225–260.
11. Kikuzawa, K. and Umeki, K. (1996). Effect of canopy structure on degree of asymmetry of competition in two forest stands in northern Japan. *Ann. Bot.*, **77**, 565–571.
12. Kubota, Y. (1995). Effects of disturbance regime and stand stratification on the regeneration process of a coniferous forest, Taisetsuzan National Park, Japan. *Ecol. Res.*, **9**, 333–341.
13. Levin, S.A. and Pain, R.T. (1974). Disturbance, patch formation, and community structure. *Proc. Natl. Acad. Sci. USA*, **71**, 2744–2747.
14. Lundqvist, L, (1994). Growth and competition in partially cut sub-alpine Norway spruce forests in northern Sweden. *For. Ecol. Manage.*, **65**, 115–122.
15. Naka, K. (1982). Community dynamics of evergreen broadleaf forests in south eastern Japan. I. Wind damage trees and canopy gaps in an evergreen oak forest. *Bot. Mag. Tokyo*, **95**, 385–399.

16. Nakashizuka, T. (1982). Regeneration process of climax beech (*Fagus crenata* Blume) forests. I. Structure of a beech forest with the undergrowth of Sasa. *Jpn. J. Ecol.*, **32**, 57–67.

17. Nakashizuka, T. (1984). Regeneration process of climax beech (*Fagus crenata* Blume) forests. V. Population dynamics of beech in a regeneration process. *Jpn. J. Ecol.*, **34**, 411–419.

18. Nakashizuka, T. (1987). Regeneration dynamics of beech forests in Japan. *Vegetatio*, **69**, 169–175.

19. Ross, M.A. and Harper, J.L. (1972). Occupation of biological space during seedling establishment. *J. Ecol.*, **64**, 77–88.

20. Sato, K. and Iwasa, Y. (1993). Modelling of wave regeneration (shimagare) of subalpine *Abies* forests: population dynamics with spatial structure. *Ecology*, **74**, 1538–1550.

21. Schwinning, S. and Weiner, J. (1998). Mechanisms determining the degree of size asymmetry in competition among plants. *Oecologia*, **113**, 447–455.

22. Spies, T.A. and Franklin, J.F. (1989). Gap characteristics and vegetation response in coniferous forests of the Pacific northwest. *Ecology*, **70**, 543–545.

23. Suzuki, E., Ota, K., Igarashi, T. and Fujiwara, K. (1987). Regeneration process of coniferous forests in northern Hokkaido I. *Abies sachalinensis* forest and *Picea glehnii* forest. *Ecol. Res.*, **2**, 61–75.

24. Stoll, P., Weiner, J. and Schmid, B. (1994). Growth variation in a naturally established population of *Pinus sylvestris*. *Ecology*, **75**, 660–670.

25. Thomas, S.C. and Bazzaz, F.A. (1993). The genetic component in plant size hierarchies: norms of reaction to density in *Polygonum* species. *Ecol. Monogr.*, **63**, 231–249.

26. Wall, R. and Begon, M. (1985). Competition for fitness. *Oikos*, **44**, 356–360.

27. Weiner, J. (1990). Asymmetric competition in plant populations. *Trends Ecol. Evol.*, **5**, 360–364.

28. Weiner, J. and Thomas, S.C. (1986). Size variability and competition in plant monocultures. *Oikos*, **47**, 211–222.

29. Whitmore, T.C. (1989). Canopy gaps and the two major groups of forest trees. *Ecology*, **70**, 536–538.

30. Yokozawa, M. (1999). Size hierarchy and stability in competitive plant populations. *Bull. Math. Biol.*, **61**, 949–961.

31. Yokozawa, M. and Hara, T. (1995). Foliage profile, size structure and stem diameter - plant height relationship in crowded plant populations. *Ann. Bot.*, **76**, 271–285.

32. Yokozawa, M., Kubota, Y. and Hara, T. (1998). Effects of competition mode on spatial pattern dynamics in plant communities. *Ecol. Model.*, **106**, 1–16.

33. Yokozawa, M., Kubota, Y. and Hara, T. (1999). Effects of competition mode on the spatial pattern dynamics of wave regeneration in subalpine tree stands. *Ecol. Model.*, **118**, 73–86.

34. Zobel, M. (1992). Plant species coexistence - the role of historical, evolution and ecological factors. *Oikos*, **65**, 314–320.

Formation of a Structure of Exponentially Forking Branches with a Steady-state Amount of Current-year Shoots in a Hardwood Tree Crown

Akihiro Sumida[1] and Yuri Takai[2]

[1] Institute of Low Temperature Science, Hokkaido University. N19W8, Sapporo 060-0819 Japan
asumida@lowtem.hokudai.ac.jp
[2] Faculty of Agriculture, Gifu University

21.1 Introduction

Forking branches are a structural pattern characterizing a tree species. For a tree species, as a sessile organism, the forking structure is essential for gaining solar energy by spreading leaves in lighter spaces as quickly as possible. The structure is an outcome of a process in which a mother shoot of a branch produces multiple daughter shoots, and thus allows a tree to exponentially increase the amount of leaves available for photosynthetic production. On the other hand, in closed hardwood forests, tree crowns are so close to each other that there is very little space for "exponential" expansion of tree crowns [8]. How can the nature of branch forking, which appears to result in an exponential increase of leaf amount, be consistent with crown development in closed hardwood stands where crown expansion is limited due to a lack of available space? To answer such a question, analyses of demographic (birth and death) and morphological patterns of annual shoots (the portion of shoots elongated during a year) in tree crowns are useful, since they can clarify how the shoot population in a crown develops and is maintained [1, 2, 3, 5, 6, 7, 9, 10, 11, 12, 13]. Here, we begin by showing an example of the structural pattern of a branch as observed in the top canopy of a closed hardwood forest. We then show how the observed pattern can be formed by introducing a model simulating demographic and structural patterns of the annual shoots.

21.2 Estimation of future branch development based on a branch structure at a point in time

If the demographic pattern of annual shoots in a crown is followed over time, it would take years to complete an observation. Hence, a demographic pattern is often estimated from an intensive observation of a structural pattern of a branch at a point in time. In the case of our target species, the deciduous oak Konara (*Quercus serrata* Thunb.), the extension of the growth of current-year shoots is monopodial with rhythmical extensions; i.e., each current-year (0-year-old) shoot having multiple leaves flushes within a short emergence duration (usually within one month) at the beginning of the growing season. Once a current-year shoot is formed, it usually remains on its mother shoot until autumn when all the leaves and several older annual shoots are shed from the abscission layer formed at their base positions. The number of current-year shoots produced on a mother (1-year-old) shoot tends to increase with the increasing length of the mother shoot. Current-year shoots near the apex of a mother shoot tend to be longer, bear more leaves, and survive more years than those that emerge near the base of the mother shoot due to the strong apical dominance of Konara oaks. Surviving 0-year-old shoots become 1-year-old shoots the next year and produce new current year shoots that in turn depend on the shoot length and position on the mother shoot. Some 0-year-old shoots, mostly shorter ones, do not bear new daughter shoots and so die the following year. If we simulate the demography of annual shoots by taking into account such observable patterns, the resulting age structure of annual shoots should be similar to the actual age structure.

Konara oak distributes widely over the warm- and cool-temperate forest zones of Japan, and often becomes the dominant canopy species. From a hardwood forest, branches were randomly sampled from the middle of the top crowns of three Konara trees. The canopy height was about 20 m and the forest age around 50 years. The age of each annual shoot in the branch sample was determined from bud scales and annual rings on the cross section of the annual shoot. The lengths of each annual shoot for each age, basal position of each annual shoot on its mother shoot, and the number of leaves on each 0-year-old shoot were also recorded. Next, we show an outline of our analyses, using data from one of the branch samples.

21.3 Age structure of a branch

Initially we examined the age structure of the branch sample. The number of annual shoots for each annual-shoot age was counted, and this was then plotted against the age of annual shoots on semi-log coordinates. The relationship between the number of annual shoots (NA) and the age of the annual shoots (A) could be fairly approximated by the exponential equation (Fig. 21.1),

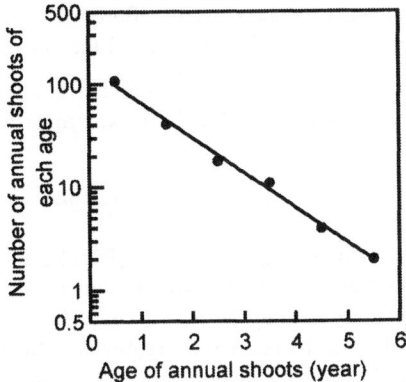

Fig. 21.1. Age structure observed for a branch sample of *Quercus serrata* as expressed by an exponential relationship between the age of annual shoots and the number of annual shoots.

$$NA = 144.3\exp(-0.782A) \qquad (21.1)$$

where $r^2 = 0.994$ and $p < 0.0001$. The value of $\exp(0.782)$ in the equation corresponds to the ratio of the number of annual shoots of a given age divided by that of the one-year older shoots. The value here was 2.19, showing that the number of annual shoots of a given age was about twice that of the one-year older shoots. As for age structure, it appeared that the number of annual shoots increased exponentially with time, but naturally the age structure is a result of the abscission of older shoots previously present on their mother shoots.

In the next section we introduce a simulation of branch-structure development and show how the observed exponential pattern of age structure can be reconstructed. Some structural patterns observed in a branch are expressed by arbitrary equations approximating the patterns.

21.4 Simulation procedure

For annual shoots older than 2 years, it is possible to estimate the number of 0-year-old shoots that had been present when their mother shoot was a 1-year-old shoot by observing leaf-scars or shoot-scars. However, the estimation is sometimes difficult because scars become obscure with age. Hence, the length and number of 0-year-old shoots produced on each 1-year-old shoot are the most reliable information obtainable from a branch sample. Therefore, to construct the simulation, only data for 0- and 1-year-old shoots on the branch sample were used. An underlying assumption in the simulation is that the coefficients of the equations used for the simulation are constant over time, meaning that there is no year-to-year difference in the process of branch development. The four major relationships used for the simulation are:

1. The relationship between the length of mother shoot (LM, cm) and the number of daughter shoots on it (ND). It was approximated by

$$[ND] = (LM - 1.09)/2.26 \qquad (21.2)$$

   where brackets show the Gaussian integer (Fig. 21.2A). This relationship shows that the number of daughter shoots can be determined simply by mother-shoot length. It should be noted that there is a critical mother-shoot length that does not produce daughter shoots the next year (arrow in Fig. 21.2A). In the present case a mother shoot shorter than 3.35 cm will not produce any daughter shoots.

2. The relationship between mother-shoot length and the length of the apical daughter shoot (shoot at the apical position on the mother shoot) ($LDapex$, cm). An apical daughter shoot tends to be the longest of all daughter shoots on a mother shoot. The relationship (Fig. 21.2B) was approximated by the exponential function,

$$LDapex = 1.25\exp(0.224LM). \qquad (21.3)$$

   As the intersection of this curve with the dotted diagonal line ($LDapex = LM$) in Fig. 21.2B shows, the apical-daughter-shoot length was the same as that of the mother shoot when the mother-shoot length was 8.6 cm. The length of a mother shoot that bears an apical daughter shoot whose length is the same as that of the mother shoot is referred to as the "recurrent length". The length of an apical daughter shoot was shorter than the mother-shoot length if the mother shoot was shorter than the recurrent length, 8.6 cm.

3. The relationship between the length of the daughter shoot and the position of the daughter shoot on the mother shoot. The daughter-shoot position was expressed by the relative distance from the apex of the mother shoot (S, %; mother-shoot length=100%), and the daughter-shoot length was represented by its length relative to its mother-shoot length (RL, %). The relationship (Fig. 21.2C) was approximated by

$$RL = 74.6S^{-0.225}. \qquad (21.4)$$

   Equations (21.3) and (21.4) are used in simulations to calculate a daughter-shoot length from a mother-shoot length.

4. The changes in the linear density of 0-year-old shoots with changes in position on a mother shoot. The daughter shoots tended to be denser near the apex of a mother shoot, and become sparser as the daughter-shoot position approached the base of the mother shoot. This observation

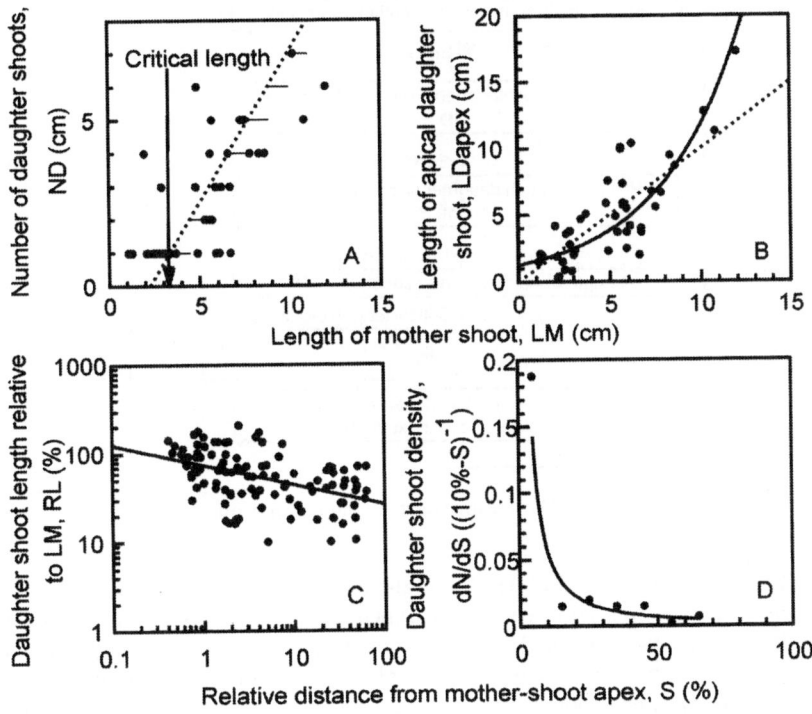

Fig. 21.2. 2A-D. Observed structural patterns of annual shoots used in the simulation. Explanation in text.

was represented by the linear density of 0-year-old shoots $(dN/dS, (10\% - S)^{-1})$ at 10% intervals of the relative position on a mother shoot $(S, \%)$ (Fig. 21.2D). It was approximated by

$$dN/dS = 1.09S^{-1.29}. \tag{21.5}$$

If the linear density is integrated from the value corresponding to the position of the apical daughter shoot (defined as $S = Sapex$) towards the base of the mother shoot (i.e., $S = 100$), the integrated value at $S = 100$ should correspond to the number of 0-year-old shoots calculated from a mother-shoot length (Eq. (21.2)). Based on this principle, we calculated the S values where the first daughter shoot, the second, the third, etc., appeared on the mother shoot.

The flowchart of the simulation is outlined in Fig. 21.3. In this simulation the length of a mother shoot determines everything regarding the daughter shoots on it; initially, only a mother shoot with a certain length was given. The length was set to be the recurrent length (=8.6 cm; see above). This is

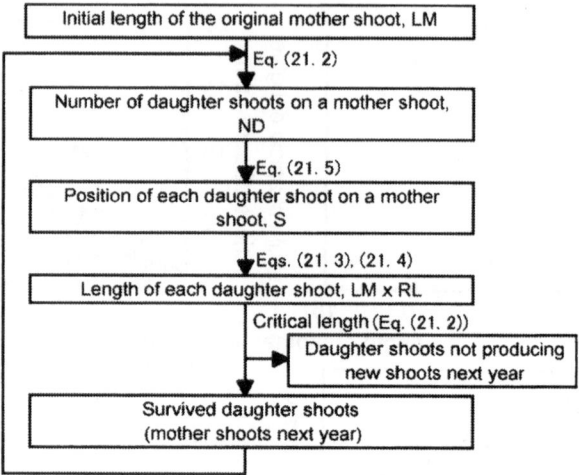

Fig. 21.3. Flowchart of the simulation

equivalent to assuming that the annual-shoot length of the main axis of a branch is always the same each year, since a mother shoot with the recurrent length always bears a daughter shoot with the recurrent length at the mother-shoot's apex. The number of daughter shoots on a mother shoot is determined by mother-shoot length from Eq. (21.2). Then the basal position of each daughter shoot on the mother shoot (Eq. (21.5)) and the length of each daughter shoot (Eqs. (21.3) and (21.4)) are automatically determined. Some of the daughter shoots may not produce new daughter shoots the following year if their lengths are shorter than the critical length for shoot production (Eq. (21.2)). Such daughter shoots are regarded as being dead the following year. If a twig (a part of the branch composed of several annual shoots) does not bear any 0-year-old shoots, this twig is also regarded as dying (dieback of a twig). Surviving 0-year-old shoots (i.e., those longer than the critical length), become mother shoots next year, each of which produces new daughter shoots the following year according to its length. Thus the developmental process of a branch can be simulated, starting with just one mother shoot in the initial year.

21.5 Simulation for branch development

Fig. 21.4 shows a two-dimensional illustration of the simulated branch structure with eight simulated years. In this illustration, the length of each annual shoot and their positions on their mother shoots are correctly drawn, but the angle of annual-shoot extension is arbitrary. In the first year, five 0-year-old shoots (one apical shoot and four lateral ones) were produced on an initial

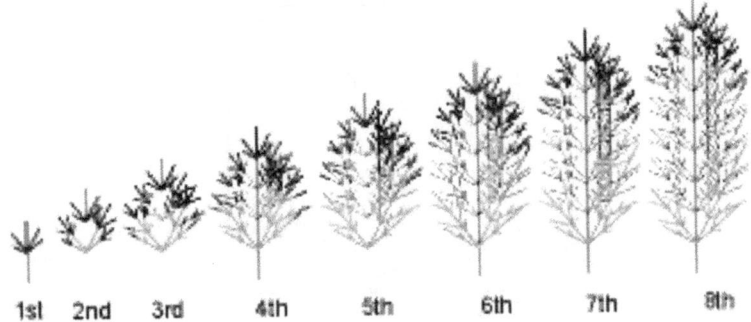

1st   2nd   3rd   4th   5th   6th   7th   8th

Fig. 21.4. Branch structure development over eight simulated years. Thick solid lines show 0-year-old shoots. Lateral twigs developing from the four lateral 0-year-old shoots of the first year are shown by thick gray lines.

mother shoot. During the first and third years, new 0-year-old shoots were produced on their respective mother shoots, while the number of 0-year-old shoots was less, and their length shorter, on lateral twigs originating from the lower shoots of the first year. In the fourth year, 0-year-old shoots were not produced on the twig originating from the lowest lateral shoot of the first year, since all the 1-year-old shoots on the twig were shorter than the critical length for shoot production (Fig. 21.2B). This means that this twig died back to the main stem. By the seventh year, the four twigs whose origins were the four lateral shoots of the first year died out, and only the apical shoot of the first year survived. On the other hand, in the sixth year, the living part of the branch (a part having one or more 0-year-old shoots) was composed of living annual shoots of six years of age. After that, the living part does not change its structure, having been raised upwards with time. This means that a cluster of foliage with a stable structure was formed.

Comparing the structures of the eight simulated years in Fig. 21.4, we see that the whole first-year structure appears in the upper part of the second-year structure, and the whole second-year structure in the upper part of the third-year structure, and so on. Hence we could say that the developmental process of the stable cluster is as if less-developed lateral twigs are added to their lower part each year until a stable cluster is formed.

21.6 Simulated increase in the number of current-year shoots and age structure

As the formation of the stable cluster suggests, the total number of current-year shoots reaches a constant value in the sixth year (Fig. 21.5, dotted line). In the seventh year, the number of 0-year-old shoots developing on the four lateral shoots of the first year died out, meaning that the four lateral twigs

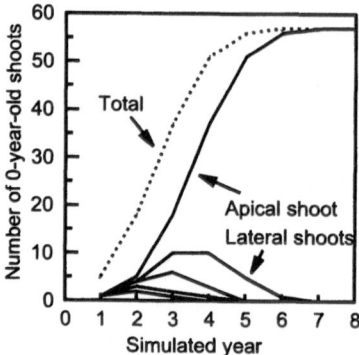

Fig. 21.5. Simulated increase in the number of 0-year-old shoots with time. The number of 0-year-old shoots on the twigs originating from the apical shoot and the four lateral shoots of the first simulated year are shown by thick lines.

died back, while the number of 0-year-old shoots developing on the apical shoot of the first year reached a constant. This indicates that the turnover time of the living part of the cluster structure is the same as the time it takes for all the lateral shoots produced in the first year to die out (six years in this case). An exponential age structure of annual shoots as in Fig. 21.1 was also reproduced with the simulation. Fig. 21.6 shows the age structure of living annual shoots in the seventh year of the simulation. Note that the exponential relationship holds for the annual shoots from zero to six years of age, which corresponds to the turnover time of the cluster. This explains why branches with an exponentially forking structure can exist in a crown with little space for increasing its size; the exponential relationship only holds in the stable cluster that does not increase in size over time. For annual shoots over six years of age, the number of surviving shoots is just one (Fig. 21.6), which had been the apical shoot of the first year.

21.7 Ecological implications of the formation of a stable cluster

The data used for the present simulation was for a branch sample taken in the middle of a crown. Hence the simulated stable cluster should be a reflection of intra-crown (i.e., inter-shoot) competition rather than inter-crown (neighborhood) competition. In forming a stable cluster, dieback of older lateral twigs would be an important process, since it suppresses the exponential increase of current-year shoots. The physiological mechanisms that cause dieback of lateral twigs are unknown, but they would be strongly affected by physical and biological conditions. If the relationships in Figs. 21.1 and 21.2 vary according to ambient physical (air temperature and humidity, light intensity, light qual-

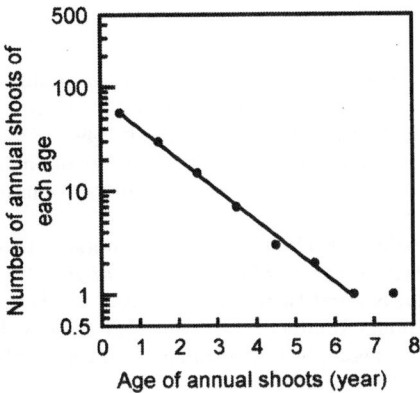

Fig. 21.6. Simulated relationship between the age of annual shoots (A) and the number of annual shoots (NA). The relationship was approximated by $NA = 79.7$ $\exp(-0.684\ A)$ ($r^2 = 0.996$, p<0.001). The slope was not significantly different from the relationship in Fig. 21.1 (p>0.05).

ity, etc.) and biological (inter-shoot competition, etc) conditions where the branch exists, then the size and turnover time of the simulated stable cluster would also vary. Once a forest canopy is closed, tree crowns need to exist with little space to spread foliage, while they have a natural characteristic that multiple current-year shoots are produced on each mother shoot when they flush new leaves. Formation of a stable cluster of branches can partly explain how such tree crowns can continue to exist in a closed canopy. In fact, if we look down on a forest from above, we can observe that a crown of a hardwood tree is composed of clusters of foliage, which can be regarded as basic components of a hardwood crown [4]. Persistence of hardwood crowns in a closed canopy may be closely related to the ability to form stable clusters of foliage.

Acknowledgments

We thank Tomohiro Mano, Asako Togashi, Chika Kanada and other students of the laboratory of Forest Ecology, Gifu University, for their assistance with fieldwork, and Mutsuki Higo, Shogo Kato, Akira Komiyama, Toshihiko Hara and Shin-Ichi Yamamoto for facilitating the study. This research was partly supported by the Ministry of Education, Science, Sports and Culture, Japan, through a Grant-in-Aid for Scientific Research (A2) (No 13356003, Yamamoto S).

References

1. Buck-Sorlin, G.H., and Bell, A.D. (2000). Crown architecture in *Quercus petraea* and *Q.robur*: the fate of buds and shoots in relation to age, position and environmental perturbation. *Forestry*, **73**, 331-349.
2. Chaar, H., Colin, F., and Collet, C. (1997). Effects of environmental factors on the shoot development of *Quercus petraea* seedlings. A methodological approach. *For Ecol Manage*, **97**, 119-131.
3. Collet, C., and Frochot, H. (1996). Effects of interspecific competition on periodic shoot elongation in oak seedlings. *Can J For Res*, **26**, 1934-1942.
4. Kira, T., Shinozaki, K., and Hozumi, K. (1969). Structure of forest canopies as related to their primary productivity. *Plant Cell Physiol*, **10**, 129-142.
5. Koike, F. (1989). Foliage-crown development and interaction in *Quercus gilva* and *Q.acuta*. *J Ecol*, **77**, 92-111.
6. Sattler, R., and Rutishauser, R. (1997). The fundamental relevance of morphology and morphogenesis to plant research. *Ann Bot*, **80**, 571-582.
7. Steingraeber, D.A., Kascht, L.J., and Franck, D.H. (1979). Variation of shoot morphology and bifurcation ratio in sugar maple (*Acer saccharum*) saplings. *Amer J Bot*, **66**, 441-445.
8. Sumida, A., Terazawa, I., Togashi, A., and Komiyama, A. (2002). Spatial arrangement of branches in relation to slope and neighbourhood competition. *Ann Bot*, **89**, 301-310.
9. Suzuki, A. (2000). Patterns of vegetative growth and reproduction in relation to branch orders: the plant as a spatially structured population. *Trees*, **14**, 329-333.
10. Suzuki, M. (2001). *Allometry and Dynamics of Current-year Shoot Populations in the Crown Development of Deciduous Trees*. Ph.D. Thesis, Graduate School of Earth Environmental Science, Hokkaido University.
11. Seleznyova, A.N., Thorp, T.G., Barnett, A.M., and Costes, E. (2002). Quantitative analysis of shoot development and branching patterns in *Actinidia*. *Ann Bot*, **89**, 471-482.
12. White, J. (1979). The plant as a metapopulation. *Ann Rev Ecol Syst*, **10**, 109-145.
13. Wilson, B.F. (1989). Tree branches as population of twigs. *Can J Bot*, **67**, 434-442.

Part VI

Spatio-Temporal Pattern Formation in
Epidemiology

Patterns in Epidemiology of Sexually Transmitted Diseases in Human Populations

Masayuki Kakehashi

Faculty of Medicine, Hiroshima University, Kasumi, Minami-ku, Hiroshima, Hiroshima 734-8551, Japan

22.1 Introduction

Infectious diseases exhibit a lot of interesting patterns when they spread in host populations. Detailed data are especially available on infectious diseases in human populations. Many typical infectious diseases of childhood, like measles and rubella, show clear seasonality with period of one year [2, 7, 8, 9, 12]. In addition to such annual patterns, they also show periodicity with period more than one year, eg., two years, five years, etc. The period of the same disease may be different in different places and in some places there may be chaotic patterns rather than periodic patterns. Moving focus from such childhood diseases to those of adulthood or adolescents, sexually transmitted diseases (or STDs for short, hereafter) also have interesting patterns although they cannot be clearly observed like childhood diseases. In most studies of pattern formation, some interesting pattern is generated similar to a shadow of some other preceding pattern. In the analysis of STDs, we may find cases where a pattern of more interest exists behind the apparently observed pattern. By the analysis of patterns observed in incidences of STDs, we can find patterns of human sexual behavior, the detail of which is usually hidden although questionnaire surveys can sometimes shed light on it. Some patterns of STDs are common to infectious diseases of childhood. The growth of the number of infected people is exponential at the beginning phase of spread. Another typical pattern is one observed in incidences according to the age of hosts. We first briefly show descriptive epidemiology of STDs in Japan below. The data are collected on the outpatients who visited urological, dermatological, obstetric or gynecological clinics in 1999. After reviewing some observed patterns in epidemiology, we try to explain these patterns by mathematical models of infectious diseases.

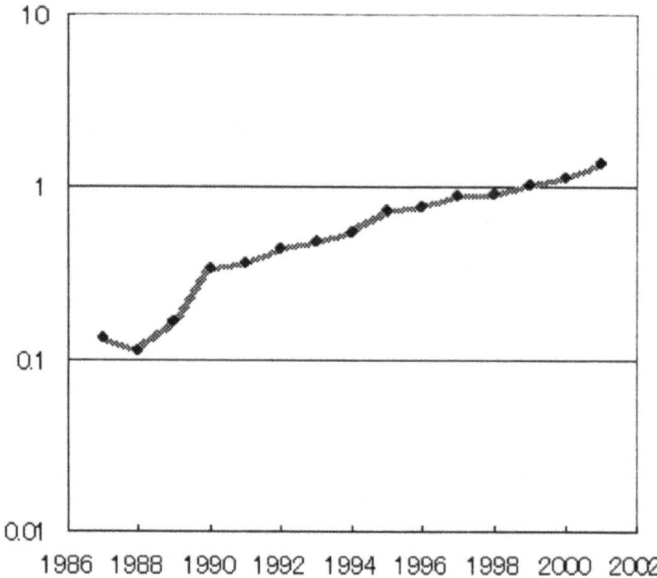

Fig. 22.1. Rate of HIV positive blood donors in Japan 1987-2001 (per 100,000)

22.2 Descriptive epidemiology of sexually transmitted diseases (STDs) in Japan

22.2.1 Exponential growth of infected population

Fig. 22.1 shows the temporal pattern observed in the rates of HIV positive blood donors (per 100,000 individuals) during 1987-2001. The rates are approximately in a straight line except at the very beginning phase. The growth is almost exactly exponential because the vertical axis is logarithmically scaled. The growth rate is about 15% per year. Similar values of the growth rates are also obtained from another source of data, i.e., the national surveillance of HIV infected individuals and AIDS cases in Japan. The exponential growth is considered to be a simple consequence of the fact that the fixed number of newly infected cases are transmitted HIV by one infected individual. The rate of growth is determined by the actual situation of the society according to the degree of sexual activity.

22.2.2 Patterns in incidence by sex, age, region and disease

To clarify the exact situation of the spread of STDs, data were collected from outpatients of clinics in eight prefectures in Japan in 1999. The diseases

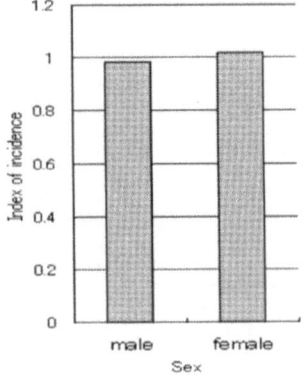

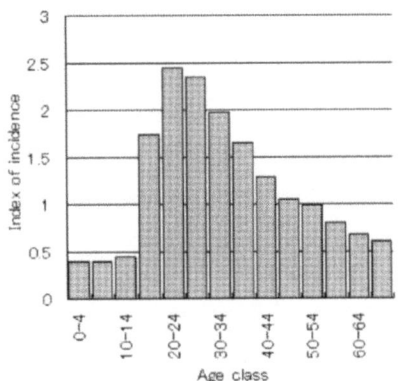

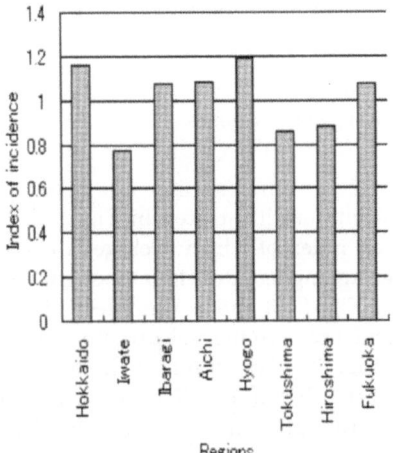

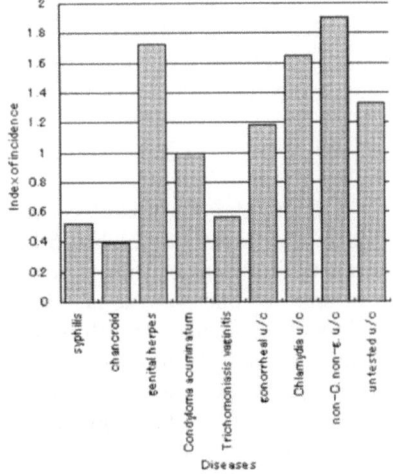

Fig. 22.4. Incidence of STDs in different prefectures.

Fig. 22.5. Incidence of different STDs.

were syphilis, chancroid, genital herpes, Condyloma acuminatum, Trichomoniasis vaginitis, gonorrheal urethritis and cervicitis, Chlamydia urethritis and cervicitis, non-Chlamydia and non-gonorrheal urethritis and cervicitis, and untested urethritis and cervicitis. To examine the effect of factors on incidence more clearly, we carried out analysis based on a linear regression model

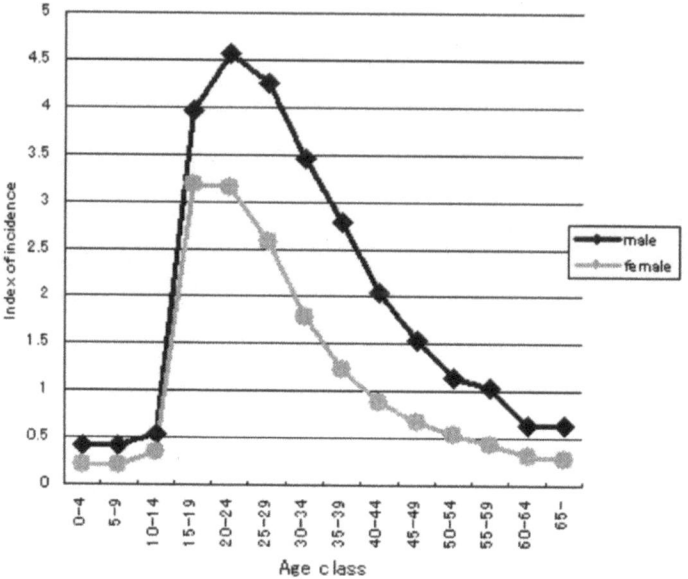

Fig. 22.6. Effect of sex and age on incidence

similar to the analysis of variance. The results are illustrated in Figs. 22.2 - 22.5. The incidence is shown in terms of an index of which average is unity. In Fig. 22.2, we see that incidence is higher among females than among males in spite of our traditional view on STDs. Fig. 22.3 shows the effect of age class on incidences and illustrates that incidences are highest around age 25, the age class of the most sexually active. Fig. 22.4 is a graph of the effect of regions. We see there is some regional variation but it is difficult to tell what characteristics of the regions are connected to higher incidence. In Fig. 22.5, the incidence of different diseases is illustrated. More interesting patterns can be observed if we examine incidence by more than one factor simultaneously (or see the effect of interactions, in terms of analysis of variance). Fig. 22.6 shows the effect of sex and age class on incidence. We see the peak in females is located at a little younger age class than that for males. Fig. 22.7 shows the effect of sex and disease to illustrate the sex that has higher incidence is quite different among different STDs. But we have to notice that this may not be the exact incidence because the data is based on the rate of access to clinics. The rate of access must depend on the degree of pain or symptoms of the disease, which may be different in different sexes even if the disease is the same.

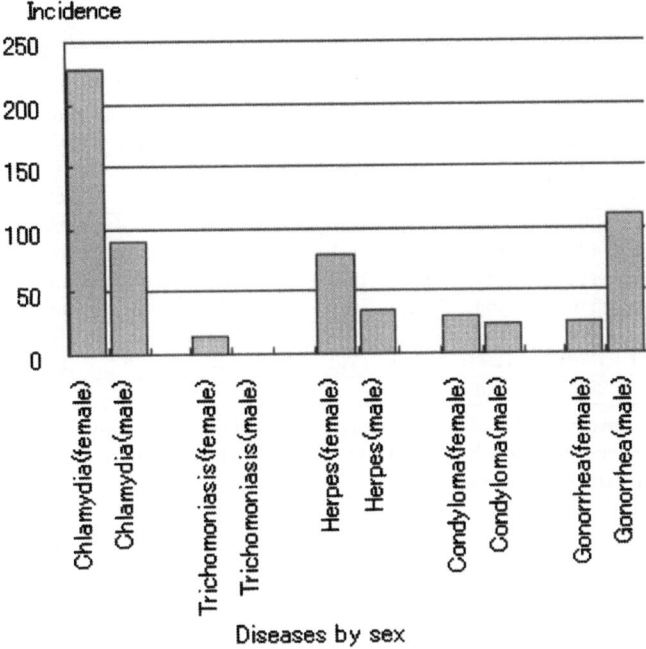

Fig. 22.7. Effect of sex and disease on incidence

The data on sexual behavior is important to explain the observed patterns in descriptive epidemiology. In general it is difficult to obtain this kind of data because of the difficulty of collection. Fortunately we now have data on human sexual behavior provided by the Sex Survey Japan. The data in Table 1 shows estimated frequency of having sex classified by sex and age class. The frequency is higher in younger age classes in both sexes. There are also data on the frequency of individuals who have sex with casual partners. It is interesting to note that there is a little gap between the average frequencies of having sex by males and females although they must be the same if the data is exactly correct.

Frequency (times per year)		1 /M	1 /M	2-3 /M	1 /W	2-3 /W	4+ /w	
age class	number answered	6	12	30	50	125	250	average
18 - 24	216	127	19.6	14.2	18.9	18.1	24.4	3.2	33.0
25 - 34	345	269	18.2	16.7	29	24.2	10.4	0.4	29.5
35 - 44	420	374	24.9	18.4	34.2	14.7	7.2	0	27.0
45 - 54	535	451	22	21.3	33	18.2	3.8	0.6	24.6
55 +	246	189	24.4	19	34.9	15.3	3.2	1	21.8
Sum (male)	1762	1410	22.1	18.7	31.6	18	7.7	0.8	26.9
18 - 24	205	109	15.5	11.9	28.4	25.7	17.4	0.9	25.4
25 - 34	402	325	21.9	15.1	29.8	22.2	8.6	0.6	28.6
35 - 44	419	351	20.3	16.2	32.8	21.7	6.8	0.9	29.0
45 - 54	510	364	29.6	19.2	30.5	14.3	3	0.6	18.3
55 +	264	149	31.5	24.3	26.2	13.4	2	0	12.3
Sum (male)	1800	1298	24.2	17.3	30.3	19.1	6.5	0.7	23.1
Total	3562	2708							

Table 22.1. Heterogeneity of sexual activity level: The Sex Survey Japan Frequencies (less than one per month, etc) are transformed into numbers and averages are calculated for each sex and age class.

22.3 Explanation of the patterns by mathematical models

22.3.1 Simple mathematical models of sexually transmitted diseases (STDs)

Simple SIS models are often used for STDs because immunity is usually not taken into account. Moreover Macdonald type transition of infection is commonly adopted in STD models because transition is considered to be proportional to the rate of infected individuals rather than to the number (or density) of infected individual as is in mass-action type transition models. Whether or not the disease can spread in the host population is determined by the basic reproductive number (R_0), the number of secondary infected individuals produced by a single primary infected individual. In the case of mass-action type models, R_0 is a product of the number of contacts per individual per unit time, the duration of time being infectious, and the number of accessible susceptible individuals. In contrast, in the Macdonald type models, R_0 is a product of only the number of contacts per individual per unit time and the duration of time being infectious. In other words, the number of accessible susceptible individuals is unity in the case of STDs. Due to this fact, STDs are considered to have been under strong selection pressure to have a long duration of infectious period from an evolutionary viewpoint. To realize long duration of infectiousness, STDs may have evolved to be less virulent in comparison with other directly transmitted diseases. This can explain the avirulent nature of most STDs [13].

To fully model STDs, it is necessary to incorporate male and female variables in the model. In the model below, the numbers (or densities) of susceptible and infected individuals are represented by S and I, respectively. The subscripts, m and f, represent male and female, respectively.

$$\frac{dS_m}{dt} = \lambda_m - \mu_m S_m - \beta_m c_m \frac{I_f}{S_f + I_f} S_m + f_m I_m \qquad (22.1)$$

$$\frac{dS_f}{dt} = \lambda_f - \mu_f S_f - \beta_f c_f \frac{I_m}{S_m + I_m} S_f + f_f I_f \qquad (22.2)$$

$$\frac{dI_m}{dt} = -\mu_m I_m + \beta_m c_m \frac{I_f}{S_f + I_f} S_m - f_m I_m \qquad (22.3)$$

and

$$\frac{dI_f}{dt} = -\mu_f I_f + \beta_f c_f \frac{I_m}{S_m + I_m} S_f - f_f I_f. \qquad (22.4)$$

In this model, λ represents a birth rate and μ represents a mortality rate. The parameter f represents the rate of recovery from the disease and β and c represent transition probability per action and the rate of sexual contact per unit time, respectively. Here again the subscripts, m and f, represent male and female, respectively. Additional mortality due to the disease is neglected. According to this model, the prevalence rates of the disease in male and female subpopulations at the equilibrium are given as follows:

$$\frac{I_m}{N_m} = \frac{\beta_m c_m \beta_f c_f - (\mu_m + f_m)(\mu_f + f_f)}{\beta_m c_m (\mu_f + f_f + \beta_f c_f)} \qquad (22.5)$$

and

$$\frac{I_f}{N_f} = \frac{\beta_m c_m \beta_f c_f - (\mu_m + f_m)(\mu_f + f_f)}{\beta_f c_f (\mu_m + f_m + \beta_m c_m)}. \qquad (22.6)$$

This result explains why different incidence rates can be realized by the different rates of transitions and recovery in different sexes.

22.3.2 The characteristics of human population: The existence of temporarily stable couples

To construct more realistic models for human population, pair formation models have been proposed since Dietz [3, 4]. In pair formation models, one male and one female form a couple for a certain duration and sexual contacts are restricted within the couple. By incorporating the formation of pairs, the rate of spread is reported to become smaller than in the models without pair formation. But the model becomes more complex due to the increased number of variables. Unfortunately the restriction of sexual partner within the pair is not always perfect in most human population. Waldstatter [14] has already proposed a pair formation model with commercial sex workers (CSWs here-after). Kakehashi [10] applied such a pair formation model with CSWs for the actual Japanese society by fitting the involved parameters as closely to the real values as possible. Some parameters that are difficult to determine are assigned plausible ranges and sensitivity analyses were carried out. In the analyses, the focus is put on critical transmission probability. If the actual transmission probability per sexual contact is above the critical transmission probability, then HIV can spread in the population. The critical transmission probability was more sensitive to contact rate to CSWs than to pair formation rate. Finding Sensitive parameters helps focus on which information should be collected in more detail and may be important in preventive actions. It was concluded that the contribution by the transmission via CSWs is very large because there would be no spread of HIV if there were no CSWs.

Even in such a more complex model like a pair formation model with CSWs, the initial spread of the disease shows exponential growth in the initial phase. The growth rate of infected population is calculated as the largest eigenvalue of the Jacobian matrix of the dynamics evaluated at the equilibrium without the disease. If the initial growth rate of infected population has been obtained, it provides further information on actual parameter values.

Another important fact about sexual behavior in human population is heterogeneity. Some people may be very sexually active and change their partners more often than others. Moreover some individuals have access to CSWs whereas others do not. As described before, the Sex Survey Japan revealed some information on heterogeneity. To incorporate such heterogeneity, we extended the pair formation model with CSWs to a model with different sexual activities. Using the model without heterogeneity, the saturating prevalence that would be finally obtained in the host population was calculated around 10% (within the range between 5-15%) under the condition that the early phase exponential growth rate is about 15% per year [11]. But from the model with heterogeneity, there is a possibility that the saturating level can be much smaller. This model can also be used to examine quantitatively the effect of condom use to prevent the spread of HIV.

22.3.3 The pattern of age specific incidence

Another interesting pattern is age specific incidence of STDs. To discuss the effect of age, the model should be an age specific one. Age specific models of STDs have been intensively analyzed by Anderson and his colleagues [1, 5, 6, 15]. In their models, difference in sexual activity level is also involved in addition to age structure. They clarified the importance of mixing pattern where more active males select more active females or else less active females are taken into account. Mixing pattern had a large influence in the way disease spreads. To explain age specific pattern of incidence, we used a simple age specific model with no difference in sexual activity level:

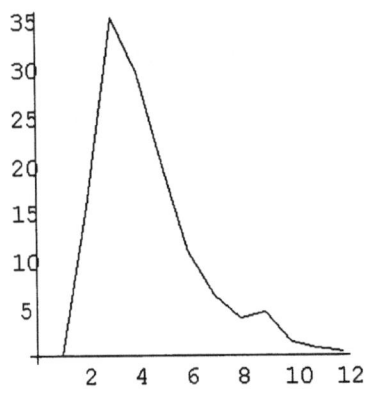

Fig. 22.8. Age specific pattern of male Chlamydia incidence: Observed

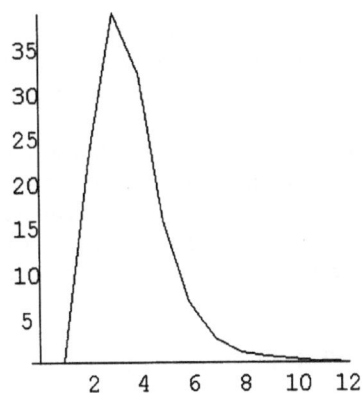

Fig. 22.9. Age specific pattern of male Chlamydia incidence: predicted

$$\frac{dS(\sigma, i)}{dt} = m_{i-1}S(\sigma, i-1) - m_i S(\sigma, i) - \mu_i S(\sigma, i) + fI(\sigma, i)$$

$$-\beta(\sigma) \sum_{j=1}^{n} c_{ji}(\sigma) \frac{I(\sigma', j)}{N(\sigma', j)} S(\sigma, i) \quad (22.7)$$

and

$$\frac{dI(\sigma, i)}{dt} = m_{i-1}I(\sigma, i-1) - m_i I(\sigma, i) - (\mu_i + \delta_i)I(\sigma, i) - fI(\sigma, i)$$

$$+\beta(\sigma) \sum_{j=1}^{n} c_{ji}(\sigma') \frac{I(\sigma', j)}{N(\sigma', j)} S(\sigma, i). \quad (22.8)$$

In the model above, $S(\sigma, i)$ and $I(\sigma, i)$ represent the number of susceptible and infected individuals of sex σ (male or female) and age class i, respectively. We calculated equilibrium prevalence level at each sex and age class. In addition, $N(\sigma, i) = S(\sigma, I) + I(\sigma, i)$ and we put $\mu_i = 0$ and $m_i = m$ for all i. By putting the rates of contact between male and female (c_{ji}) to have a peak where males and females are of the same age class or females are a little younger than males, the observed age specific pattern of male Chlamydia incidence was reproduced by the model (Figs. 22.8 - 22.9).

22.3.4 Regional variation

The incidences of STDs show regional variation like other infectious diseases. If the level of sexual activity commonly determines the level of incidences of multiple STDs, there must exist a unique score for each region that represents the level of incidence of STDs. Yamaguchi and Kakehashi [16] calculated such a unique score and statistically analyzed the relationship between the score and socioeconomic factors. The socioeconomic factors in the analysis consisted of demographic factors, industrial/commercial factors, factors of lifestyle, educational factors, the levels of medical resources and services, the level of welfare supply, traffic factors, and the consumption of durable consumer goods, etc., in total nearly 200 variables. They found that no single or combined factors that determine the spread of STDs in a temporally stable way could be obtained although some variables explained the data very well in each year. They concluded that the spread of STDs cannot be explained very well by commonly used socioeconomic indices.

22.4 Discussion

We have briefly reviewed remarkable patterns in epidemiology of sexually transmitted diseases and seen how they can be explained by mathematical models. Simple mathematical models considering at most sex, age and the level of sexual activity can explain basic patterns observed in human populations. But the analyses shown here are only the beginning of more intensive analyses. Moreover human sexual behavior has a very complex aspect and there are still many unresolved questions. Making use of questionnaire survey and health statistical data, in addition to the use of mathematical models, we would like to carry out further analyses.

Today HIV/AIDS is a threat to all over the world. In the selection of most effective preventive policy, mathematical models can play an important role. Using mathematical models to simulate the spread of infectious diseases is not only a challenge of theoretical interest but also of practical importance.

Acknowledgements

The author is grateful to Professor T. Sekimura for giving him the opportunity to talk about this topic. He is also grateful to Professor T. Usui and Dr. M. Shigeta at Hiroshima University for helping in using the STD data. This work has partly supported by Grant-in-Aid from the Ministry of Education, Culture, Sports, Science and Technology of Japan.

References

1. Anderson, R. M. and Garnett, G. P. (2000). Mathematical models of the transmission and control of sexually transmitted diseases. *Sexually Transmitted Diseases*, **27**(10): 636-643.
2. Bolker, B. M. and Grenfell, B. T. (1993). Chaos and biological complexity in measles dynamics. *Proc. R. Soc. Lond. B* **251**(1330): 75-81.
3. Dietz, K. (1988). On the transmission dynamics of HIV. *Math. Biosci.* **90**: 397-414.

4. Dietz, K. (1988). The dynamics of spread of HIV infection in the heterosexual populations. In: *Statistical Analysis and Mathematical Modelling of AIDS*. (J. C. Jager and E. J. Ruitenberg, eds.), Oxford University Press.

5. Garnett, G. P. and Anderson, R. M. (1993). Factors controlling the spread of HIV in heterosexual communities in developing countries: patterns of mixing between different age and sexual activity classes. *Phil. Trans. R. Soc. Lond. B* **342**: 137-159.

6. Garnett, G. P., and Anderson, R. M. (1995). Strategies for limiting the spread of HIV in developing countries: conclusions based on studies of the transmission dynamics of the virus. *Journal of Acquired Immune Deficiency Syndrome and Human Retrovirology* **9**: 500-513.

7. Grenfell, B. T. (1992). Chance and chaos in measles dynamics. *Journal of Royal Statistical Society*, **B54**: 383-398.

8. Grenfell, B. T. and Dobson, A. P. (1995). *Ecology of Infectious Diseases in Natural Populations*. Cambridge University Press.

9. Grenfell, B. T., Kleczkowski, A., Gilligan, C. A. and Bolker, B. M. (1995). Spatial heterogeneity, nonlinear dynamics and chaos in infectious diseases. *Statistical Methods in Medical Research*, 4(2): 160-83.

10. Kakehashi, M. (1998). A mathematical analysis of the spread of HIV/AIDS in Japan. *IMA Journal of Mathematics Applied in Medicine and Biology* **15**: 299-311.

11. Kakehashi, M. (2000). Validity of simple pair formation model for HIV spread with realistic parameter setting. *Mathematical Population Studies* 8(3): 279-292.

12. Keeling, M. and Grenfell, B. T. (1999). Stochastic dynamics and a power law for measles variability. *Phil. Trans. R. Soc. Lond. B* **354**: 769-776.

13. Sheldon, B. C. and Read, A. F. (1997). Comparative biology and disease ecology. *Trends in Evolution and Ecology* 12(2): 43-44.

14. Waldstätter, R. (1989). Pair formation in sexually-transmitted disease. In: *Mathematical and Statistical Approaches to AIDS Epidemiology*. (C. Castillo-Chavez, ed.) *Lecture Notes in Biomathematics* **83**. Springer-Verlag.

15. Williams, J. R. and Anderson, R. M. (1994). Mathematical models of the transmission dynamics of Human Immunodeficiency Virus in England and Wales: Mixing between different risk groups. *Journal of Royal Statististical Society A* **157**: 69-87.

16. Yamaguchi, F. and Kakehashi, M. (2002). A prefectural index of the spread of sexually transmitted diseases and the related socioeconomic factors. *Kosei no Shihyo*. **49**(10): 24-30 (In Japanese).

Backward Bifurcation in a Model for Vector Transmitted Disease

Hisashi Inaba

Department of Mathematical Sciences, University of Tokyo,
3-8-1 Komaba Meguro-ku, Tokyo 153-8914 Japan
E-mail: inaba@ms.u-tokyo.ac.jp

23.1 Introduction

In mathematical models for the spread of infectious diseases, it is well known that there is a threshold phenomenon: if the basic reproduction number R_0 is greater than one, the disease can invade into the susceptible host community, whereas it cannot if R_0 is less than one. The basic reproduction number is the average number of secondary cases produced by one infectious individual during its total infective period, in a population that is in the disease-free steady state (see [1, 3]).

Mathematically speaking, the threshold phenomenon implies that the disease-free steady state is locally asymptotically stable if $R_0 < 1$, it is unstable for $R_0 > 1$ and at $R_0 = 1$ an endemic steady state bifurcates from the disease-free steady state. In most epidemic models investigated so far, the bifurcation at $R_0 = 1$ is forward (supercritical), that is, for $R_0 < 1$ there is no endemic steady state and the disease-free steady state is globally asymptotically stable, and for $R_0 > 1$ the bifurcated endemic steady state is unique and locally stable at least as long as it is small enough (near the bifurcation point), though we do not know what will happen for the endemic steady state away from the bifurcation point. It may lose its stability in various ways [2].

On the other hand, recently some authors have found epidemic models leading to backward bifurcations (subcritical bifurcations) at $R_0 = 1$ and have stressed important consequences of this for the control of infectious disease (see [5, 7, 8]).

In this note, we present a new epidemic mechanism leading to backward bifurcation of endemic steady states. We consider here a simple epidemic model for vector transmitted disease. The basic model can be seen as a simplified version of models for Chagas disease developed in [6, 9, 10, 11]. Chagas disease, or *American trypanosomiasis*, is caused by the protozoan parasite *Trypanosoma cruiz* and transmitted by blood-feeding triatomine bugs, is a chronic, frequently fatal infection that is common in Latin America. There is so far no effective medical treatment and no vaccine. Though Chagas disease can be also transmitted by blood transfusion, organ transplants and the placenta (vertical transmission from mother to babies), here we only consider vector transmission and neglect variable infectivity since our purpose is not to develop a realistic model for Chagas disease, but to show a possible mechanism leading to backward bifurcation as simply as possible.

23.2 The basic model

Let us consider a fatal disease transmitted by a vector population. We divide the host population into two groups: $S(t)$ denotes the density of uninfected, but susceptible host population, $I(t)$ denotes the density of infected hosts, and t denotes time. The infected population is removed with the extra death rate $\gamma > 0$ due to disease. Then the total size of the host population is given by $T(t) := S(t) + I(t)$. Let $M(t)$ denote the density of susceptible vectors at time t, $V(t)$ the density of infected vectors at time t and $U(t) := M(t) + V(t)$ the total density of vectors. Let b_1 (b_2) be the birth rate of host population (vectors), μ_1 (μ_2) the natural death rate of host population (vectors). For the vector population, we assume that there is no extra death rate due to infection.

For modeling vector transmitted disease, the most important ingredient is the formula for the force of transmission. Here we adopt the most simple assumption that is very common among traditional modeling of vector transmitted diseases such as malaria. The reader may refer to [4] for another type of assumption on the force of infection.

Let a be the number of bites per vector per unit time and c_1 be the proportion of infected bites that give rise to infection from infected vector to susceptible host. For simplicity, we assume that a is a constant. Then the V vector makes $ac_1 V$ infectious bites of which a fraction S/T are on susceptible hosts. That is, the number of new host infections per unit time by vector transmission is given by $ac_1 V(S/T)$.

Hence the force of infection (the probability per unit of time for a susceptible to become infected) for the host population, denoted by $\lambda_1(t)$, is given by

$$\lambda_1(t) = \alpha V(t)/T(t), \tag{23.2.1}$$

where $\alpha := ac_1$. By using a similar argument, we know that the force of infection for the vector population, denoted by $\lambda_2(t)$, is given by

$$\lambda_2(t) = \beta I(t)/T(t), \tag{23.2.2}$$

where $\beta := ac_2$ and c_2 denotes the proportion of bites that give rise to infection from infected host to susceptible vector.

Under the above assumption, our basic model for a vector transmitted disease can be formulated as follows:

$$dS(t)/dt = b_1 - (\mu_1 + \lambda_1(t))S(t), \tag{23.2.3a}$$

$$dI(t)/dt = \lambda_1(t)S(t) - (\mu_1 + \gamma)I(t), \tag{23.2.3b}$$

$$dM(t)/dt = b_2 - (\mu_2 + \lambda_2(t))M(t), \tag{23.2.3c}$$

$$dV(t)/dt = \lambda_2(t)M(t) - \mu_2 V(t). \tag{23.2.3d}$$

If we add the two equations (23.2.3c) and (23.2.3d) for the vector population, we obtain a single equation for the total vector population $U = M + V$ as

$$dU(t)/dt = b_2 - \mu_2 U(t). \tag{23.2.4}$$

It is easily seen that $U^* := b_2/\mu_2$ is a globally stable steady state. In the following, we assume in advance that $M(t) + V(t) = b_2/\mu_2$ for all $t \geq 0$. Then it is easy to see that the solution of (23.2.3a)-(23.2.3d) starting from the region

$$\Omega := \{(S, I, M, V) \in \mathbf{R}_+^4 : 0 \leq S + I \leq b_1/\mu_1, M + V = b_2/\mu_2\}$$

stays in Ω for all $t > 0$, that is, Ω is positively invariant with respect to the flow generated by the basic system (23.2.3).

The system (23.2.3a)-(23.2.3d) has a trivial steady state (the disease-free steady state)

$$(S^*, I^*, M^*, V^*) = (b_1/\mu_1, 0, b_2/\mu_2, 0).$$

The dynamics of the initial invasion phase is described by the linearized system at the disease-free steady state. At the disease-free steady state, we can formulate the linearized system as follows:

$$\begin{pmatrix} \dot{I} \\ \dot{V} \end{pmatrix} = \begin{pmatrix} -(\mu_1 + \gamma) & \alpha \\ \beta b_2 \mu_1/(b_1 \mu_2) & -\mu_2 \end{pmatrix} \begin{pmatrix} I \\ V \end{pmatrix}. \tag{23.2.5}$$

By calculating the eigenvalues, we can conclude that the disease-free steady state is locally asymptotically stable if $R_0 < 1$, whereas it is unstable if $R_0 > 1$, where the basic reproduction number R_0 is calculated as

$$R_0 = \frac{\alpha \beta b_2 \mu_1}{b_1 \mu_2^2 (\mu_1 + \gamma)}. \tag{23.2.6}$$

By using the basic reproduction number, we can formulate the threshold condition for this vector transmitted disease as follows:

Proposition 1. *If $R_0 < 1$, then the disease-free steady state is locally asymptotically stable, and if $R_0 > 1$, it is unstable. Moreover, if*

$$R_0 < \frac{\mu_1}{\mu_1 + \gamma}, \tag{23.2.7}$$

then the disease-free steady state is globally asymptotically stable.

Proof. The eigenvalues of the coefficient matrix of (23.2.5) are given by the solution of the quadratic equation

$$\lambda^2 + (\mu_1 + \mu_2 + \gamma)\lambda + (\mu_1 + \gamma)\mu_2(1 - R_0) = 0,$$

where R_0 is given by (23.2.6). If $R_0 < 1$, all eigenvalues have negative real parts, and if $R_0 > 1$ one eigenvalue is positive. Therefore the first part of this statement is obvious. Let us prove the second part. It follows from (23.2.3a) and (23.2.3b) that

$$dT(t)/dt = b_1 - \mu_1 T(t) - \gamma I(t) \geq b_1 - (\mu_1 + \gamma)T(t).$$

Then we obtain

$$T(t) \geq \frac{b_1}{\mu_1 + \gamma} + (T(0) - \frac{b_1}{\mu_1 + \gamma})e^{-(\mu_1+\gamma)t}.$$

Hence if $T(0) \geq b_1/(\mu_1 + \gamma)$, it follows that $T(t) \geq b_1/(\mu_1 + \gamma)$ for all $t \geq 0$. Even if $T(0) < b_1/(\mu_1+\gamma)$, the condition $T(t) \geq b_1/(\mu_1+\gamma)$ will be satisfied as time goes to infinity, so without loss of generality, we can assume that $T(0) \geq b_1/(\mu_1+\gamma)$ holds in advance. Hence we can assume that $1/T(t) \leq (\mu_1+\gamma)/b_1$.

Next let us define functions $B(t)$ and $C(t)$ as follows:

$$B(t) := \lambda_1(t)S(t), \quad C(t) =: \lambda_2(t)M(t),$$

that is, $B(t)$ denotes the number of newly infected hosts at time t and $C(t)$ is the number of newly infected vectors at time t. From (23.2.3), we have the following expressions:

$$S(t) = \frac{b_1}{\mu_1} + (S(0) - \frac{b_1}{\mu_1})e^{-\mu_1 t} - \int_0^t e^{-\mu_1(t-s)}B(s)ds, \tag{23.2.8}$$

$$I(t) = I(0)e^{-(\mu_1+\gamma)t} + \int_0^t e^{-(\mu_1+\gamma)(t-s)}B(s)ds, \tag{23.2.9}$$

$$V(t) = V(0)e^{-\mu_2 t} + \int_0^t e^{-\mu_2(t-s)}C(s)ds. \tag{23.2.10}$$

From these we obtain a system of Volterra integral equations:

$$B(t) = \alpha\frac{S(t)}{T(t)}\left(V(0)e^{-\mu_2 t} + \int_0^t e^{-\mu_2(t-s)}C(s)ds\right), \tag{23.2.11}$$

$$C(t) = \beta\frac{M(t)}{T(t)}\left(I(0)e^{-(\mu_1+\gamma)t} + \int_0^t e^{-(\mu_1+\gamma)(t-s)}B(s)ds\right). \tag{23.2.12}$$

By using inequalities $S/T \leq 1$ and $M/T \leq (\mu_1 + \gamma)b_2/(b_1\mu_2)$ and inserting (23.2.12) into (23.2.11), it follows that

$$B(t) \leq G(t) + \int_0^t K(s)B(t - s)ds, \tag{23.2.13}$$

where

$$K(s) := \frac{\alpha\beta(\mu_1 + \gamma)b_2}{b_1\mu_2} \int_0^s e^{-\mu_2(s-\sigma)}e^{-(\mu_1+\gamma)\sigma}d\sigma,$$

$$G(t) := \alpha V(0)e^{-\mu_2 t} + \frac{\alpha\beta(\mu_1 + \gamma)b_2}{b_1\mu_2}I(0) \int_0^t e^{-\mu_2(t-s)}e^{-(\mu_1+\gamma)s}ds.$$

Therefore we conclude that $\lim_{t\to\infty} B(t) = 0$ if

$$\int_0^\infty K(s)ds = \frac{\alpha\beta b_2}{b_1\mu_2^2} = R_0\frac{\mu_1 + \gamma}{\mu_1} < 1.$$

Then (23.2.7) is a sufficient condition for global stability of the disease-free steady state.

23.3 Backward bifurcation of endemic steady states

Let us denote by S^*, I^*, M^* and V^* the stationary states of $S(t)$, $I(t)$, $M(t)$ and $V(t)$ respectively, and let λ_j^*, $(j = 1, 2)$ be the force of infection corresponding to the stationary state. Then we have

$$\begin{cases} 0 = b_1 - \lambda_1^* S^* - \mu_1 S^*, \\ 0 = \lambda_1^* S^* - (\mu_1 + \gamma)I^*, \\ 0 = b_2 - \lambda_2^* M^* - \mu_2 M^*, \\ 0 = \lambda_2^* M^* - \mu_2 V^*, \end{cases} \tag{23.3.1}$$

and it is easy to obtain the expression for the steady state as follows:

$$(S^*, I^*, M^*, V^*) = \left(\frac{b_1}{\mu_1 + \lambda_1^*}, \frac{b_1\lambda_1^*}{(\mu_1 + \lambda_1^*)(\mu_1 + \gamma)}, \frac{b_2}{\mu_2 + \lambda_2^*}, \frac{\lambda_2^* b_2}{\mu_2(\mu_2 + \lambda_2^*)}\right). \tag{23.3.2}$$

From relations $\lambda_1^* = \alpha V^*/(S^* + I^*)$ and $\lambda_2^* = \beta I^*/(S^* + I^*)$, we obtain

$$\lambda_1^* = \frac{\alpha b_2}{\mu_2 b_1}\frac{(\mu_1 + \lambda_1^*)(\mu_1 + \gamma)}{\lambda_1^* + \mu_1 + \gamma}\frac{\lambda_2^*}{\mu_2 + \lambda_2^*}, \tag{23.3.3}$$

$$\lambda_2^* = \frac{\lambda_1^* \beta}{\mu_1 + \gamma + \lambda_1^*}. \tag{23.3.4}$$

From (23.3.3) and (23.3.4), we can derive the equation satisfied by λ_1^* as

$$\lambda_1^* = \frac{\alpha\beta b_2}{\mu_2 b_1}\frac{(\mu_1 + \lambda_1^*)(\mu_1 + \gamma)}{\lambda_1^* + \mu_1 + \gamma}\frac{\lambda_1^*}{\mu_2(\mu_1 + \gamma) + \lambda_1^*(\mu_2 + \beta)} \tag{23.3.5}$$

$$= \lambda_1^* R(\lambda_1^*),$$

where $R(\lambda)$ is the function defined by

$$R(\lambda) := \frac{\alpha\beta b_2}{\mu_2 b_1}\frac{(\mu_1 + \lambda)(\mu_1 + \gamma)}{\lambda + \mu_1 + \gamma}\frac{1}{\mu_2(\mu_1 + \gamma) + \lambda(\mu_2 + \beta)}.$$

By using the expression for the steady states, we can decompose $R(\lambda_1^*)$ as

$$R(\lambda_1^*) = \left[\alpha \cdot \frac{1}{\mu_2} \cdot \frac{S^*}{T^*}\right] \cdot \left[\frac{1}{\mu_1 + \gamma} \cdot \beta \cdot \frac{U^*}{T^*} \cdot \frac{M^*}{U^*}\right],$$

which shows that $R(\lambda)$ is the product of the average number of infected hosts produced by an infected vector during its infective period and the average number of infected vectors produced by an infected host during its infective period at the endemic steady state. That is, $R(\lambda)$ is no other than the reproduction number at the endemic steady state with the force of infection λ. We can prove the following bifurcation result for the endemic steady states:

Proposition 2. *Let us define the numbers D and G as*

$$D := \mu_1^2 - (\mu_1 + \gamma)G, \quad G := \mu_1 - \frac{\gamma\mu_2}{\mu_2 + \beta}.$$

If $G \geq 0$, then there exists an unique endemic steady state if and only if $R_0 > 1$. If $G < 0$, then the following hold:

(1) If $R_0 \geq 1$, there exists only one endemic steady state.
(2) If $R_0 < 1$ and $R(-\mu_1 + \sqrt{D}) > 1$, then there exist two endemic steady states.
(3) If $R_0 < 1$ and $R(-\mu_1 + \sqrt{D}) = 1$, then there exists only one endemic steady state.
(4) If $R_0 < 1$ and $R(-\mu_1 + \sqrt{D}) < 1$, then there is no endemic steady state.

Proof.   Observe that

$$R(\lambda) = \frac{\alpha\beta b_2(\mu_1 + \gamma)}{\mu_2 b_1} \left(1 - \frac{\gamma}{\lambda + \mu_1 + \gamma}\right) \frac{1}{\mu_2(\mu_1 + \gamma) + \lambda(\mu_2 + \beta)},$$

so

$$R'(\lambda) = \frac{\alpha\beta b_2(\mu_1 + \gamma)}{\mu_2 b_1} \frac{-f(\lambda)}{(\mu_1 + \gamma + \lambda)^2(\mu_2(\mu_1 + \gamma) + \lambda(\mu_2 + \beta))^2},$$

where

$$f(\lambda) = (\mu_2 + \beta)(\lambda^2 + 2\mu_1\lambda + (\mu_1 + \gamma)G).$$

If $G \geq 0$, then $R'(\lambda) \leq 0$ for all $\lambda > 0$, so $R(\lambda)$ is monotone decreasing for $\lambda > 0$. Since

$$R(0) = R_0 = \frac{\alpha\beta b_2\mu_1}{b_1\mu_2^2(\mu_1 + \gamma)}, \quad \lim_{\lambda \to \infty} R(\lambda) = 0,$$

we know that the characteristic equation $R(\lambda) = 1$ has only one positive root if and only if $R_0 > 1$. On the other hand, if $G < 0$, then we have $D > 0$ and

$$R'(\lambda) = -\frac{\alpha\beta b_2(\mu_1 + \gamma)}{\mu_2 b_1} \frac{(\lambda + \mu_1 + \sqrt{D})(\lambda + \mu_1 - \sqrt{D})}{(\mu_1 + \gamma + \lambda)^2(\mu_2(\mu_1 + \gamma) + \lambda(\mu_2 + \beta))^2}.$$

Hence we know that $R(\lambda)$ has a unimodal pattern and it attains the maximum value at $\lambda = -\mu_1 + \sqrt{D} > 0$. Then the above statement follows immediately.

Corollary 1. *The backward bifurcation does not occur for our model if the extra death rate γ due to the disease is zero.*

From the above propositions, we know that a backward bifurcation of endemic steady states could occur at $R_0 = 1$ under the condition $G < 0$, but we do not yet know what kind of parameter set satisfies the condition $R(-\mu_1 + \sqrt{D}) \geq 1$. However, observe that

$$R'(0) = -\frac{\alpha\beta b_2(\mu_2 + \beta)G}{\mu_2^3 b_1(\mu_1 + \gamma)^2}.$$

Hence if $G < 0$, then $R'(0) > 0$, and it is easy to see that if $R_0 < 1$ and $|R_0 - 1|$ is small enough, there exist two endemic steady states.

On the other hand, note that the equation $R(\lambda) = 1$ can be reduced to a quadratic equation

$$g(\lambda) := (\mu_2 + \beta)\lambda^2 + H\lambda + (\mu_1 + \gamma)^2\mu_2(1 - R_0) = 0, \tag{23.3.6}$$

where H is given by

$$H := (\mu_1 + \gamma)(2\mu_2 + \beta - \frac{\alpha\beta b_2}{\mu_2 b_1}).$$

Since $g'(0) = H$ and $g(0) = (\mu_1 + \gamma)^2\mu_2(1 - R_0)$, by graphical consideration we can easily obtain the following formulation:

Proposition 3. *Let us define the discriminant E of (23.3.6) as*

$$E := H^2 - 4(\mu_1 + \gamma)^2(\mu_2 + \beta)\mu_2(1 - R_0).$$

Then the following hold:

(1) If $R_0 > 1$, there is only one endemic steady state.
(2) If $R_0 = 1$, there is only one endemic steady state when $H < 0$, and there is no endemic steady state when $H \geq 0$.
(3) If $R_0 < 1$, there is no endemic steady state when $H \geq 0$. If $H < 0$, there are two endemic steady states when $E > 0$, there is only one endemic steady state when $E = 0$ and there is no endemic steady state when $E < 0$.

Corollary 2. *Suppose that*

$$2 + \frac{\beta}{\mu_2} < \frac{\alpha\beta b_2}{\mu_2^2 b_1} < 2 + \frac{\beta}{\mu_2} + 2\sqrt{1 + \frac{\beta}{\mu_2}}. \tag{23.3.7}$$

Then the backward bifurcation occurs at $R_0 = 1$.

Proof. First observe that under condition (23.3.7) we have

$$H = \mu_2(\mu_1 + \gamma)(2 + \frac{\beta}{\mu_2} - \frac{\alpha\beta b_2}{\mu_2^2 b_1}) < 0. \tag{23.3.8}$$

Let us choose R_0 as a bifurcation parameter. Consider γ as a function of R_0 and let us fix the other parameters. Observe that

$$\gamma = \frac{\alpha\beta b_2 \mu_1}{b_1 \mu_2^2} \frac{1}{R_0} - \mu_1. \tag{23.3.9}$$

Then R_0 moves from zero to $\alpha\beta b_2/(b_1 \mu_2^2)$, γ decreases monotonically from ∞ to zero. Moreover note that $E = 0$ when $R_0 = R_0^*$, where

$$R_0^* := 1 - \frac{\mu_2}{4(\mu_2 + \beta)} (2 + \frac{\beta}{\mu_2} - \frac{\alpha\beta b_2}{\mu_2^2 b_1})^2,$$

which is positive under the condition (23.3.7). Then we obtain the following table:

R_0	0		R_0^*		1		$\alpha\beta b_2/(\mu_2^2 b_1)$
γ	∞				$\mu_1(\alpha\beta b_2/(\mu_2^2 b_1) - 1)$		0
E	$E < 0$	0			$E > 0$		

Therefore we know that for $R_0^* < R_0 < 1$, there exist two endemic steady states, and the backward bifurcation occurs at $R_0 = 1$.

23.4 Discussion

We have developed a mathematical model for vector transmitted disease with fatality, and calculated the basic reproduction number R_0 to show that the disease can invade into the susceptible population and an unique endemic steady state exists if $R_0 > 1$, whereas the disease dies out if R_0 is small enough. We also proved that depending on the parameters, the backward bifurcation of an endemic steady state can occur, so even if $R_0 < 1$, there could exist endemic steady states. In our modeling, existence of disease-induced death rate plays a crucial role for the existence of a backward bifurcation.

The presence of a backward bifurcation has important practical consequences for the control of infectious diseases. If the bifurcation of the endemic state at $R_0 = 1$ is a forward one, the size of the infected population will be approximately proportional to the difference $|R_0 - 1|$. On the other hand, in a system with a backward bifurcation, the endemic steady state that exists for R_0 just above one could have a large infectious population, so the result of R_0 rising above one would be a drastic change in the number of infecteds. Conversely, reducing R_0 back below one would not eradicate the disease, as long as its reduction is not sufficient. That is, if the disease is already endemic, in order to eradicate the disease, we have to reduce the basic reproduction number so far that it enters the region where the disease-free steady state is globally asymptotically stable and there is no endemic steady state.

In this note, we have only discussed existence of backward bifurcation of endemic steady states; to consider their stability is a future problem. From the general principle of subcritically bifurcated steady state solutions, we would expect that if $R_0 < 1$ and there are two endemic steady states, then the steady state corresponding to the smaller force of infection would be unstable, and the other steady state would be locally stable. This kind of stability argument will be shown in a separate paper (see [6]).

References

1. Anderson, R. M. and May, R. M. (1991). *Infectious Diseases of Humans: Dynamics and Control*, Oxford UP, Oxford.
2. Cha, Y., Iannelli, M. and Milner, F. A. (2000). Stability change of an epidemic model, *Dyn. Sys. Appl.* **9**: 361-376.
3. Diekmann, O. and Heesterbeek, J.A.P. (2000). *Mathematical Epidemiology of Infectious Diseases: Model Building, Analysis and Interpretation*, John Wiley and Sons, Chichester.
4. Esteva, L. and Matias, M. (2001). A model for vector transmitted diseases with saturation incidence, *J. Biol. Sys.* **9**(4): 235-245.
5. Hadeler, K. P. and van den Driessche, P. (1997). Backward bifurcation in epidemic control, *Math. Biosci.* **146**: 15-35.
6. Inaba, H. and Sekine, H. (2002). A mathematical model for Chagas disease with infection-age-dependent infectivity, submitted.
7. Kribs-Zaleta, C. M. and Velasco-Hernández, J. X. (2000). A simple vaccination model with multiple endemic states, *Math. Biosci.* **164**: 183-201.
8. Kribs-Zaleta, C. M. and Martcheva, M. (2002). Vaccination strategies and backward bifurcation in an age-since-infection structured model, *Math. Biosci.* **177/178**: 317-332.
9. Sekine, H. (2002). *A Model for Chagas Disease Continuously Depending on the Duration of Infection*, M.A.Thesis, University of Tokyo. [in Japanese]
10. Velasco-Hernández, J. X. (1991). An epidemiological model for the dynamics of Chagas' disease, *Biosystem* **26**: 127-134.
11. Velasco-Hernández, J. X. (1994). A model for Chagas disease involving transmission by vectors and blood transfusion, *Theor. Popul. Biol.* **46**: 1-31.

Part VII

Morphogenesis and Pattern Formation in
Medicine

Mathematical Modelling of Solid Tumour Growth: Applications of Pre-pattern Formation

Mark A.J. Chaplain[1], Mahadevan Ganesh[2], Ivan G. Graham[3] and Georgios Lolas[1]

[1] The SIMBIOS Centre, Division of Mathematics, University of Dundee, Dundee DD1 4HN, Scotland U.K.
   E-mail: chaplain@maths.dundee.ac.uk, glolas@maths.dundee.ac.uk
[2] School of Mathematics, University of New South Wales, Sydney NSW 2052, Australia
   E-mail: ganesh@maths.unsw.edu.au
[3] Department of Mathematical Sciences, University of Bath, Bath BA2 7AY, U.K.
   E-mail: igg@maths.bath.ac.uk

24.1 Introduction

The year 2002 saw both the 50th anniversary of Turing's seminal paper on morphogenesis [33], and the 30th anniversary of Gierer and Meinhardt's equally important paper concerning activator-inhibitor theory [9]. These two papers have had a huge influence on the application of reaction-diffusion pre-pattern theory as a mechanism to describe spatio-temporal pattern formation in many biological systems. Specific applications of the theory (to name but a few) can be found in processes in developmental biology, population biology, ecology and interacting chemical systems. It is not our intention in this chapter to discuss the range of applications – for a comprehensive account of the theory and references to the many other applications, the interested reader is referred to the books [17, 22]. Instead, here we apply reaction-diffusion pre-pattern theory to a specific problem on a spherical domain, that of a growing avascular solid tumour. We also suggest actual chemicals known to be produced by tumours (autocrine growth factors) which could give rise to the pre-patterns and examine their relevance in the light of clinical and experimental observations.

From a general mathematical perspective, the model we present in this chapter is concerned with examining reaction-diffusion systems on the surface of the unit sphere $S = \{\mathbf{x} \in \mathbb{R}^3 : |\mathbf{x}| = 1\}$. The generic reaction-diffusion system which we will analyse (numerically) in this chapter may be written:

$$\mathbf{u}_t = D\Delta_*\mathbf{u} + \mathbf{f}(\mathbf{u}) \qquad (24.1)$$

on the space-time domain $(\mathbf{x}, t) \in S \times [0, \infty)$, where $(u_1, \ldots, u_s)^T = \mathbf{u} = \mathbf{u}(\mathbf{x}, t)$ is a vector (for example, of chemical concentrations such as growth activating and inhibiting factors), $D = \mathrm{diag}\{d_1, \ldots, d_s\}$ is a diagonal matrix of positive diffusion coefficients, Δ_* is the Laplace-Beltrami operator:

$$\Delta_* u = \frac{1}{\sin \theta} \left\{ \frac{\partial}{\partial \theta} \left(\sin \theta \frac{\partial u}{\partial \theta} \right) + \frac{1}{\sin \theta} \frac{\partial^2 u}{\partial \phi^2} \right\} ,$$

and $\mathbf{f} : \mathbb{R}^s \to \mathbb{R}^s$ is a (nonlinear) autonomous vector-valued function representing the reaction kinetics.

For our purposes in this chapter we restrict to the special two-species case

$$\begin{aligned} u_t &= \Delta_* u + \gamma f(u, v), \\ v_t &= d\Delta_* v + \gamma g(u, v), \end{aligned} \tag{24.2}$$

with d, γ given positive parameters and f, g given functions, but in fact the method applies equally well to any number of chemical species. In general system (24.1) arises naturally in studies of pre-pattern formation in biological systems [21, 22]. There it is of interest to study the stability of spatially homogeneous steady states of (24.1) with respect to the diffusion represented by $D\Delta_* \mathbf{u}$, and in particular to identify spatial patterns which evolve in practice from unstable (in the Turing sense) homogeneous steady states. In the next section we describe the application of pre-pattern theory to solid tumour growth and present some simulations of our system. A more detailed description of the background biology, modelling and numerical technique used in this chapter can be found in [4].

24.2 The rôle of pre-pattern theory in solid tumour growth and invasion

24.2.1 Application to a spherical tumour

Solid tumours are known to progress through two distinct phases of growth - the avascular phase and the vascular phase. During the former growth phase the tumour remains in a diffusion-limited, dormant state while during the latter growth phase, invasion and metastasis may take place. The initial avascular growth phase can be studied in the laboratory by culturing cancer cells in the form of three-dimensional *multicell spheroids* ([20, 30] and references therein). It is well known that these spheroids, whether grown from established tumour cell lines or actual *in vivo* tumour specimens, possess growth kinetics which are very similar to *in vivo* tumours. Typically, these avascular nodules may grow to a few millimetres in diameter depending on the cell types and the culture conditions used, although carcinoma *in vivo* may reach dormancy at a smaller size of between $250 - 500$ μm. Cells towards the centre of the spheroid, being deprived of vital nutrients, die and give rise to a necrotic core. Proliferating cells can be found in the outer three to five cell layers, that is, *essentially on the surface of the tumour*. Lying between these two regions is a layer of quiescent cells, a proportion of which can be recruited into the outer layer of proliferating cells. Much experimental data has been gathered on the internal architecture of spheroids, and studies regarding the distribution of vital nutrients (for example, oxygen) and metabolites within the spheroids have been carried out [8, 35].

The transition from the dormant avascular state to the vascular state, wherein the tumour possesses the ability to invade surrounding tissue and metastasise to distant parts of the body, depends upon its ability to induce new blood vessels from the surrounding tissue to grow and eventually connect with the tumour. This permits vascular growth to take place. It is during this stage of growth that the insidious process of invasion takes place. In certain types of cancer, for example, carcinomas arising within an organ, this process typically consists of columns of cells projecting from the central mass of cells and extending into the surrounding tissue area and the local spread of these carcinomas often assumes an irregular jagged shape.

Prior to successful completion of angiogenesis, the avascular tumour, although dormant (or quasi-dormant) with regard to its growth, is still very much in a "dynamic state of equilibrium", with cell birth and proliferation in balance with cell loss and death. The cancer cells are also known to produce and secrete a variety of growth-activating and growth-inhibiting chemicals [12, 24].

In formulating our mathematical model we take into account certain important experimental/biological observations from multicell spheroid studies and make other reasonable mathematical assumptions, namely:

- We assume that the tumour is perfectly spherical in shape and that it has grown in a radially symmetric manner.
- We assume that the tumour has reached its diffusion-limited avascular maximum size and consists of a large internal necrotic core surrounded by a thin layer of proliferating cells at the surface. The thin layer of live cells essentially defines the surface of the solid tumour.
- Experimental results have demonstrated that tumour cells secrete both growth-inhibiting and growth-activating chemicals in an autocrine manner [24] and that the balance and interaction between these factors play an important role in the development and progression of tumours [7, 19, 24, 25, 27].
- Transforming growth factor betas (TGF-βs) constitute a family of local mediators that regulate the proliferation and functions of many cell types. Indeed TGF-βs have an identified effect of specifically suppressing tumour cell proliferation in many types of cancers [14, 18, 26, 36], including carcinomas.
- TGF-βs also are known to induce apoptosis (cell death) in carcinoma cells [36] and can stimulate the synthesis of the extracellular matrix and equally

importantly the tumour stroma. They have therefore been implicated in controlling cancer invasion [1].

- There is also much evidence to demonstrate that many types of tumour cells (including carcinoma cells) also secrete a variety of growth-activating factors. For example, epidermal growth factor (EGF) and transforming growth factor-α (TGF-α) [37]; basic fibroblast growth factor (bFGF) [32]; platelet-derived growth factor (PDGF) [34]; insulin-like growth factor (IGF) [25]; interleukin-1α (IL-1α) [11] and granulocyte colony-stimulating factor (G-CSF) [19].
- We assume that the production of the growth activating and growth inhibitory factors is restricted to the thin layer of live, proliferating cells at the tumour surface.
- Not only can we identify specific growth inhibitors and activators (as opposed to generic chemicals), but there is direct experimental evidence that in tumour cell lines these chemicals interact and modulate the effect of each other [10, 16].

In addition to the above experimental observations, it is also well-known that the timescale of a growing tumour is very much slower than the diffusion timescale of chemicals. Any chemical which is produced by the tumour cells will therefore diffuse and reach a steady-state distribution within its domain on a much faster timescale than the growing tumour itself. We therefore consider the possibility of the development of a genuine heterogeneous chemical pre-pattern on the surface of a solid tumour which takes place prior to successful angiogenesis. This chemical pre-pattern predisposes cells in certain regions on the surface of the tumour (that is, in regions where the concentration of the growth-activating factor is high) to invasion and subsequently facilitates the vascular, invasive growth. Such cellular heterogeneity in tumours is well documented [2, 13, 23, 29].

The mathematical model we propose consists of a system of reaction-diffusion equations on the surface of a sphere (that is, the tumour surface), modelling the interaction of the growth-activating (u) and growth-inhibiting (v) chemicals which are produced by the tumour cells. The specific system we consider is given by:

$$u_t = \Delta_* u + \gamma(a - u + u^2 v),$$

$$v_t = d\Delta_* v + \gamma(b - u^2 v),$$

(24.3)

where, as before, Δ_* is the Laplace-Beltrami operator, d, γ, a, b are positive constants. Using the numerical scheme developed in [4], we can solve the above system on the surface of a sphere with a set of parameter values (see legend in Fig. 24.1) which satisfy the conditions for Turing-instability and we can obtain spatially heterogeneous steady-state distributions of the two chemicals on the surface of the tumour (cf. [3, 4]). These can be seen in Fig. 24.1.

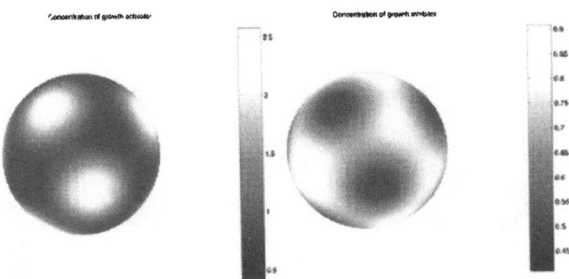

Fig. 24.1. Steady-state, spatially heterogeneous distributions of the chemical concentration profiles u (growth-activator) and v (growth-inhibitor) over the surface of a multicell spheroid. Initial conditions were taken as small perturbations around the spatially-homogeneous steady-state. Parameter values: $d = 25, \gamma = 100, a = 0.2, b = 1$.

It is a well-known feature of solid tumours such as carcinomas that they invade the surrounding local tissue with columns of cells projecting outward from the central mass. We suggest that while a solid tumour is in its avascular, dormant state, a steady-state chemical pre-pattern is set up. This is biologically feasible given the difference in timescales between the tumour growth rate and the diffusion rate of the chemicals [3]. Once angiogenesis takes place and the tumour becomes vascularized, tumour cells which are located on the surface in regions of high concentrations of the growth promoting factor will be stimulated into proliferating faster and begin to invade the local tissue through increased migration. A chemical pre-pattern of this type is also consistent with the observation that tumours can directly manipulate their local environment by secretion of the growth factors. Thus this chemical pre-pattern will not only predispose the tumour cells to higher proliferation and increased mobility but will also directly affect the local surrounding tissue as well, thus facilitating invasion of the tissue by the cells [1, 14].

24.2.2 Model extension: Application to a growing spherical tumour

The results of the previous section were obtained by considering the reaction-diffusion system on a domain of fixed size, that is, the surface of the *unit sphere*. The fact that a tumour grows on a much slower timescale than the diffusion of the chemicals enabled a genuine chemical pre-pattern to form. The model, as described, is therefore most applicable when applied to a solid tumour which has already reached its diffusion-limited avascular size. However, in the case of smaller tumours which are still growing, growth promoting and growth inhibiting chemicals will still be produced by the tumour cells. These chemicals will reach a steady-state distribution (on a faster timescale than the tumour growth rate) and a pre-pattern will be formed. If the tumour

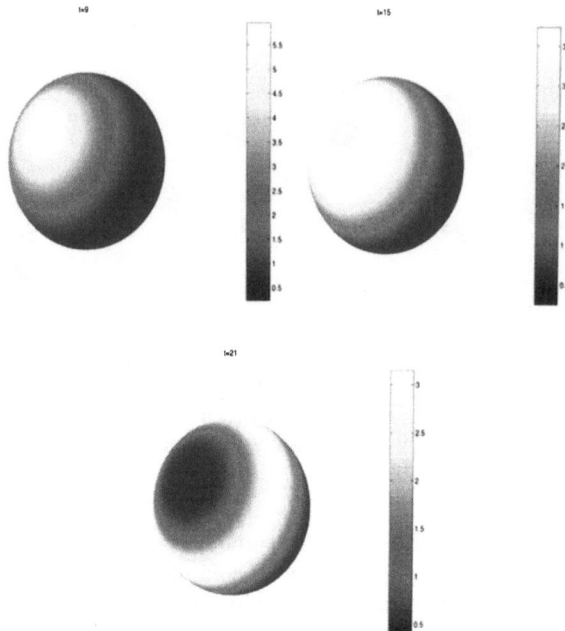

Fig. 24.2. Plots of the concentration profiles of the growth activating chemical, u, over the surface of the multicell spheroid at times $t = 9, 15$ and 21 in the situation of a growing tumour. As time increases the spatially heterogeneous pattern evolves with the changing chemical concentration profiles on the surface of the tumour. Parameter values $a = 0.2, b = 1, \gamma = 5, d = 100$.

is not at the stage of its growth where invasion of the tissue occurs, then it will continue to grow, the chemicals will form a new pre-pattern (on a faster timescale) and so on. Thus a more appropriate and realistic way to model the distribution of the chemicals on the surface of a growing tumour would be to consider the reaction-diffusion system on a growing, time-dependent domain.

We therefore now consider the application of the results of the previous section to the case of the growing domain described above. The reaction-diffusion system is therefore considered on the domain $S(t)$, the surface of the sphere of radius $R(t)$, that is, $S(t) = \{\mathbf{x} \in \mathbb{R}^3 : |\mathbf{x}| = R(t)\}$. The reaction-diffusion system on $S(t)$ (cf. [4]) is then

$$u_t = \frac{1}{[R(t)]^2} \Delta_* u + \gamma(a - u + u^2 v) , \tag{24.4}$$

$$v_t = \frac{d}{[R(t)]^2} \Delta_* v + \gamma(b - u^2 v) , \tag{24.5}$$

which is to be solved for functions u, v of θ, ϕ and t (cf. the formulation of [5]).

It is possible to prescribe in detail the specific growth law of an avascular tumour and then couple the ODE modelling this to (24.4) and (24.5). However, since we are interested only in qualitative results in this paper, it is sufficient to consider monotonically increasing functions of time for $R(t)$, and here we restrict to the case $R(t) = 1 + \alpha t$, $\alpha > 0$, representing linear growth. We solved (24.4) and (24.5) using our numerical scheme with $\alpha = 0.1$. The actual spatio-temporal distributions of the growth activating chemical on the surface of the tumour are given in Fig. 24.2. Clearly one can see that the spatial pattern generated is heterogeneous and changes with time (a similar result is seen for the growth-inhibiting chemical). The results of these numerical simulations give a predictive insight into the "dynamic activity" which occurs during the growth of solid tumours and are consistent with the experimentally and clinically observed proliferative heterogeneity of cancer cells in solid tumours [2, 8, 13, 23, 29].

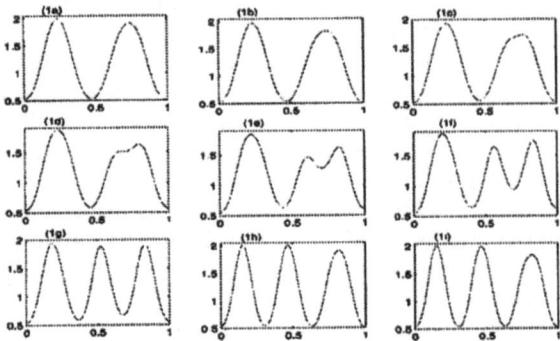

Fig. 24.3. Numerical simulation of (24.3) on a growing domain. The results show the dynamic evolution of a pattern in a 1-dimensional growing domain $x = [0, s(t)]$ where growth occurs only at the right hand boundary $x = s(t)$. The initial pattern shows "two stripes" with a third "stripe" being added once the domain has grown large enough. As the domain continues to grow further stripes are added one-by-one at the right hand boundary. See [15] for full details.

24.3 Discussion and conclusions

In this paper we have studied a system of reaction-diffusion equations on the surface of a sphere. We have applied the pre-pattern theory (Turing-type models) of reaction-diffusion systems to a novel biological (pathological) problem –

that of the growth of solid tumours, for example, carcinomas – and, moreover, have suggested a number of specific chemicals which may be involved in this process. Finally, we have studied the system of reaction-diffusion equations on a growing domain using a moving-boundary formulation. This formulation models the dynamic process of tumour growth more realistically.

We have also shown in this paper that the spatially heterogeneous chemical pre-patterns which arise on the surface of a sphere may be an important process occurring in solid tumour growth and may help to explain certain clinically and experimentally observed phenomena in carcinoma and multi-cell spheroids, that is, the heterogeneous distributions of proliferating cells in carcinoma and multicell spheroids and the characteristic invasive patterns of these cancers. Of course there are many other factors and processes which are involved in tumour growth, for example, the distribution of nutrient supply to the cancer cells. These are also very important and we certainly do not claim that the results of the model provide a complete answer to the problem of cancer growth and invasion but rather may be an important part of the complex overall mechanisms governing solid tumour growth (cf. [24, 25, 27, 31]).

The application of our system of reaction-diffusion equations to a growing, spherical domain has enabled us to model more realistically an actual growing solid tumour and we believe that the results of the numerical simulations of Sec. 24.2.2 are highly consistent with *in vitro* experimentally observed proliferative heterogeneity of cancer cells in solid tumours at all stages of their development [2, 8, 13, 23, 29]. This aspect of reaction-diffusion theory (applications on growing domains) is attracting a good deal of interest and is providing a better explanation for many hitherto unexplained aspects of pattern formation (see, for example, [5, 6, 15]). Figure 24.3 shows how a pattern emerges dynamically in a 1-dimensional growing domain $x = [0, s(t)]$, which is growing only at the boundary $x = s(t)$ [15].

Finally, the results of the model suggest that some degree of control or regulation of cancer invasion may be possible through manipulation of the levels of growth factors as has already been suggested experimentally [1, 14], that is, it may be possible to intervene with the growth factor kinetics in such a way as to ensure that, even if one cannot halt the growth of a cancer, one may be able to prevent the highly heterogeneous distributions of proliferating cells from occurring [7, 19, 28]. Given that a solid tumour can be detected at an early enough stage in its development (for example, the avascular stage), this fact alone may prevent the irregular spread of columns of cancer cells into the surrounding tissue and may reduce the likelihood of the secondary spread of the disease.

References

1. Albo, D., Berger, D.H., Wang, T.N., Xu, X.L., Rothman, V. and Tuszynski, G.P. (1997). Thrombospondin-1 and transforming-growth-factor-beta1 promote

breast tumor cell invasion through up-regulation of the plasminogen/plasmin system. *Surgery*, **122**, 493–499

2. Becciolini, A., Balzi, M., Barbarisi, M., Faraoni, P., Biggeri, A. and Potten, C.S. (1997). 3H-thymidine labelling index (TLI) as a marker of tumour growth heterogeneity: evaluation in human solid carcinomas. *Cell Prolif.* **30**, 117–126

3. Chaplain, M.A.J. (1995). Reaction-diffusion prepatterning and its potential role in tumor invasion. *J. Biol. Sys.*, **3**, 929–936

4. Chaplain, M.A.J., Ganesh, M. and Graham, I.G. (2001). Spatio-temporal pattern formation on spherical surfaces: numerical simulation and application to solid tumour growth. *J. Math. Biol.*, **42**, 387–423

5. Crampin, E.J., Gaffney, E.A. and Maini, P.K. (1999). Reaction and diffusion on growing domains: Scenarios for robust pattern formation. *Bull. Math. Biol.*, **61**, 1093-1120

6. Crampin, E.J., Hackborn, W.W. and Maini, P.K. (2002). Pattern formation in reaction-diffusion models with nonuniform domain growth. *Bull. Math. Biol.*, **64**, 747–769

7. Ethier, S.P. (1995). Growth factor synthesis and human breast cancer progression. *J. Natl. Cancer Inst.*, **87**, 964–973

8. Freyer, J.P. and Sutherland, R.M. (1986). Proliferative and clonogenic heterogeneity of cells from EMT6/Ro multicellular spheroids induced by the glucose and oxygen supply. *Cancer Res.*, **46**, 3513–3520

9. Gierer, A. and Meinhardt, H. (1972). A theory of biological pattern formation. *Kybernetik*, **12**, 30–39

10. Hata, A., Shi, Y.G. and Massagué, J. (1998). TGF-β signaling and cancer: structural and functional consequences of mutations in Smads. *Molecular Medicine Today*, **4**, 257–262

11. Ito, R., Kitadai, Y., Kyo, E., Yokozaki, H., Yasui, W., Yamashita, U., Nikai, H. and Tahara, E. (1993). Interleukin 1α acts as an autocrine growth stimulator for human gastric carcinoma cells. *Cancer Res.*, **53**, 4102–4106

12. Iversen, O.H. (1991). The hunt for endogenous growth-inhibitory and or tumor suppression factors - their role in physiological and pathological growth-regulation. *Adv. Cancer Res.*, **57**, 413–453

13. Jannink, I., Risberg, B., Vandiest, P.J., and Baak, J.P.A. (1996). Heterogeneity of mitotic-activity in breast-cancer. *Histopathol.*, **29**, 421–428

14. Keski-Oja, J., Postlethwaite, A.E. and Moses, H.L. (1988). Transforming growth factors and the regulation of malignant cell growth and invasion. *Cancer Invest.*, **6**, 705–724

15. Lolas, G. (1999). *Spatio-temporal Pattern Formation and Reaction Diffusion Equations*. MSc Thesis, University of Dundee, Dundee

16. Massagué, J. (1998). TGFβ signal transduction. *Annu. Rev. Biochem.*, **67**, 753–791

17. Meinhardt, H. (1982). *Models of Biological Pattern Formation*. Academic Press, London

18. Moses, M.L., Yang, E.Y. and Pietenpol, J.A. (1990). TGF-β stimulation and inhibition of cell proliferation: new mechanistic insights. *Cell*, **63**, 245–247

19. Mueller, M.M., Herold-Mende, C.C., Riede, D., Lange, M., Steiner, H.-H. and Fusenig, N.E. (1999). Autocrine growth regulation by granulocyte colony-stimulating factor and granulocyte macrophage colony-stimulating factor in human gliomas with tumor progression. *Am. J. Pathol.*, **155**, 1557–1567

20. Mueller-Klieser, W. (1987). Multicellular spheroids: A review on cellular aggregates in cancer research. *J. Cancer Res. Clin. Oncol.*, **113**, 101–122
21. Murray, J.D. (1982). Parameter space for Turing instability in reaction diffusion mechanisms: a comparison of models. *J. theor. Biol.*. **98**, 143–163
22. Murray, J.D. (1993). *Mathematical Biology* (Second Edition). Springer-Verlag, London
23. Palmqvist, R., Oberg, A., Bergstrom, C., Rutegard, J.N., Zackrisson, B. and Stenling, R. (1998). Systematic heterogeneity and prognostic significance of cell proliferation in colorectal cancer. *Br. J. Cancer*, **77**, 917–925
24. Pusztai, L., Lewis, C.E. and Yap, E. (eds.). (1996). *Cell Proliferation in Cancer: Regulatory Mechanisms of Neoplastic Cell Growth.* Oxford University Press, Oxford
25. Quinn, K.A., Treston, A.M., Unsworth, E.J., Miller, M.-J., Vos, M., Grimley, C., Battey, J., Mulshine, J.L. and Cuttitta, F. (1996). Insulin-like growth factor expression in human cancer cell lines. *J. Biol. Chem.*, **271**, 11477–11483
26. Rahimi, N., Tremblay, E., McAdam, L., Roberts, A. and Elliott, B. (1998). Autocrine secretion of TGF-beta 1 and TGF-beta 2 by pre-adipocytes and adipocytes: A potent negative regulator of adipocyte differentiation and proliferation of mammary carcinoma cells. *In Vitro Cell. Dev. Biol. Animal*, **34**, 412–420
27. Rosfjord, E.C. and Dickson, R.B. (1999). Growth factors, apoptosis and survival of mammary epithelial cells. *J. Mammary Gland Biol. Neoplasia*, **4**, 229–237
28. Rozengurt, E. (1999). Autocrine loops, signal transduction and cell cycle abnormalities in the molecular biology of lung cancer. *Curr. Opin. Oncol.*, **11**, 116–122
29. Sessa, F., Bonato, M., Bisoni, D., Bosi, F. and Capella, C. (1997). Evidence of a wide heterogeneity in cancer cell population in gallbladder adenocarcinomas. *Lab. Invest.*, **76**, 860
30. Sutherland, R.M. (1988). Cell and environment interactions in tumor microregions: the multicell spheroid model. *Science* **240**, 177–184
31. Tahara, E., Yasui W. and Yokozaki, H. (1996). Abnormal growth factor networks in neoplasia, chapter 6, pp. 133-153, in: L. Pusztai, C.E. Lewis and E. Yap (eds.). *Cell Proliferation in Cancer: Regulatory Mechanisms of Neoplastic Cell Growth.* Oxford University Press, Oxford
32. Takahashi, J.A., Mori, H., Fukumoto, M., Igarashi, K., Jaye, M., Oda, Y., Kikuchi, H. and Hatanaka, M. (1990). Gene expression of fibroblast growth factors in human gliomas and meningiomas: demonstration of cellular source of basic fibroblast growth factor mRNA and peptide in tumor tissues. *Proc. Natl. Acad. Sci. USA*, **87**, 5710–5714
33. Turing, A.M. (1952). The chemical basis of morphogenesis. *Phil. Trans. Roy. Soc. Lond.*, **B237**, 37–72
34. Westermark, B. and Heldin, C.-H. (1991). Platelet-derived growth factor in autocrine transformation. *Cancer Res.*, **51**, 5087–5092
35. Wibe, E., Lindmo, T. and Kaalhus, O. (1981). Cell kinetic characteristics in different parts of multicellular spheroids of human origin. *Cell Tissue Kinet.*, **14**, 639–651
36. Yanagihara, K. and Tsumuraya, M. (1992). Transforming growth factor $\beta 1$ induces apoptotic cell death in cultured human gastric carcinoma cells. *Cancer Res.*, **52**, 4042–4045

37. Yoshida, K., Kyo, E., Tsujino, T., Sano, T., Niimoto, M., and Tahara, E. (1990). Expression of epidermal growth factor, transforming growth factor-α and their receptor genes in human carcinomas: implication for autocrine growth. *Cancer Res.*, **81**, 43–51

The Formation of Branching Systems in Human Organs

Ryuji Takaki[1], Hiroko Kitaoka[1], Tetsuji Nishioka[1], Hironobu Kaneko[1], Masaniri Nara[1], and Hideo Shimizu[2]

[1] Tokyo University of Agriculture and Technology, Koganei, Tokyo 184-8588, Japan
[2] Shonan-Kamakura Hospital, Kamakura, Kanagawa 247-8533, Japan

Summary. In this chapter computer algorithms for the formation of branching systems in human organs are proposed and results of simulations are shown. As specific examples of branching systems the lung airway, the liver blood vessels and the liver capillary network are chosen. In setting up algorithms, functions of these organs and anatomical data are considered, where the former is applied for rules to determine details of structures and the latter mainly for boundary conditions. The resulting computer-generated structures show a good resemblance with real organs, and some quantitative comparisons with observations are made to support the present algorithms.

Keywords: human organs, computer simulation, bifurcation, lung, airway, liver, blood vessel, capillary

25.1 Introduction

Numerical construction of human organs is a big challenge, in the sense that the organs are highly complex systems and that they provide us with a basic problem of the relation between forms and functions. There is also an expectation that a human body constructed artificially within a computer will serve as an efficient tool for basic studies in physiology and medicine. The present authors began this activity a few years ago when one of the present authors (Kitaoka), who is a medical doctor, proposed a collaboration with another (Takaki). This kind of work is essentially interdisciplinary because it is possible only through joint work by those from both mechanical and medical sciences.

There are some motivations for this work. First, the National Laboratory of Medicine (USA) has a large database called the "Visible Human", which includes complete anatomical data from a real human body obtained by CT technology and it is arailable for reference through the internet [12]. This data are used for education and simulation of surgery. This project reminds us of an idea that the accumulation of data itself is valuable because it will stimulate a range of research in the future. On the other hand, numerically constructed organ structures are a different kind of useful data, which will serve as models for basic studies of organ functions.

Secondly, cancers in some parts of the human body are difficult to detect. Since X-ray photos are 2D projections of 3D structures and important signs are often disturbed, lung cancer has the worst detection rate. A better method is an analysis of 3D CT-data, but there arises the problem of how to analyze a large quantity of 3D data. A model organ, if it is constructed and is deformed to simulate pathological states, will be useful for developing a new method of analyzing 3D data.

Thirdly, there is the fundamental and common question of how shapes of human organs are realized. Motivated by these, the present authors decided to construct branching structures of organs within a computer by setting up suitable algorithms. We began with the lung airway [7] and the labyrinth-like structure of acinus connected to each terminal of airway [8]. Recently we began simulating the liver blood vessels. The reason why we began with the lung airway is that it is a tree-like branching system formed in a free space filled with liquid (amniotic fluid). Therefore, morphogenesis of this structure will not be affected much by other parts of body, hence a relatively simple algorithm will lead to a successful result.

25.2 Simulation of lung airway

Simulation of the lung airway has been carried out in [4, 9, 11], but their models do not consider 3D structure of the branching system. A model treating 3D structure is proposed by the present authors [7]. In this model an algorithm is set up according to the following requirements connected to functions of the lung:

(i) The airway supplies air to all parts in the lung equally, hence terminals should be distributed uniformly.
(ii) For an effective gas supply the flow rate of each branch should match the volume of its basin (the region to which the branch supplies air) (see Fig. 25.1).
(iii) Total energy loss due to air viscosity should be minimized.
(iv) For narrow branches diffusion is more effective than convection, hence branches should terminate at a certain flow rate, where terminals are connected to acini.
(v) The location and shape of the main trunk and shape of the outer boundary are determined as boundary conditions, hence they should be fixed based on anatomical data.

Note that the structure of acini has been considered separately and is not discussed here [8].

The algorithm for 3D tree structure of airways with the above requirements is composed of the following two kinds of rules, geometrical and fluid-dynamical. In particular, the following formulae are used in setting up fluid-dynamical rules.

In dichotomous branching a relation between the parent diameter d_0 to those of the daughters d_1, d_2 (see Fig. 25.2) is given by

$$d_0^m = d_1^m + d_2^m. \tag{25.1}$$

Morphometric data gave $n = 2.6$ or 2.7 for arteries and $n = 2.4 \sim 2.9$ for airways. The theoretical value from an energy minimum principle is $r = 3$ [3]. By introducing a flow dividing ratio r $(0 < r \leq 0.5)$, so that the flow rates in daughters are r and $1 - r$, Eq. (25.1) is rewritten as

$$d_1 = d_0 r^{1/n},$$
$$d_2 = d_0 (1 - r)^{1/n}. \tag{25.2}$$

The relation between the branching angle and the flow rate was obtained theoretically in [1, 5], which, after combination with Eq. (25.2) becomes:

$$\cos \theta_1 = \frac{1 + r^{4/n} - (1 - r)^{4/n}}{2r^{2/n}},$$

$$\cos \theta_2 = \frac{1 + (1 - r)^{4/n} - r^{4/n}}{2(1 - r)^{2/n}}. \tag{25.3}$$

This formula is supported qualitatively by observations [2].

Now, the geometrical rules (1-5) and fluid-dynamical rules (6-9) are listed as follows:

Rule 1: Branching is dichotomous.

Rule 2: A parent branch and the two daughter branches lie on the same plane (branching plane).

Rule 3: The length of each daughter branch is three times its diameter (based on anatomical data).

Rule 4: The branching plane is perpendicular to the preceding branching plane.

Rule 5: The basin of a parent branch is divided into two daughter basins by a plane (space dividing plane), which is perpendicular to the branching plane.

Rule 6: The flow rate is conserved after branching.

Rule 7: The flow dividing ratio r is equal to the volume dividing ratio of the daughter basins.

Rule 8: The diameters and branching angles of the two daughters are determined by Eqs. (25.2), (25.3).

Rule 9: Branches terminate if flow rates are less than a threshold or if they come out of their basins.

R. Takaki et al.

298

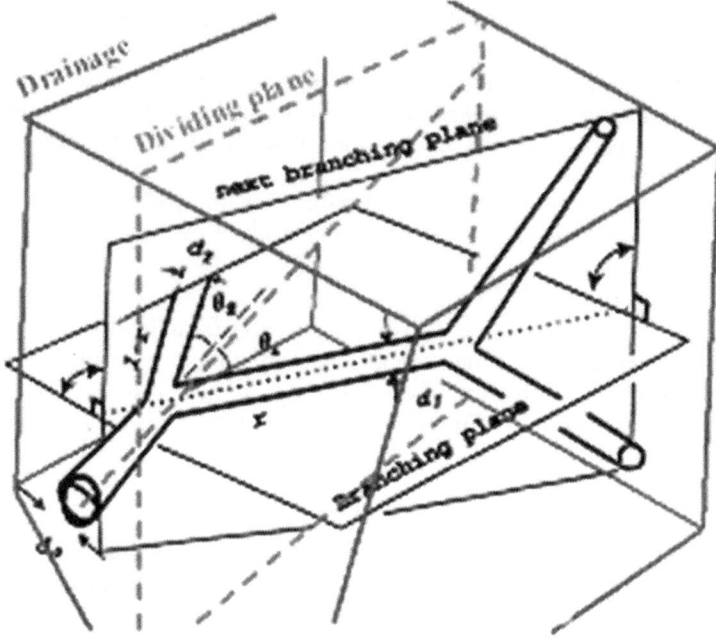

Fig. 25.1. Schematic diagram showing the relation between branching and division of a basin.

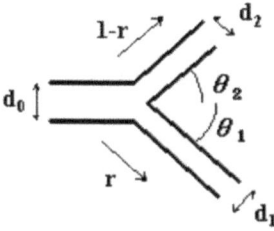

Fig. 25.2. Schematic diagram showing the relationship between the parameters at the instant of branching.

Figure 25.3 shows a computer-generated image of an airway. This model resembles the real airway remarkably. Some quantitative data are estimated from the present results. It has about $27,000$ terminals, and mean generation numbers 17.6 ± 3.4 down to the terminals. If the diameter of the main trunk is assumed to be $18 \ mm$, the mean diameter of terminals, the lung volume, the total airway volume and the mean volume of acini become 0.48 ± 0.06

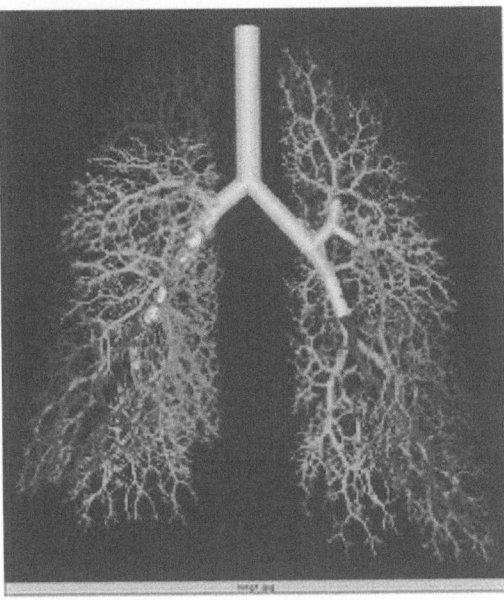

Fig. 25.3. Result of a simulation of a lung as a branched airway.

mm, $6,388$ ml, 174 ml and 0.211 ± 0.076 ml, respectively. These values are consistent with observations.

25.3 Construction of liver blood vessels

The main function of the liver is to detoxify poisonous materials which have been eaten. The inlet blood vessel coming from the digestive organs is called a portal vein, whose entrance is located at the lower side of the liver. The outlet vessel is called a hepatic vein and exits the liver at its rear side. Between the terminals of the portal vein and the hepatic vein capillary networks, called sinusoids, are developed. In this section the algorithm and result of simulations for the combination of the portal vein and the hepatic vein are explained.

According to the observation in [13] tip parts of the portal vein and the hepatic vein are separated by an almost uniform distance (about 350 μm). This fact suggests that the simplest configuration of the blood vessels in their numerical simulation would be a branching system on a cubic grid, where one of both kinds of veins goes along the intermediate positions of the grid as shown in Fig. 25.4.

The main points of the present algorithm used to construct a branching system are summarized below, and the algorithm for both of the two veins is set up as shown in Fig. 25.5

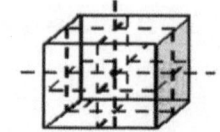

Solid: Grid for portal vein

Dashed: Grid for hepatic vein

Fig. 25.4. Schematic diagram showing the cubic grids used for constructing liver blood vessels.

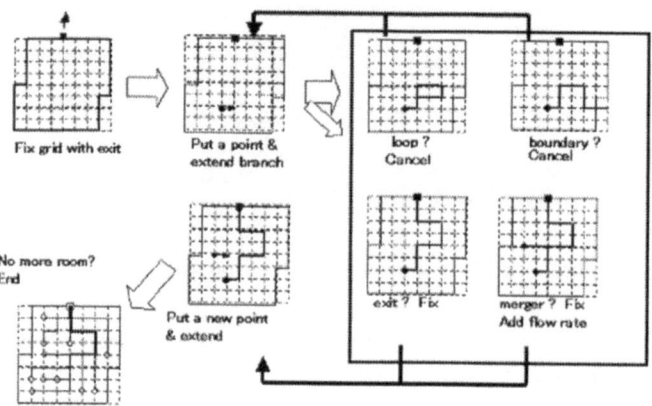

Fig. 25.5. Schematic diagram showing the algorithm used for the formation of 3D liver blood vessels.

(1) Both vessels are formed on cubic grids.

(2) An outer boundary is fixed according to anatomical data.

(3) An entrance and an exit are fixed according to anatomical data.

(4) Branches are formed beginning with tip points. When a branch meets another the two merge to form a thicker branch, as is done in the simulation of rivers.

(5) The type of branching is not necessarily dichotomous.

The simple application of this algorithm led to an unsatisfactory result of a structure which was not similar to the real one. Therefore, the tendency of a branch direction to the exit is introduced through a new parameter $C1$. Figure 25.6 shows the results of simulations with various choices of $C1$ in the

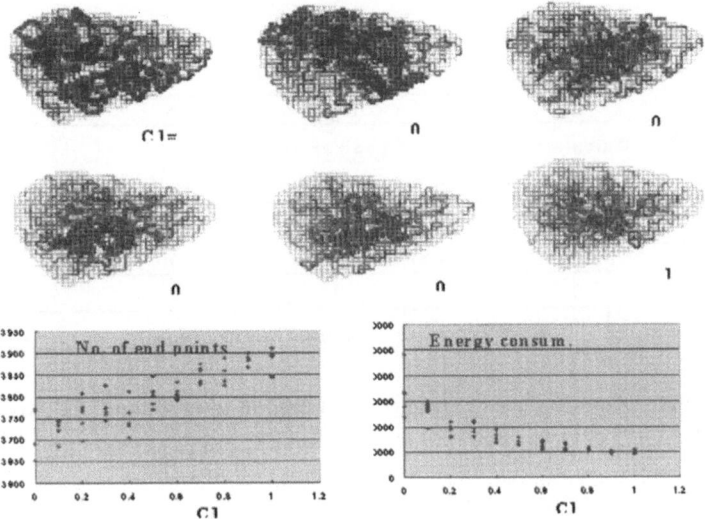

Fig. 25.6. Results of a simulation of liver blood vessels according to the algorithm shown in Fig.25.5. $C1$ is a parameter for the probability preference to the direction to the exit (from left to right, $C1=0$(no preference), 0.2, 0.4, 0.6, 0.8, 1.0).

upper row. For evaluation of the resulting systems, the number of tip points and the viscous energy consumption in the whole system (a flow rate 1 is given at every tip point) are computed, as shown in the lower row. This figure suggests that the maximum preference of direction ($C1 = 1$) gives the best result. Quantitative comparison with observations has not yet been made.

25.4 Construction of liver capillary

In simulating liver capillaries the following characteristics of the liver tissue should be kept in mind. First, capillaries should occupy the space of the liver in the optimum way so that they have sufficient blood flow of blood. Second, capillaries should be distributed uniformly in the liver and touch all cells. Third, the capillaries form a 3D network, while the group of liver cells also forms a network. Both networks are dual in the sense that one constitutes the outer part of another.

In order to construct such a capillary system the algorithm shown in Fig. 25.7 is set up. It is based on the idea that an efficient network is constructed by eliminating branches with small flow rates. This algorithm has already been applied successfully by [6].

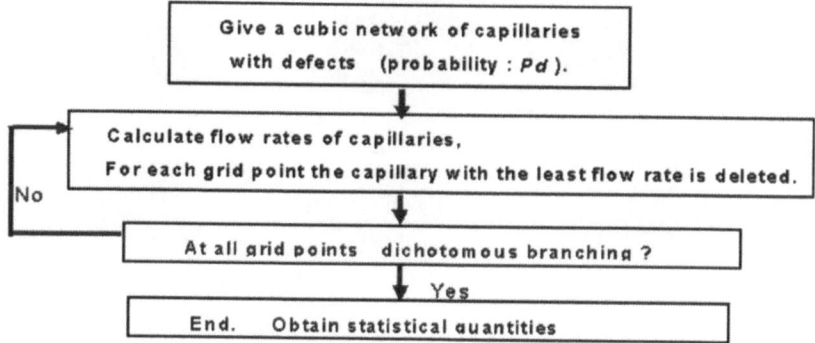

Fig. 25.7. Diagram showing the algorithm used for constructing the liver capillary system.

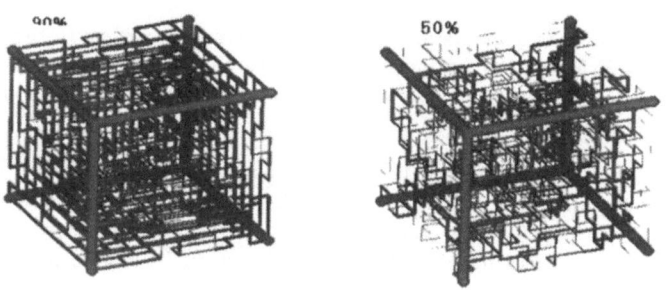

Fig. 25.8. Results of simulations of liver capillaries. 90% (left) or 50%(right) of initial branches are randomly chosen and eliminated. The six thick edges of the cubic are the portal vein and the hepatic vein. The thickness of the branches indicates the flow rate.

As a computational space $10 \times 10 \times 10$ grids are arranged in a cubic region. Three edges of the cube are assigned as an upstream vessel (portal vein) and three edges as a downstream vessel (hepatic vein). The flow rate in each branch is calculated by giving a high (low) blood pressure in the upstream (downstream) vessel. A method developed in electrical engineering is applied to calculate the flow rates.

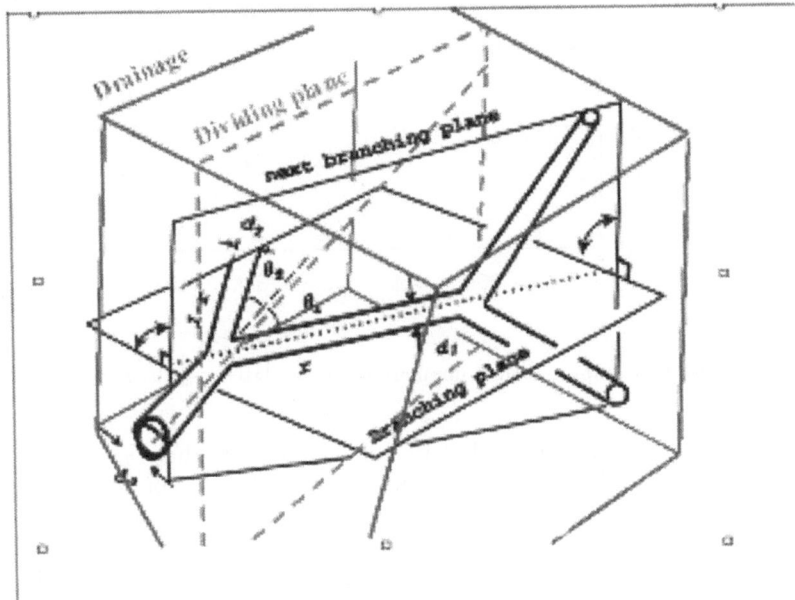

Fig. 25.9. Histograms of number of branches which are in contact with one cell.

The resulting capillary networks having initial defect rates of 90% and 50% respectively are shown in Fig. 25.8, which have only dichotomous branching at each grid point. In this figure the flow rate in each branch is indicated by thickness of the branch (in the original figure the pressure was indicated by color).

Histograms of the numbers of edges making contact with one liver cell (one grid region in the computational space is considered as a cell) are shown in Fig. 25.9 for several values of initial defect rate. For defect rates between 70% and 99% most cells are in contact with 4 to 6 branches, which means that one of the requirements (capillaries are distributed uniformly) is satisfied.

Comparison of the simulated networks with real liver capillaries was made by computing the Betti number of the networks. The Betti number is defined as follows in terms of the numbers of edges and vertices:

$$N_{Betti} = N_{edge} - N_{vertex} + 1. \tag{25.4}$$

Computed Betti numbers based on this formula are shown in Fig. 25.10. As is seen from this figure, N_{Betti} is nearly constant with a value of about 440 for initial defect rates between 50% and 99%, and below 50% it decreases (about 220 at 30%).

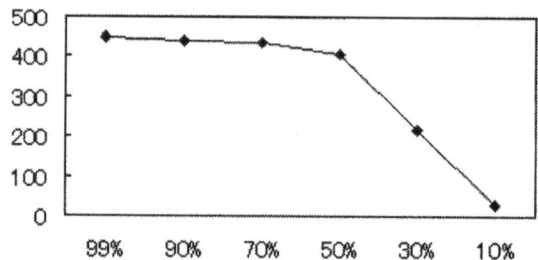

Fig. 25.10. Computed Betti numbers from simulated capillary networks.

From this constancy of the Betti number we can see that the value of 440 is a rather universal value for networks with dichotomous branching and an optimized flow distribution.

Measurement of Betti number was made by one of the present authors (Shimizu) from a specimen of $200 \times 200 \times 80 \ \mu m^3$ [10]. He obtained the following values for normal and pathological cases (cirrhosis):

$$N_{Betti} = \begin{cases} 181 \pm 24 \ (normal) \\ 85 \pm 19 \ \ (cirrhosis). \end{cases} \tag{25.5}$$

Since the computation is made without any dimension of scale, we must introduce a dimension for comparison with these data. Considering the fact that a typical size for a liver cell is 20 μm, the specimen for observation corresponds to a grid with $10 \times 10 \times 4$ points, which is $\frac{2}{5}$ of the grid used for simulation. Then, from the computed values 440 (99%-50%) and 220 (30%) we obtain values 176 and 88, respectively, by multiplying this fraction. These values agree well with the above data.

This agreement suggests that the Betti number in the normal state of a real liver is the result of the optimization of a dichotomous network system, and that an extreme elimination of branches at the beginning (owing to some pathological condition) results in the reduction of the Betti number to a value of half (or less) its original size.

25.5 Concluding remarks

From the above results we can conclude that algorithms to simulate human organs with several rules, which are given partly from required functions of organs and partly from anatomical data, can produce realistic organs. The results of the simulations are acceptable because they look quite similar to real organs and because some quantitative agreements are obtained.

Applications of these simulations seem to be possible, especially for the simulation of pathological states, which will be considered in future.

In conclusion, the term "computational anatomy" is proposed for the construction of organs within a computer. It is expected to open up a wide range of research fields.

References

1. Kamiya, A., Togawa, T., and Matsumoto, A. (1970). Optimum structure of branching system of blood vessels. *Med. Electron. and Bioeng.*, **8**, 136 (in japanese).
2. Thurlbeck, A., and Horsfield, K. (1980). Branching angles in the bronchial tree related to order of branching, *Respir. Physiol.*, **41**, 173-181.
3. Murray, C.D. (1926). The physiological principle of minimum work. I., *Proc. Nat'l. Acad. Sci. USA*, **12**, 207-214.
4. Weibel, E.R. (1963). *Morphometry of the Human Lung*, Academic, New York. .
5. Uylings, H.B.M. (1977). Optimization of diameters and bifurcation angles in lung and vascular tree structures, *Bull. Math. Biol.*, **39**, 509-519.
6. Honda, H. (1997). Formation of the branching pattern of blood vessels in the wall of the avian yolk sac studied by a computer simulation, Develop. *Growth Differ.*, **39**, 581-589.
7. Kitaoka, H., Takaki, R., and Suki, B. (1999). A Three-Dimensional Model of the Human Air way Tree, *J.Appl.Physiol.*, **87**, 2207-2217.
8. Kitaoka, H., Tamura, S., and Takaki, R. (2000). A three-dimensional model of the human pulmonary acinus, *J.Appl.Physiol.*, **88**, 2260-2268.
9. Kitaoka, H., and Suki, B. (1997). Branching design of the bronchial tree based on a diameter-flow relationship. *J.Appl.Physiol.*, **82**, 968-976.
10. Shimizu, H., and Yokoyama, T. (1993). Three-dimensional structural changes of hepatic sinusoids in cirrhosis providing an increase in vascular resistance of portal hypertension. *Acta Pathologica Japonica*, **43**, , 625-634.
11. Horsfield, K., Dart, G, Olson, D.E., Filley, G.F., and Cumming, G. (1971). Models of the human bronchial tree, *J.Appl.Physiol.*, **31**, 202-217.
12. National Laboratory of Medicine:
    http://www.nlm.nih.gov/research/visible/photos.html
13. Takahashi, T., and Chiba, T. (1990). Three-D computational geometry: the "Form" of vascular tree as expressed by the distribution of distance in the space, its organ difference and significance in blood flow, *Science on Form*, Proc. 2nd ISSF, ed. S. Ishizaka, KTK Sci. Publ., 17-30

Computer Simulation on Morphogenesis of Colonic Neoplasia

Sasuke Miyazima[1], Kenzo Ono[2] and Tomomasa Nagamine[3]

[1] Department of Bioscience and Biotechnology, Chubu University, Kasugai,Aichi 487-8501, Japan.
   e-mail: miyazima@isc.chubu.ac.jp
[2] Department of Pathology, Tosei General Hospital, Nishioiwakecho 160, Seto, Aichi 489-0065, Japan.
[3] Department Of Human Science, Shiga Cultural Junior college, Fuse 29, Yokaichi, Shiga 527-0081, Japan.

26.1 Introduction

There are many serious problems facing human beings which must be overcome in a short time, for example, population growth, food and energy problems including CO^2 reduction, diseases such as AIDS and various types of cancer as well as wars. Cancer is one of worst things that humans cannot control at the moment, even though a great number of scientists are researching it. Now one of the important actions against cancer is to observe it as early as possible or predict it. If we could find it at the initial stage, most patients could recover.

It would be impossible to observe tumor morphogenesis in vivo with the passage of time and also it is difficult to get an objective indicator of cancer from observation, i.e., how quickly the cancer grows. Even an amateur can say if cells are well or if they are completely invaded by cancer. In this study we simplified the histological architecture of human colons by computer graphics and simulated growth of colonic neoplasia under conditions as close as possible to natural conditions. We set up 5 main parameters to reproduce tumor cell growth in this simulation ; 1) ability to proliferate which is characterised by the doubling time, 2) contact inhibition between tumor cells, 3) nutritional factors, measured by the distance from the nearest blood vessel, 4) competition to gain space, and 5) cell adhesion for tumor proliferation. Among these 5 conditions, the first 4 directly affect cell proliferation. By various combinations of these conditions, we could elucidate what factors are essential for forming the shape of tumor tissue. Our attempt will serve a new tool to analyze the course of tumor morphogenesis and will give a strict indicator of tumor status.

26.2 Characteristic points of neoplasm

An abnormal mass of tissue makes the exceeding growth, is uncoordinated with that of the normal tissues and persists in the same excessive manner after cessation of the stimuli which evoked the change. Abnormal mass is purposeless, preys on the host and is virtually autonomous. At first we can classify the cell growth into epithelial and non-epithelial proliferation, each of which is classified into two cases, benign and malignant. Typical cancers, and characteristic properties of benign and malignant neoplasia are given in Tables 26.1 26.2.

Table 26.1. Classification of neoplasia

Epithelial	↗	Benign	Adenoma etc.
	↘	Malignant	Carcinoma
Non-epithelial	↗	Benign	Leiomyoma etc.
	↘	Malignant	Sarcoma

* Others : Hematopoietic, Neuroectodermal tumors

Table 26.2. Characteristic properties of benign neoplasia vs. malignant neoplasia from the point of view of gross and microscopic findings

		Benign	Malignant
Gross findings	Growth rate	slow	rapid
	Growth style	exposive	invasive
	Boundary	clear	unclear
	Adhesion	-	+
	Metastasis	-	+
	Recurrence	rare	often
	Risk of life	rare	high

		Benign	Malignant
Microscopic findings	Hemorrhage & Necrosis	rare	often
	Atypia	slightly	markedly
	Nucleus	small	large
	Nucleoles	small	large
	N/C ratio	small	large
	Differentiation	well	poorer
	Mitosis	rare	many

26.3 Grade of differentiation, atypia and polarity

It is very important to know the grade of differentiation, i.e., degree of similarity of tumor cells to normal cells and grade of atypia, i.e., degree of difference of tumor cells with normal cells. These two grades are complementary properties to each other. Another important factor is polarity. In a secretary cell, the secretary side and the basal side are morphologically defined. In the secretary side, there are microvilli, and the nucleus usually localizes on the basal side. This kind of ordered structure is called "cellular polarity".

Here it is important to describe histological types of epithelial neoplasia of the large intestine. Examples of benign epithelial neoplasia include tubular, tubulovillous and villous adenoma. On the other hand there are adenocarcinoma, mucous carcinoma, signet-ring cell carcinoma, squamous cell carcinoma, adenosquamous carcinoma and others. Typical pictures of normal tissue and some types of carcinoma are shown in Figs. 26.1 - 26.3.

At first we show the normal tissue of the large intestine in Fig.26.1, and some typical tumor tissues are shown in Figs. 26.2 - 26.3. There are many other tumors such as adenoma, carcinoma and mixed patterns of adenoma and carcinoma, and we refer the reader to appropriate textbooks [1].

It is natural that normal cells are created by the normal proliferative process. On the other hand, there are two routes from normal cell to carcinoma cell. The first route is a change from normal to adenoma cell and the succeeding change from it to carcinoma cell at the proliferative zone. The other route is from normal to carcinoma cell directly. Here we take into account of the high ability of proliferation of carcinoma cells in the growth process at the proliferative zone. The normal cell can make a daughter cell only at a site along a choosen direction and the adenoma cell can choose one site of four neighbors, however the carcinoma cell can grow into any directions in our cubic lattice. Therefore the numbers of growth directions are assumed to be 1, 4 and 8 for normal, adenoma and carcinoma cells respectively.

26.4 Methods of simulation

The model is programmed in Java under the following conditions:

1) Change of normal cells into neoplastic cells occurs only in cells of the normal proliferative zone.
2) The carcinoma cells were mainly derived from the adenoma cells or the carcinoma cells.
3) The mutation rate of changing from normal to adenoma cells, from normal to carcinoma cells (de novo carcinoma), from adenoma to carcinoma cells or from carcinoma to poorer differentiated carcinoma cells was set at 0.00005 (the mutation rate is actually 10^{-15} but we changed the rate to a much larger number for shorter calculation).

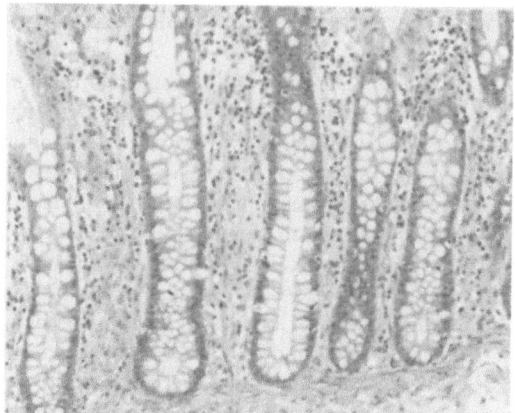

Fig. 26.1. Normal tissue of large intestine.

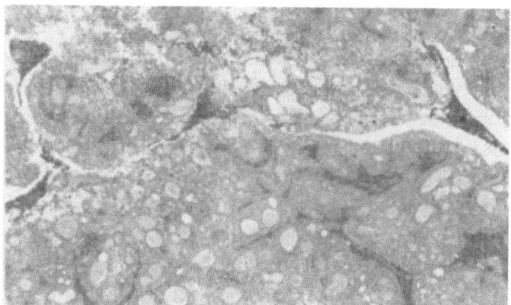

Fig. 26.2. Adenocarcinoma of large intestine, which is moderately differentiated.

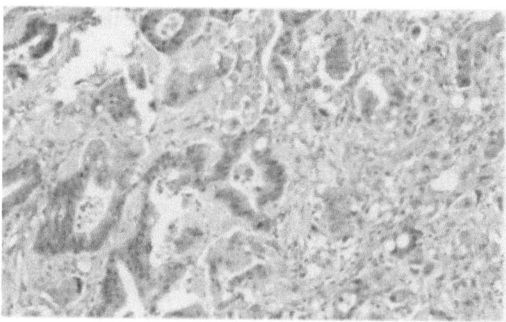

Fig. 26.3. Adenocarcinoma of large intestine, which is a mixed state of poorly differentiated (left) and moderately differentiated parts (right).

4) The movement of the normal and the adenoma cells was restricted to the basement membrane.
5) The adenoma and carcinoma cells had doubling time of 30 and 10 steps respectively. Mutated carcinoma cells to a poorer differentiation state had a doubling time of 5 steps.
6) The properties of cells were basically inherited from the parent cells.
7) The doubling time of the neoplastic cells is affected by mutation, contact inhibition between the neighboring cells, and distance from blood vessel (nutritional factor).
8) The carcinoma cells compete for space and the probability of the proliferation decreases when cells density increases.
9) Every tumor cell has attachable surfaces from 0 to 6 surfaces. If at least one surface between two cells, which are going to attach, is attachable, they are attached. If not, they are detached.

For cell growth, we assume conditions shown in Table 26.3.

Table 26.3. Parameters used in the present simulation

	Normal Cell	Adenoma Cell	Carcinoma Cell
Ability of Proliferation	low	$\longleftrightarrow$	high
Doubling Time	long	$\longleftrightarrow$	short
Directions of Growth (in 2-dimensions)	1	2	4
On Basement Membrane	yes	yes	no
Process Formation	—	+	+

Probability of formation of processes or depressions in adenoma is set up to 0.3 under the following conditions:

1) adenoma cells are always fixed on basement membrane, and
2) selection of process or depression is decided under probability of 0.5.

Factors associated with proliferation and morphogenesis of neoplastic cells are nutrient factors, distance from blood vessels, cell adhesion, polarity, contact inhibition, and competition for space.

26.5 Conclusion and discussion

In the present study our simulations indicate that the morphogenesis of colon carcinoma deeply depends on the number of attachable surfaces of the carcinoma cells as shown in Figs. 26.4-26.6. These examples with attachable surface number 1, 3 and 5, respectively, show the strong effect on the patterns.

Number of attachable surfaces: 1

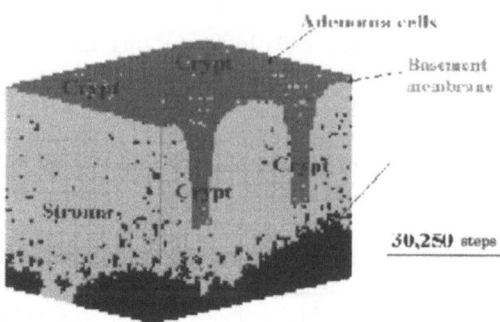

Fig. 26.4. The number of attachable surfaces is 1, and a poorly differentiated pattern is obtained.

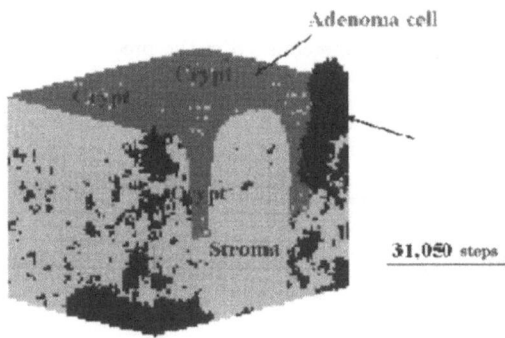

Fig. 26.5. The number of attachable surfaces is 5, and a moderately differentiated pattern is obtained.

The other 4 factors affect only the speed of cell proliferation. The conclusion comes from the simulation results that the carcinoma tissue consisted of the cell with 6 attachable surfaces formed homogenous cell cluster, and cells without attachable surface distributed sparsely in the stroma. Therefore the morphological features formed by these two types of cells were similar to poorly differentiated adenocarcinoma. On the other hand, the cells with 3 to 4 attachable surfaces had a tendency to form gland-like and cribriform-like structures similar to moderately differentiated adenocarcinoma in the real colon.

For future investigation we are planning to extend our model by adding biological phenomena more precisely. We think that our simulation system

Number of attachable surfaces:5

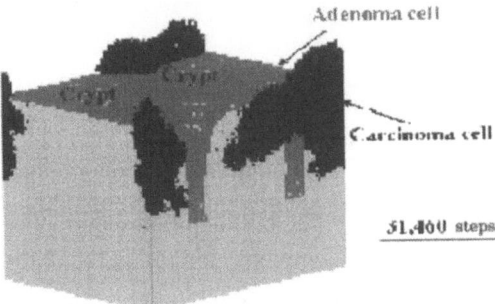

Fig. 26.6. The number of attachable surfaces is 3, and a poorly differentiated pattern is obtained.

could be a new tool for morphological analysis on tumor pathology (so to speak "Theoretical Pathology") and this method is also useful to make clear what biological problems remain to be elucidated.

Main problems to be solved for the simulation:

1) How do the adenoma cells or carcinoma cells connect with the normal cells? What is the physical strength of the binding between two types of cells?
2) How are the processes or depressions formed in adenoma? What factor decides if the proliferating cells protrude or depress?
3) What is the relationship between carcinoma cells and the stroma? Is there any resistance when the carcinoma cells invade through the stroma?
4) How are the glandular spaces in adenocarcinoma made? Do the spaces originate from the normal crypt spaces or are they made by exclusion of stromal substances, or both?
5) How long can the carcinoma cells live?

References

1. Cotran, R. Z., Kumar, V. and Colllins, T. (1999). eds. The gastrointestinal tract. In *Robbins : Pathologic Basis of Disease.* 6th ed. Saunders, Philadelphia. pp.826-835.
2. Eden, M. (1991). *Dynamics of Fractal Surfaces* Vicsek, World Scientific, Singapore.
3. Japanese Society for Cancer of the Colon and Rectum (1998). *Japanese Classification of Colorectal Carcinoma.* 1st ed., Kanehara Shuppan Inc. Tokyo.

The Effects of Cell Adhesion on Solid Tumour Geometry

Alexander R. A. Anderson

Department of Mathematics,
University of Dundee,
Dundee DD1 4HN,
Scotland
E-mail: anderson@maths.dundee.ac.uk

27.1 Introduction

The development of a primary solid tumour (e.g. a carcinoma) begins with a single normal cell becoming transformed as a result of mutations in certain key genes. This transformed cell differs from a normal one in several ways, one of the most notable being its escape from the body's homeostatic mechanisms, leading to inappropriate proliferation. An individual tumour cell has the potential, over successive divisions, to develop into a cluster (or nodule) of tumour cells. Further growth and proliferation leads to the development of an avascular tumour consisting of approximately 10^6 cells. This cannot grow any further, owing to its dependence on diffusion as the only means of receiving nutrients and removing waste products. For any further development to occur the tumour must initiate angiogenesis - the recruitment of blood vessels. Once angiogenesis is complete, the blood network can supply the tumour with the nutrients it needs to grow further. There is now also the possibility of tumour cells finding their way into the circulation and being deposited at distant sites in the body, resulting in metastases (secondary tumours).

Central to the invasive process are the molecules that facilitate interactions between cells and between cells and the extracellular matrix (ECM), know as cell adhesion molecules. A common feature of cell adhesion molecules is the ability to function as a molecular bridge between an external ligand and the cytoskeleton within the cell [7]. Over the past few years, it has become clear that receptors that mediate cell adhesion do not just affect cell migration, since occupancy of cell-surface receptors results in the initiation of signal-transduction pathways that regulate many aspects of cell function including transcription, proliferation, differentiation, cytoskeletal organisation and receptor activation [7, 11].

A crucial part of the invasive/metastatic process is the ability of the cancer cells to degrade the surrounding tissue or ECM [15]. This is a complex mixture of macromolecules (MM), some of which, like the collagens, play a structural role while others, such as laminin, fibronectin and vitronectin, are important for cell adhesion, spreading and motility. We note that all of these macromolecules are *bound* within the tissue i.e. they are non-diffusible. The ECM can also sequester growth factors and itself be degraded to release fragments which can have growth-promoting activity. Thus, while ECM may have to be physically removed in order to allow a tumour to spread, its degradation may, in addition, have biological effects on tumour cells.

A number of *matrix degradative enzymes* (MDEs) such as the *plasminogen activator* (PA) system and the large family of *matrix metalloproteinases* (MMPs) have been described [18] and both of these have been repeatedly implicated in tumour invasion and metastasis. In addition to opening migratory pathways, MDEs can alter cell adhesion properties regulated through several classes of cell surface receptors. These receptors, including cadherins, CD-44, integrins, and receptors for fibronectin, laminin, and vitronectin, negatively regulate cell motility and growth through cell-cell and cell-matrix interactions [15]. Therefore, proteolytic degradation of receptor and/or ECM components could release tumour cells from these constraints. Recent studies have shown that CD-44 mediates the attachment of cells to various MM. In fact invasion of human glioma cells has been inhibited by antibodies against CD-44 [12]. It is therefore important for any model that considers tumour invasion to include both cell-cell and cell-matrix interactions.

Tumour heterogeneity at the genetic level is well known. The so-called "Guardian of the Genome", the p53 gene is widely considered as a precursor to much wider genetic variation [13]. The p53 protein links three cellular functions: proliferation, death and DNA repair. In normal cells, p53 blocks proliferation and enables damaged DNA to be repaired. If DNA repair is incomplete, then apoptosis is initiated and the cell dies. Loss of p53 function (e.g. through mutation) allows for the propagation of damaged DNA to daughter cells [13]. As a step towards the inclusion of true tumour heterogeneity we shall consider a tumour that has phenotypic heterogeneity, with p53 being the only specific gene considered. Its effect is to simply allow the genetic mutations to begin and to of course allow the tumour cell to survive. The tumour cell phenotype will be defined here by the level of the cell's aggressiveness, i.e. a combination of its cell-cell adhesiveness, proliferation, degradation and migration rates (further details will be discussed below).

The aim of this chapter is to examine the effects of tumour cell heterogeneity upon the overall spatial structure of the tumour and to discuss the importance of the roles of cell-cell and cell-matrix interactions.

27.2 The PDE model of invasion

We will base our mathematical model on generic solid tumour growth, which we will assume has just been vascularised i.e. a blood supply has been established. We choose to focus on four key variables involved in tumour cell invasion, thereby producing a minimal model, namely; tumour cell density (denoted by n), MDE concentration (denoted by m), MM concentration (denoted by f) and oxygen concentration (denoted by c). Initially we define a system of coupled nonlinear partial differential equations to model tumour invasion of surrounding tissue and use these as the basis for the *Hybrid Discrete-Continuum* (HDC) technique [4].

The complete system of equations describing the interactions of the tumour cells, MM, MDEs and oxygen is

$$\frac{\partial n}{\partial t} = \overbrace{D_n \nabla^2 n}^{random\ motility} - \overbrace{\chi \nabla \cdot (n \nabla f)}^{haptotaxis},$$

$$\frac{\partial f}{\partial t} = - \overbrace{\delta m f}^{degradation},$$

$$\frac{\partial m}{\partial t} = \overbrace{D_m \nabla^2 m}^{diffusion} + \overbrace{\mu n}^{production} - \overbrace{\lambda m}^{decay},$$

$$\frac{\partial c}{\partial t} = \overbrace{D_c \nabla^2 c}^{diffusion} + \overbrace{\beta f}^{production} - \overbrace{\gamma n}^{uptake} - \overbrace{\alpha c}^{decay},$$

(27.1)

where D_n, D_m and D_c are the tumour cell, MDE and oxygen diffusion coefficients respectively, χ the haptotaxis coefficient and δ, μ, λ, β, γ and α are positive constants. Since this model represents a development of the work by Anderson *et al.* [3], we will not discuss its derivation here. However, some explanation should be given for the inclusion of oxygen in the model. Oxygen is assumed to diffuse into the MM, decay naturally and be consumed by the tumour. For simplicity oxygen production is proportional to the MM density. This is a crude way of modelling an angiogenic oxygen supply (see [2] for a more appropriate way of modelling the angiogenic network). We should also note that the cell-matrix adhesion is modelled here by the use of haptotaxis in the cell equation i.e. directed movement up gradients of MM (see [3] for more details).

This system is considered to hold on some square spatial domain Ω (a region of tissue) with appropriate initial conditions for each variable. We assume that the MM, oxygen, tumour cells and consequently the MDEs, remain within the domain of tissue under consideration and therefore no-flux boundary conditions are imposed on $\partial\Omega$, the boundary of Ω. See [3] for full details of the model derivation.

27.3 The HDC model

The HDC technique (see [1, 2, 3, 4, 5]) will be used to follow the path of
an individual tumour cell and first of all involves discretising (using standard
finite-difference methods) the system of partial differential equations (27.1).
We then use the resulting coefficients of the five-point finite-difference stencil
to generate the probabilities of movement of an individual cell in response to
its local milieu (see Appendix of [3] for the full discrete system). For clarity
we only consider the tumour cell equation,

$$n_{i,j}^{q+1} = n_{i,j}^q P_0 + n_{i+1,j}^q P_1 + n_{i-1,j}^q P_2 + n_{i,j+1}^q P_3 + n_{i,j-1}^q P_4. \qquad (27.2)$$

The coefficient P_0, which is proportional to the probability of no movement,
has the form,

$$P_0 = 1 - \frac{4kD_n}{h^2} - \frac{k\gamma}{h^2}\left(f_{i+1,j}^q + f_{i-1,j}^q - 4f_{i,j}^q + f_{i,j+1}^q + f_{i,j-1}^q\right),$$

and the coefficients P_1, P_2, P_3 and P_4, which are proportional to the probabilities of moving left, right, up and down, respectively, have the forms,

$$P_1 = \frac{kd_n}{h^2} - \frac{k\gamma}{4h^2}\left[f_{i+1,j}^q - f_{i-1,j}^q\right], P_2 = \frac{kd_n}{h^2} + \frac{k\gamma}{4h^2}\left[f_{i+1,j}^q - f_{i-1,j}^q\right],$$

$$P_3 = \frac{kd_n}{h^2} - \frac{k\gamma}{4h^2}\left[f_{i,j+1}^q - f_{i,j-1}^q\right], P_4 = \frac{kd_n}{h^2} + \frac{k\gamma}{4h^2}\left[f_{i,j+1}^q - f_{i,j-1}^q\right],$$

where the subscripts specify the location on the grid and the superscripts the
time steps. That is $x = ih$, $y = jh$ and $t = qk$ where i, j, k, q and h are
positive parameters.

The central assumption in the HDC technique is that the five coefficients
P_0 to P_4 are proportional to the probabilities of the tumour cell being stationary (P_0) or moving left (P_1), right (P_2), up (P_3) or down (P_4). From the
above probabilities we see that if there were no MM the values of P_1 to P_4
would be equal, with P_0 smaller (or larger, depending on the precise values
chosen for the space and time steps) i.e. there is no bias in any one direction
and the tumour cell is less (more) likely to be stationary - approximating an
unbiased random walk. However, if there are gradients in the MM, haptotaxis
contributes to the migration process and the coefficients P_0 to P_4 will become
biased towards the direction of increased MM concentration. The motion of
an individual cell is therefore governed by its interactions with the matrix
macromolecules in its local environment. Of course the motion will also be
modified by interactions with other tumour cells.

27.4 Individual-based processes

We shall now discuss in detail the processes each tumour cell will experience
as it migrates through the MM, driven by the movement probabilities defined
in the previous section.

Life Cycle: At each time step a tumour cell will initially check if it can move with regards to cell-cell adhesion restrictions (see the next paragraph for criteria), if it can then the movement probabilities (above) are calculated and the cell is moved. A check is then made if the cell should die (see death paragraph for criteria) or not. If it should not die, its age is increased and a check to see if it has reached proliferation age is performed. If it has not reached this age then it starts the whole loop again. If proliferation age has been reached then a check is made to see if the criteria for proliferation are satisfied (see proliferation paragraph for details). If proliferation criteria are not met then the cell becomes quiescent. If they are satisfied then we check to see if this mitosis results in a p53 mutation. If a p53 mutation does not result then death occurs, if it does then further mutations are possible (see mutation paragraph for details). This whole process is repeated at each time step of the simulation.

Cell-Cell Adhesion: To model cell-cell adhesion explicitly we assume each cell has its own internal adhesion value (A_i) i.e. the number of neighbours to which it will preferentially adhere . We therefore examine the number of external neighbours each cell has (A_e) and if $A_e \geq A_i$ then the cell is allowed to migrate, otherwise it remains stationary. Whilst this is a somewhat crude way of modelling cell adhesion, it does capture some features of cell-cell adhesion e.g. certain cells are more likely to bind to others and in so doing restrict their own ability to migrate.

Death: For the tumour cell to survive it requires sufficient oxygen, since some tumour cells have been found to survive in very poorly oxygenated environments, we make the assumption that the concentration has to drop to 0.05 non-dimensional units (where 1 would be the initial concentration) for cell death to occur. This assumption is also applied to quiescent tumour cells. The space that dead cells occupy becomes available to new cells as soon as they die. Cell death can also occur if the cell does not undergo a p53 mutation after the first proliferation and undergoing we assume that each cell has a probability $P_{p53} = 0.1$ of a p53 mutation. As with all the mutations considered here, they can only occur as a result of proliferation.

Proliferation: In our model we assume that each individual cell has the capacity for proliferation and will produce two daughter cells provided: (i) the parent cell has reached maturity ($8\ hrs - 16\ hrs$, see mutation section) and (ii) there is sufficient space surrounding the parent cell for the two new daughter cells to occupy. In order to satisfy condition (ii), we assume that one daughter cell replaces the parent cell and the other daughter cell will move to any one of the parent cell's four orthogonal neighbours that is empty. If more than one of the neighbouring grid points is empty then the new cell position is chosen randomly from these points. If no empty neighbours exist then the cell becomes quiescent and proliferation is delayed until space becomes available. Quiescent cells are assumed to be non-motile and consume half the oxygen of normal cells.

Mutation: Initially all cells are assumed to have wild-type p53 i.e. they are non-mutated. After a p53 mutation occurs the genome becomes unstable and is then open to many more mutations. So after the p53 mutation has occurred, the cell now has predefined phenotypic traits that define its behaviour. Here we shall consider 100 randomly defined phenotypes, each phenotype has an equal probability of being selected. We define each phenotype to be a set of parameter values that describe the behaviour of the cell expressing it. Therefore a particular phenotype will have a randomly selected proliferation age ($Phen_{age} = 8 - 16$ hrs), O_2 consumption ($Phen_{O_2} = \gamma - 4\gamma$), MDE production ($Phen_{mde} = \mu - 4\mu$), haptotaxtis coefficient ($Phen_{taxis} = \chi - 4\chi$) and adhesion value ($Phen_A = 0 - 3$). In most cases the range of values each parameter can take to define the phenotype were chosen to represent biologically realistic limits. Following the first p53 mutation a cell is assigned the values of one of the hundred randomly selected phenotypes and for each subsequent proliferation there is a small probability (P_{mutat}) of further mutations occurring which will lead to another randomly selected phenotype and so on.

Production/Degradation: Since we are modelling individual tumour cells we must consider MDE production at the level of a single cell. In the continuum model (27.1) we have MDE production as being proportional to the tumour cell density. Now MDE is only produced at a grid point if a tumour cell is occupying that grid point. Since we have no precise parameter estimates for this production rate, we take $n = 1$ in the discrete form of the MDE equation when a tumour cell is occupying the current location and take $n = 0$ otherwise. Similarly for O_2 uptake, we take $n = 1$ (since ω is scaled as per cell) in the discrete form of the oxygen equation when a cell is consuming oxygen at the current location and $n = 0$ otherwise.

27.5 HDC model simulation results and discussion

The following simulations were carried out on a 400×400 grid, which is a discretisation of the unit square $[0, 1] \times [0, 1]$, with a space step of $h = 0.0025$ and a time step of $k = 0.0005$. No flux boundary conditions were imposed on the square grid, restricting the tumour cells, MDE, MM and oxygen to within the grid. Initially, 50 tumour cells are centred around $(0.5, 0.5)$ with a random age between 0 $hrs - 16$ hrs, the MDE concentration was zero throughout the domain ($m(x, y) = 0$) and the oxygen concentration was taken to be one ($c(x, y) = 1$). We consider the effects, upon tumour invasion, of two different MM initial distributions: (i) homogeneous ($f(x, y) = 1$) and (ii) random ($0 \leq f(x, y) \leq 1$). For clarity we shall label the resulting tumour cell distributions as (i) homogeneous tumour and (ii) random tumour. The non-dimensional parameter values used in all the following simulations are $d_n = 0.0005$, $D_m = 0.0005$, $D_c = 0.5$, $\chi = 0.01$, $\delta = 50$, $\mu = 1$, $\lambda = 0$, $\beta = 0.5$, $\gamma = 0.57$ and $\alpha = 0.025$. We also take the phenotype mutation probability to

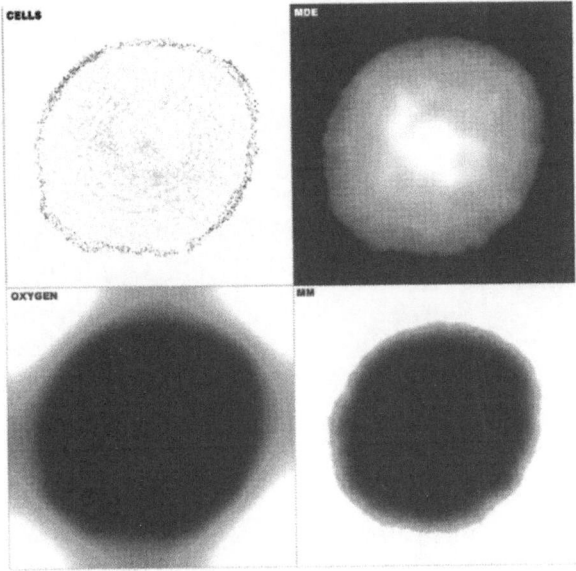

Fig. 27.1. Homogeneous tumour results, spatial distribution of tumour cells, MDE, MM and oxygen (clockwise) at time $t = 200$ units (i.e. 200 generations, approximately 133 days). Dark gray colouration of the tumour cells represents cells with a cell-cell adhesion value, $A_i = 0$ and light gray represents dead cells. For the MDE, MM and oxygen concentration, the gray colour map is used i.e. white=high concentration, black=low concentration and gray is in between.

be $P_{mutat} = 0.1$. Other values were considered and produced similar results but for shorter or longer times depending on whether the probability was smaller or larger.

Figure 27.1 shows the simulation results for each of the four variables for the the homogeneous initial MM distribution at $t = 200$ time units. The tumour cell distribution shows a dead central region with a thin proliferating boundary. The dark gray cells which have survived and continue to proliferate have the least adhesive phenotype i.e. they have an adhesion value $A_i = 0$. In fact two phenotypes dominate the tumour population and survive the full length of the simulation. Both of these phenotypes have a zero cell adhesion value and have short proliferation age (8.9 hrs and 9.2 hrs, respectively) as well as high haptotaxis coefficients (3.9χ and 2.4χ, respectively). Interestingly, one of these phenotypes is the most dominant phenotype, always being expressed by the largest fraction of cells in the tumour population, and has the shortest proliferation age, highest haptotaxis coefficient and no cell-cell adhesion restrictions. In some sense, through random mutation the most aggressive phenotype has been naturally selected.

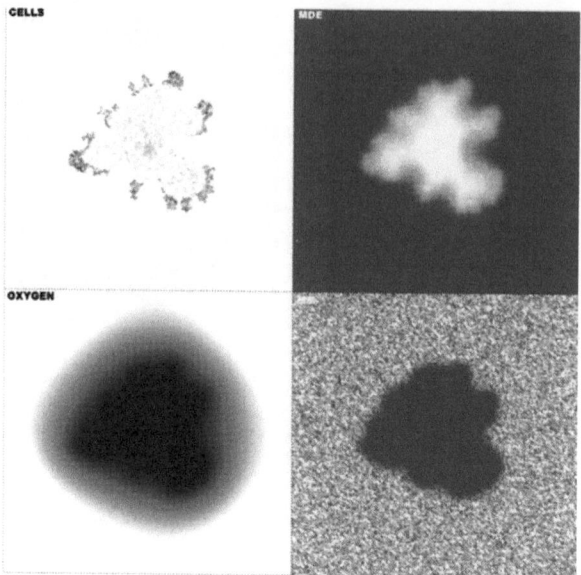

Fig. 27.2. Random tumour results: spatial distribution of tumour cells, MDE, MM and oxygen (clockwise) at time $t = 200$ units (i.e. 200 generations, approximately 133 days). Colouration as in Fig. 27.1.

Since the cells were invading through an initially homogeneous distribution of MM, it is perhaps not too surprising that this has produced the most invasive and symmetric tumour (this symmetry is also seen in all three of the other variables). The homogeneous tumour has also produced the largest number of individual cells, due to the faster invasion rate which gives access to empty space for proliferation leading to further invasion. The faster invasion is mainly driven by the cell-matrix interactions via haptotaxis, giving directed motion towards higher concentrations of MM. Since the least adhesive cells on the boundary also have a large haptotaxis coefficient they can exploit the sharp change in gradient the most.

Figure 27.2 shows the simulation results for an initially random MM distribution. The tumour cell distribution is now radically different geometrically, showing a more fingered morphology, although it still consists of dead cells and a thin proliferating outer layer of cells. These dark gray cells are mainly from one phenotype which has dominated the tumour population for the whole of the simulation. This phenotype has an adhesion value $A_i = 0$, a proliferation age of 8.2 hrs and a haptotaxis coefficient of 2.6χ. Again one of the most aggressive phenotypes has been naturally selected to lead the way for invasion at the boundary of the tumour. Since these cells have no cell-cell adhesion dependence, their migration is mainly driven by haptotaxis via the local MM gradients and it is these local gradients that ultimately define the tumour geometry.

Given that both simulations use the same parameters, with the exception of the MM initial distributions, these results illustrate the importance of tumour cell-matrix interactions in aiding or hindering the migration of individual cells that define the tumour geometry. This result is dependent on the fact that the tumour cells must mutate to an aggressive phenotype in order to exploit the changes in local MM gradients. If the cell-cell adhesion parameter were to be fixed for all phenotypes (e.g. $A_i = 3$), then roughly the same small circular tumour would result for both MM distributions. However, it is known that tumour cells lose the ability (via mutations) to recognise cell-cell adhesion molecules at an early stage in their development e.g. E-cadherin is lost at an early stage of breast carcinogenesis and N-CAM loss in gliomas is associated with a high probability of metastasis [16]. Therefore the results from the random tumour (Fig. 27.2) perhaps represent the most realistic simulation, where the bias from the cell-matrix interactions is purely driven by local degradation of MM and subsequent creation of gradients which lead to directed migration of the tumour cells.

The fact that the resulting tumour cell population consists of living cells with only one or two phenotypes might be surprising due to the random nature of the mutations. However, it seems logical to assume that it will be the most aggressive phenotypes that dominate the tumour population. Where aggressiveness would be defined as those phenotypes that have low proliferation age, zero cell-cell adhesion and a large haptotaxis coefficient.

In conclusion, whilst cell-cell interactions are important at the early stages of the tumour's development, subsequent loss of cell-cell adhesion molecules (via mutation) results in a tumour that is dominated by cell-matrix interactions. Therefore, these results predict that local tumour cell-matrix interactions are ultimately what define the overall geometry of the tumour and not the cell-cell interactions.

Much more work needs to be done in examining the range of behaviour that this model can display. In particular the manner in which oxygen production is modelled needs to be refined perhaps by modelling more accurately the angiogenic network. There is certainly a need for a sensitivity analysis of the parameters, especially the mutation probabilities and the production/uptake parameters that are not known. Nonetheless, the results presented above show the importance of considering both cell-cell and cell-matrix interactions in a model of tumour invasion and perhaps the strong dependence of the tumour geometry on local cell-matrix interactions points a way for future cancer research and treatment.

Acknowledgements: Dr. A.R.A. Anderson is supported by a Personal Research Fellowship from the Royal Society of Edinburgh and an equipment grant from the Royal Society of London and would like to thank Prof. Mark Lewis for inspiring the random mutation idea.

References

1. Anderson, A. R. A., Sleeman, B. D., Young, I. M. and Griffiths, B. S. (1997). Nematode movement along a chemical gradient in a structurally heterogeneous environment: II. Theory, *Fundam. appl. Nematol.*, **20**, 165-172.

2. Anderson, A. R. A. and Chaplain, M. A. J. (1998). Continuous and discrete mathematical models of tumour-induced angiogenesis, *Bull. Math. Biol.*, **60**, 857-899.

3. Anderson, A. R. A., Chaplain, M. A. J., Newman, E. L., Steele, R. J. C. and Thompson, A. M. (2000), Mathematical Modelling of Tumour Invasion and Metastasis, *J. Theoret. Med.*, **2**, 129-154.

4. Anderson, A. R. A. (2003). A hybrid discrete-continuum technique for individual based migration models, in *Polymer and Cell Dynamics*, eds. Alt, W., Chaplain, M., Griebel, M., Lenz, J., Birkhauser.

5. Anderson, A. R. A. and Pitcairn, A. (2003). Application of the hybrid discrete-continuum technique, in *Polymer and Cell Dynamics*, eds. Alt, W., Chaplain, M., Griebel, M., Lenz, J., Birkhauser.

6. Bray, D. (1992). *Cell Movements*, Garland Publishing, New York.

7. Burridge, K. and Chrzanowska-Wodnicka, M. (1996). Focal adhesions, contractability, and signalling, *Annu. Rev. Cell Dev. Biol.*, **12**, 463-518.

8. Calabresi, P. and Schein, P. S., eds. (1993). *Medical Oncology*, 2nd ed., McGraw-Hill, New York.

9. Casciari, J. J., Sotirchos, S. V., and Sutherland, R. M. (1992). Variation in tumour cell growth rates and metabolism with oxygen-concentration, glucose-concentration and extracellular pH, *J. Cell Physiol.*, **151**, 386-394.

10. Hotary, K., Allen, E., Punturieri, A., Yana, I. and Weiss, S. J. (2000). Regulation of cell invasion and morphogenesis in a 3-dimensional type I collagen matrix by membrane-type metalloproteinases 1, 2 and 3, *J. Cell Biol.*, **149**, 1309-1323.

11. Hynes, R. O. (1992). Integrins: versatility, modulation, and signalling in cell adhesion, *Cell*, **69**, 11-25.

12. Koochekpour, S., Pilkington, G. J. and Merzak, A. (1995). Hyaluronic acid/CD44H interaction induces cell detachment and stimulates migration and invasion of human glioma cells in vitro, *Intl. J. Cancer*, **63**, 450-454.

13. Lane, D. P. (1994). The regulation of p53 function. Steiner Award Lecture, *Int. J. Cancer*, **57**, 623-627.

14. Sherwood, L. (2001). *Human Physiology: From Cells to Systems*, 4th ed., Brooks/Cole, California.

15. Stetler-Stevenson, W. G., Aznavoorian, S. and Liotta, L. A. (1993). Tumor cell interactions with the extracellular matrix during invasion and metastasis, *Ann. Rev. Cell Biol.*, **9**, 541-573.

16. Takeichi, M. (1993). Cadherins in cancer: implications for invasion and metastasis, *Curr. Opin. Cell Biol.*, **5**, 806-811.

17. Terranova, V. P., Diflorio, R., Lyall, R. M., Hic, S., Friesel, R. and Maciag, T. (1985). Human endothelial cells are chemotactic to endothelial cell growth factor and heparin, *J. Cell Biol.*, **101**, 2330-2334.

18. Thorgeirsson, U. P., Lindsay, C. K., Cottam, D. W. and Gomez, Daniel E. (1994). Tumor invasion, proteolysis, and angiogenesis, *J. Neuro-Oncology*, **18**, 89-103.

Diversity of Biological Patterns in the Fossil Record and Their Meaning in Morphological Evolution

Part VIII

Diversity of Biological Species in Tree Zone: Renewal and Tree Thinning in Meteorological Conditions

Pattern Formation and Function in Palaeobiology

Enrico Savazzi

Hagelgränd 8, 75646 Uppsala, Sweden
e-mail: `enrico@savazzi.net`

28.1 Introduction

In the study of biological organisms, the perspective of palaeobiologists is radically different from that of most biologists. For the latter, verifying a hypothesis usually involves experiments that alter the properties of an organism (and/or of its immediate surroundings), with the purpose of observing directly the effects of these changes on the organism itself. Although palaeobiologists are aware that they are studying the properties of living organisms, they lack the possibility of performing such direct experiments on fossils. For this reason, palaeobiologists are intrinsically more willing than biologists to seek indirect evidence and alternative methods of investigation. Thus, palaeobiologists have developed and used an array of indirect practical and conceptual methods for studying the function and development of morphological characters in fossils [23, 30].

This difference in attitude is especially evident when palaeobiologists extend their studies to recent organisms, thus overlapping the domain of biologists. In these instances, biologists often have been critical of the indirect methods accepted by palaeobiologists. The latter, on the other hand, have argued that their methods, besides being useful with fossils, should be used also on recent organisms (this debate often has been carried out during peer-review and informal communication, but is scarcely evident from published literature). The strongest point that palaeobiologists can make in this context is that the results obtained with indirect methods applied to living organisms can be compared with the results obtained with direct techniques, thus providing an objective test of the validity and limitations of palaeobiological methods [30].

Until recently, developmental biologists have been forced to rely on indirect methods more often than other bioscientists. For instance, much of the theoretical work on reaction-diffusion systems was carried out before methods were available to study morphogens and their associated genetic machinery (and indeed, before the very existence of morphogens was proved). Thus, developmental biologists are in a position to appreciate indirect and inferential methods of investigation. This paper discusses three examples of problems in pattern formation and function that have been studied by palaeobiologists, and the conceptual methods used for their investigation.

28.2 Materials

Repositories of illustrated specimens are specified in the figure captions. Where no repository is given, the material is in the possession of the author.

28.2.1 Terrace-sculptures in brachiopods

Terrace-patterns are sets of subparallel ribs present on the external surface of invertebrate exoskeletons. Individual terraces are asymmetrical in cross-section, and have a steep side and a gently sloping opposite side. Savazzi (this volume) provided a summary of the literature on terrace-patterns in fossil and Recent invertebrates. In the present paper, it may be remembered that terrace patterns occur in organisms that grow either by marginal accretion (Mollusca, Brachiopoda) or by moulting (Arthropoda). Thus, in spite of the similarity in appearance, these patterns are produced by very different morphogenetic programs, and their similarity can be attributed to convergent evolution due to functional requirements [27, 33].

Terraces have found a few technological applications as well. For instance, a terrace-like pattern of scales has been used in cross-country skis (Fig. 28.1A) in which they prevent backslippage without hindering movement in the forward direction. Terraced nails (Fig. 28.1B) are hammered easily into a resilient material like wood, but are much harder to pull out in the opposite direction. In burrowers that perform a push-pull sequence of movements within a loose substrate, terraces are functional by providing a low friction in the burrowing direction and a higher friction in the opposite direction. Preventing backslippage by increased friction in the backward direction increases the efficiency of the burrowing process. This is accomplished by orienting terraces roughly perpendicularly to the burrowing direction and with the steep sides facing away from this direction. In organisms with other life habits, terraces have a different adaptive significance (see Savazzi, this volume, and references therein), albeit also based on asymmetrical friction.

The Lingulidae s.l. (Ordovician-Recent) are inarticulate (or lingulate) brachiopods living in a vertical orientation within mucus-lined burrows in loose sediment. An extensible pedicle is anchored at the bottom of the burrow, and can withdraw the shell deep into the sediment. Lingulids that are exposed by erosion can reburrow obliquely downward with the shell foremost and the pedicle trailing behind [26]. After reaching a suitable depth, burrowing lingulids turn upwards and continue to burrow upwards until they emerge at the

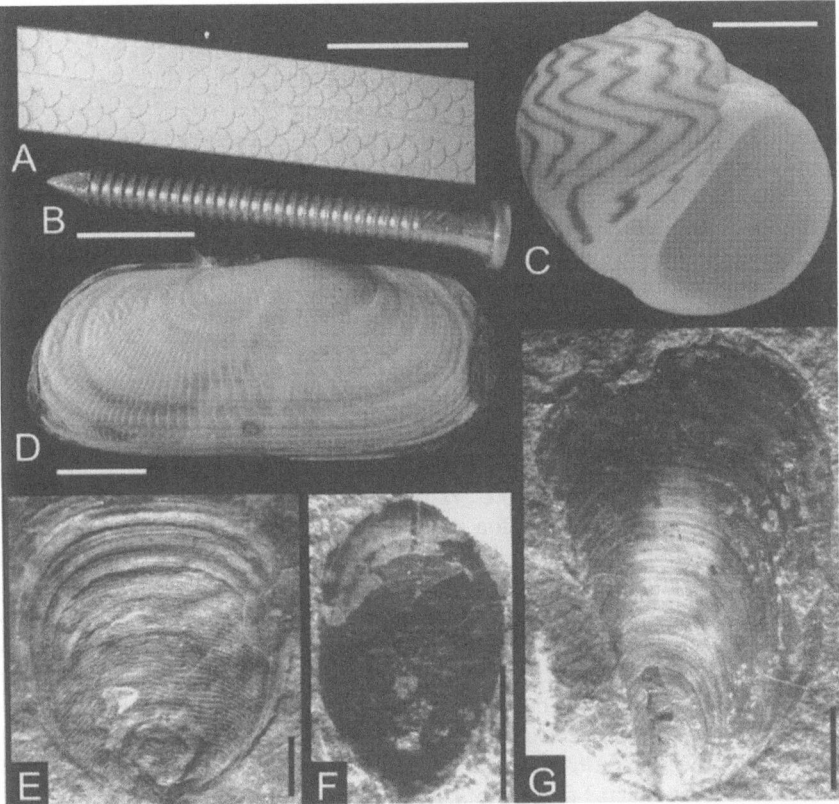

Fig. 28.1. **A**. Terrace-like scales on the sole of a cross-country ski. **B**. Terraced nail. **C**. Natica fulminea, East of Belair, Dakar, Senegal. **D**. Solecurtus divaricatus, Japan. **E**. Westonia stoneana, Ordovician, Mazomani, Wisconsin, USA, divaricate terrace-pattern (SI = Smithsonian Institution, Washington DC, USA). **F**. Westonia ollus, Ordovician, Newton, New Jersey, USA (SI). **G**. Lingula (s.l.) punctata, Ordovician, New York, USA (New York State Museum, New York, USA). Scale bars represent 100mm (**A**), 10mm (**B**), 5mm (**C**), 2mm (**E-G**).

surface of the sediment. In this way, the pedicle, which trails behind the shell and does not participate actively in the burrowing process, becomes located underneath the shell (see above references).

The Obolidae (Cambrian-Ordovician) are a family of inarticulate brachiopods differing from the Lingulidae in internal anatomical characters (inferred mostly from muscle-scars). Some Obolidae are similar in shell shape and size to Lingulidae (Fig. 28.1E-F), and are found in sandy to silty sediments. They likely had an infaunal life habit comparable in life-position and general traits with that of the Lingulidae. However, the Obolidae likely had a shorter pedicle and a preference for higher energy environments [24]. Several of these obolids with lingulid-like shells possess terrace sculptures (Fig. 28.1E-F), as well as other types of sculptures compararable in geometry to terraces [24]. Recent lingulids lack sculptures, but an Ordovician species of Lingula s.l. does possess a terrace pattern comparable to that of obolids (Fig. 28.1G).

These patterns are produced by travelling waves moving along the edge of the mantle, with each wave building a terrace-line. Shell growth at the mantle-edge records this activity pattern in time as shell sculpture. Morphogenetic programs involving travelling waves along an essentially one-dimensional mantle-edge produce characteristic divaricate colour- or sculpture-patterns [8, 9, 10, 33], which are common among bivalves (Fig. 28.1D) and, to a lesser extent [36], gastropods (Fig. 28.1C).

In bivalves, divaricate terrace patterns have been shown to be functional as burrowing sculptures, and have evolved multiple times in several families [27, 32, 33, 35]. Divaricate patterns can provide a better approximation than growth-conformable patterns (i.e., radial or commarginal sculpture; e.g. [27], Fig. 2) to the optimal orientation required by terraces (i.e., perpendicular to the burrowing direction). In several bivalves, the speed with which waves travel along the mantle edge varies along the shell margin, in order to further optimise the pattern. A similar phenomenon is observed in obolids with divaricate terraces. For instance, in Fig. 28.1F the travelling speed of waves is highest in the central region of the shell margin, thus providing an almost commarginal terrace in this region, and decreases along the shell sides, where the terraces become more oblique to the shell margin in order to remain properly oriented to the hypothesised burrowing direction. In Figs. 28.1E and 28.1G, several points along the central portion of the mantle edge fired simultaneously, producing multiple V-shaped terraces that cancelled out each other where they came into contact.

This can be an alternative pathway to optimisation, since, on a larger scale, several small V-shaped terraces placed side-by-side average to a single straight terrace. Along the sides of the shell, the travelling waves continue uninterrupted, providing the required obliquity. These optimisations to comply with the requirements of burrowing sculptures further indicate a burrowing function for these features.

Since the terraces in the Ordovician Lingula punctata (Fig. 28.1G) have the steep faces oriented toward the pedicle, the burrowing direction must have been such that the anterior commissure was foremost with the pedicle trailing behind the shell. This indicates that the burrowing mechanism of the Lingulidae evolved in the early Palaeozoic, and has not changed since.

The terraces in obolids, instead, have the steep faces oriented away from the pedicle, and so do the other types of burrowing sculptures in this family. So much evidence is available on the significance of terraces in burrowing organisms (see above references; Savazzi, this volume, and references therein), that it is unlikely that the terraces in obolids were functional in a different context. Therefore, the orientation of terraces in this family can be interpreted only by accepting that the burrowing mechanism in this family was different to that in lingulids, i.e., that burrowing took place with the pedicle lowermost and the shell trailing behind, and that making a U-turn upwards halfway during the burrowing process therefore was unnecessary in this group.

In this example, the study of terrace patterns allows us to reconstruct the behaviour of extinct organisms. In turn, behavioural characters may provide evidence relevant to the reconstruction of evolutionary relationships. In the case of obolids, they strengthen other evidence [24] pointing to the fact that this family is not closely related to lingulids, and that instead it evolved from non-burrowing stocks into burrowers independently of lingulids, achieving a shell shape similar to that of lingulids through a partly similar (albeit not identical) mode of life [24].

28.2.2 Control of shell-shape in gastropods

The role of periodic sculpture

Most gastropod molluscs possess helicospirally coiled shells, in which the last whorl adheres to the immediately preceding one (see Fig. 28.1C and Fig. 28.2). In Mesozoic to recent gastropods, a strong sculpture often is built at the shell aperture, in the form of a rib or a thickened varix The repetition of this process in time leads in many gastropods to a periodicity in the shell sculpture. In several instances, the number of these sculptural elements on each shell-whorl is constant, and sculptural elements on adjacent whorls are placed at precise reciprocal positions (e.g. Fig. 28.2A). This is called synchronised sculpture (Savazzi and Sasaki, in press).

Two alternative mechanisms can produce a synchronised sculpture. A combination of precise time-keeping, coupled with a precisely controlled rate of shell growth, can yield the observed regularity. Alternatively, a feedback from the sculpture located on the preceding whorl can be used to trigger the construction of new sculptural elements. The second morphogenetic program looks especially attractive, because it does not require a precise biological clock and control of shell secretion. In addition, it would provide regulatory properties, allowing the sculpture pattern to recover from accidental or predator-inflicted shell damage.

The two mechanisms should produce different results in the presence of repaired shell damage. Such damage is common in gastropods, and this provides us with "natural experiments" that allow us to decide which mechanism is at work, without requiring complex experiments to directly study shell morphogenesis in living gastropods.

The sum of the evidence shows that sculpture patterns disrupted by damage of the shell aperture generally reacquire their regular arrangement quickly (Savazzi and Sasaki, in press). In Fig. 28.2B, shell growth along each whorl took place from right to left and obliquely downward, and successive varices

Fig. 28.2. A-B. Epitonium sp.,Tateyama,Chiba Prefecture, Japan (Tokyo University Museum, Tokyo, Japan). **C-D.** Melanella martini, NW Australia (SMNH = Swedish Museum of Natural History, Stockholm). The arrow in (**D**) indicates repaired shell damage. **E.** Melanella sp., Recent, Catarman, Cebu, the Philippines. **F.** Colubraria strepta, Philippines. **G.** Distorsio kurzi, Philippines. **H.** Distorsio reticulata, Philippines, sectioned to show columella. Scale bars represent 10mm.

were built in this order. On the last whorl, an unusually thick varix (bottom left) disrupted the pattern, which was repaired immediately afterwards by building a correctly placed varix in correspondence with a varix on the preceding whorl. This took place in spite of the fact that the spacing from the preceding, abnormal varix is narrower than usual. This normal varix was subsequently broken (probably by a predator), and its lower portion is missing. The shell aperture was repaired by a flat portion of whorl, but the varix itself was not rebuilt. The subsequent varix, nonetheless, is placed correctly with respect to the varix on the preceding whorl, in spite of the fact that this yields an unusually long spacing from the broken varix.

It must be noted that the pattern must be repaired in a space not exceeding one whorl, because only the sculpture located on the last whorl is accessible for feedback. The external surface of earlier whorls generally is outside the reach of the soft parts. In practical instances, a biological clock must be present in the construction of synchronised sculpture, in addition to tactile feedback, because new sculptural elements often are built also if corresponding elements on earlier whorls are missing. However, this clock may not be precise, and sculptures produced by this mechanism often are not placed exactly at the expected positions.

Synchronised sculpture patterns possessing only one element per whorl (base-1 patterns of Savazzi and Sasaki, in press) are especially interesting in this context, because a disruption of the synchronisation pattern should produce permanent anomalies: if the only synchronising signal per whorl is temporarily lost, there is no second chance to repair the pattern. In fact, base-1 patterns behave as predicted. The varices that constitute the synchronising signal (and apparently correspond to temporary stops in shell construction) normally are aligned (Fig. 28.2C). In Fig. 28.2D, breakage of the aperture disrupted the pattern roughly mid-way along the shell (arrow in Fig. 28.2D). The varices on subsequent whorls regained synchronisation with respect to each other, but no feedback was possible from shell portions above the disturbed whorl, and therefore the disturbance resulted in two sets of synchronised varices separated by an offset as a consequence of the damage.

Base-1 patterns also allow the construction of shells with curved coiling axes (Fig. 28.2E-F), which evolved independently in two families. Each growth stage between successive varices is curved in the same direction, and synchronisation ensures that successive growth stages are aligned at a constant angle to each other, so that the curvature becomes cumulative. In fact, an evenly curved coiling axis is possible only in association with a base-1 pattern because, in patterns with a different base (i.e. a different number of synchronisation signals per whorl) the curvature of each growth stage cancels out the curvature of preceding stages (as in Fig. 28.2G-H, for example), and on a larger scale the coiling axis averages to a straight line (observe the columella, which corresponds to the coiling axis, in the sectioned shell in Fig. 28.2H).

Fig. 28.3. A-B, D. Caracolus caracolla, Portorico (SMNH = Swedish Museum of Natural History, Stockholm). **C.** Trochus rota, Philippines. **E-F.** Cepolis ovumreguli, Cuba (SMNH). The dashed line indicates the position of the substrate relative to a clamped shell. **G.** Strophostoma anomphalus, Oligocene, Arnegg near Ulm, Germany (Department of Earth Sciences, Cambridge, UK). **H.** Anostoma depressum, Brazil (SMNH). **I.** Streptaxis bulbulus, Vietnam (SMNH). **J.** Urocoptis anafensis, Cuba (SMNH). Arrows point to repaired shell damage. Scale bars represent 10mm (**A-F, J**), 5mm (**G, I**).

The role of whorl geometry

In coiled gastropod shells, successive whorls must be built at appropriate reciprocal positions in order to achieve a given shell geometry. This geometry is adaptive in several respects, so it can be expected that its control by the organism is usually strict. In fact, overall shell geometry is quite constant in most species (typically, more so than other morphological characters like colour pattern, sculpture and adult shell size). In a growing gastropod shell, the position of the whorls with respect to each other is determined by the position maintained by the aperture with respect to the preceding shell-whorl, since shell material is secreted in annular increments around the aperture. The soft parts present in the aperture region are suitably placed to gain a feedback from the preceding whorl by tactile feeling of its surface in proximity of the aperture. It has been proposed [6] that a peripheral keel on the preceding whorl, or alternatively, a flat region of the preceding whorl between the umbilical region and a peripheral keel, are used as feedback signals in a "road-holding" morphogenetic program that keeps the aperture and current whorl in a suitable position during growth. In fact, the suture of the current whorl often is located in correspondence of a peripheral keel on the preceding whorl (Fig. 28.3A). Experiments with artificially modified shells of terrestrial gastropods [3] seem to confirm the idea of such a feedback. A paper by Morita [13] complements this study by proposing additional or alternative factors that may be involved in controlling shell geometry.

Anomalous gastropod shells also seem to confirm the importance of the peripheral keel as a morphogenetic cue. In Fig. 28.3B-D, damage to the shell margin caused the morphogenetic program to derail, and the whorl to be built in an abnormal position. Often, contact with the peripheral keel is re-established, and the coiling pattern repaired (Fig. 28.3B-C).

In Fig. 28.3D, damage to the shell margin apparently was accompanied by damage to the soft parts, and the peripheral keel became rounded. Since the feedback signal was no longer available, the coiling pattern could not be repaired, and coiling continued in an abnormal way. Aside from feedback from a peripheral keel, coiling in gastropods may be controlled by other feedback signals (e.g. contact with the previous whorl in order to avoid an open-coiled, fragile shell) as well as intrinsic factors (e.g. the shape of the soft parts within the shell "moulding" the shell aperture and steering its growth in preferred directions), so that, in the absence of one feedback signal, coiling may not derail completely and still produces a viable, if not optimal, shell geometry. For instance, breakage of the shell margin only rarely results in complete uncoiling, or in the complete loss of contact between the earlier whorls and the current one (the latter may happen if the soft parts also are damaged; pers. obs.).

Adaptive derailing of coiling

In most terrestrial gastropods, the growth process is determinate (i.e. the shell reaches an adult size and shape and permanently stops growing; [18], and references therein). In terrestrial gastropods with a globular or discoid shell, a short portion of the last whorl preceding the adult aperture usually is bent abapically (Fig. 28.3E-F). This derailing from the earlier growth pattern is adaptive, and allows the adult aperture to clamp more tightly onto the substrate (Fig. 28.3F).

Prior to the adult stage, the position and orientation of the aperture is a compromise between a sufficiently good clamping and a continued growth

process, which does not allow the aperture to deviate substantially from a helicospiral coiling pattern because this would hinder further growth. Derailing
near the adult stage could be due to a relaxation of the feedback mechanism
that causes regular coiling, coupled with other factors that force the shell to
change its growth direction (see [13] and Morita this volume, for an example
of such factors).

In a few groups, derailing takes place in the adapical direction, and results
in an adult aperture which is bent about 90° from its previous growth direction
(Fig. 28.3G-H). The adaptive value of this geometry is unclear [18], but its
multiple homoplasy in a few families suggests it does have one.

Count-downs

The relaxation of feedback mechanisms that control the geometry of shell coiling in proximity of the adult aperture may open the way for other morphogenetic programs to take control in this region, and to produce more radically
different shell geometries. This, in turn, can favour the evolution of longer
sequences of morphogenetic steps, yielding more complex geometries. Such
sequences have been called count-downs [34], because the process must follow
a specific order of stages, and eventually ends with the attainment of the adult
shape and size (or alternatively, with a repetition of the series of steps: periodic count-downs, [34]). Examples of count-downs in terrestrial gastropods
are shown in Fig. 28.3I-J. In some cases, the steps of the count-down visibly
encompass most of the shell, and therefore affect most of its growth process
(Fig. 28.3J). In most cases, it is possible to associate an adaptive value to specialised shell morphologies resulting from count-downs (Savazzi and Sasaki,
unpublished).

28.2.3 Computer modelling of shell geometry and ornamentation

The regular coiling of mollusc shells has attracted the attention of scientists
at least since the 18th century [28]. With the advent of electronic computers
it became possible to generate graphic models of coiled shells. Already the
earliest computer models of shells [19, 20, 21, 22] incorporated essential characteristics of actual shells, like an aperture and an external surface delimiting
an internal volume.

Raup's algorithm [20] for generating coiled shell morphologies uses the z-
axis of a fixed reference frame as the coiling axis of the shell. Two parameters
are used to control the rate of outward expansion of the spiral following the
growth trajectory of the shell, and the rate of growth of the perimeter of
the shell aperture. When the parameters are correctly sized, isometric growth
results. This procedure results in planispiral shell models that coil in the
xy-plane, and is especially useful to model the regular coiling displayed by
most ectocochleate cephalopods. The addition of a parameter controlling the

Fig. 28.4. Computer-generated models of coiled shells.

translation rate of the aperture about the z-axis during growth results in helicospiral shells, as in the majority of gastropods.

Raup's algorithm allows the generation of regularly coiled (i.e., isometrically growing) shells. A few types of allometric growth can be simulated by varying the values of the parameters during growth. However, there are entire classes of shell geometries that occur in nature (e.g. most heteromorphic ammonoids) that cannot be modelled with this algorithm. Even simple straight, orthoconic shells lay beyond its possibilities. The limitations of this algorithm are due to its fundamental nature, i.e., to the fact that the coiling axis is defined *a priori*, and cannot change position during growth because the shell is fixed to the reference frame. This suggests that this algorithm does not model the processes involved in the growth and morphogenesis of biological shells that grow by marginal accretion. In spite of these limitations, Raup's algorithm can still be used when the purpose is to provide a simple descriptor of regular shell coiling, not necessarily realistic in biological terms [28].

Several algorithms have been devised to overcome the limitations of Raup's method. A group of these algorithms [1, 14, 15, 16, 17] uses a reference frame attached to the shell aperture, and the coiling axis (when present) emerges *a posteriori* as a by-product of coiling. In these algorithms, the position of the aperture is computed by applying a set of rigid transformations (rotations and translations) to the existing portion of the shell, before adding to it a new growth increment and aperture. When the parameters controlling these transformations, as well as those controlling the growth rates of the aperture and of the current growth increment, are constant, isometric growth results. These methods are successful in reproducing a broad range of isometric and allometric shell geometries, and the addition of further translations and rotations similar, in nature, to the original ones can further broaden their scope [25]. A criticism that can be made of these methods is that growth follows a "table" of arbitrary values, but the methods themselves tell us nothing of the biological mechanisms producing these values (or of whether the values and parameters do have any biological meaning). Examples of shells produced by these algorithms are shown in Fig. 28.4. Compare, for instance, Fig. 28.4B with Fig. 28.3G.

Other methods are more restricted in scope, and use specific functions to describe the changes of shell morphology that occur during anisometric growth. For instance, Illert [7] devised a method that describes the growth trajectories of coiled shells by analogy with the shapes taken by clock springs, i.e., flexible rods constrained in position and orientation at both extremities. By rotating one or both of the tie points of a clockspring, one can obtain a variety of isometrically or allometrically coiled patterns. Although some of the resulting geometries are compellingly similar to certain non-trivial coiling patterns that do occur in nature, it is not clear which relationships exist between a clockspring (which is an equilibrium shape stable in time) and an organism that builds its shells as a dynamic process, often with no visible feature corresponding to the tie points of the clockspring. Other methods for modelling the coiling patterns of shells have been described, but their scope is restricted, and they are not considered here. The shape of the aperture can be modelled independently of the coiling pattern. In particular, the former is easy to generate by choosing an arbitrary shape for the perimeter of the aperture. Unlike the mode of coiling, apertural shape does appear to be just a simple by-product of the shape of the mantle secreting the shell. Shell thickness also is simple to model, as it is sufficient to define two perimeter shapes (one for the external and one for the internal surfaces) instead of one.

It is more difficult to model the effects of the earlier portion of the shell (see above) on shell growth and geometry. This requires a volumetric or voxel-oriented rendering which, as far as I am aware, has not been used in this context. A realistic modelling of the interactions between soft mantle-tissues and rigid shell requires finite element modelling, the applications of which are rare in palaeobiology [11, 12].

The colour patterns of mollusc shells have received considerable attention from palaeobiologists and developmental biologists alike. Studies by Meinhardt [8, 9, 10] have shown that a broad variety of colour patterns and related sculpture patterns can be modelled by reaction diffusion mechanisms (including the terrace patterns discussed above). More recently, cellular automata have been used to model a seemingly equally broad range of colour patterns [5]. It may be remembered that the colour patterns on most shells are generated by an essentially one-dimensional mantle secreting a two-dimensional pattern on the shell surface through time (see also above). However, two- and (slightly) three-dimensional colour patterns produced by a two-dimensional mantle surface stationary with respect to the shell are also present in mollusc shells, and have been modelled [29].

While these studies allow the realistic modelling of virtually all colour patterns found in natural shells, the modelling of sculptures has largely lagged behind. This is due, at least in part, to the additional difficulties in modelling the complex changes in three-dimensional shape of the mantle associated with the construction of shell sculpture. In fact, the observation of the periodic sculptures in many gastropods (e.g. most Muricidae and Coralliophilidae) shows that the construction of this sculpture requires complex three-dimensional changes in the shape of the mantle, while sculpture typically has been modelled by using simple modulations in two dimensions of the apertural perimeter [4]. The latter mechanism appears to be appropriate only to the simulation of minor shell-relief. A realistic modelling of strong shell sculpture will, most probably, have to take into account the three-dimensional shape and resiliency of soft tissues, which may require the application of finite element modelling with fine, detailed and computation-intensive element meshes. Models of three-dimensional pattern formation may also be necessary to reproduce the shape of the mantle tissues involved in the construction of sculpture. Lastly, it can be noted that the field of theoretical shell morphology seems to consist of two largely isolated compartments: palaeobiologists have been pursuing models of shell coiling that attempt to be more and more faithful in biological terms, but have not shown a comparable concern about surface features like colour and sculpture, and only recently have begun to model the growth processes of shell microstructures (Ubukata, this volume and references therein). At the same time, developmental biologists, although devoting much attention to colour patterns of shells, have not demonstrated a similar interest in shell coiling. As a result, papers on shell coiling by developmental biologists and computer-graphic scientists tend to ignore the modern palaeobiological literature on this subject [4].

28.3 Concluding remarks

The above examples are representative of several palaeobiological themes that deal with pattern formation and other subjects related to developmental biology. In both disciplines there is a general willingness to accept indirect methods of investigation as valid, and the conceptual bases of these methods are partly similar in the two fields. These examples show also that palaeobiologists and developmental biologists often have been working on themes of common interest, although the awareness of each others' work probably needs to be increased.

References

1. Ackerly, S.C. (1989). Kinematics of accretionary shell growth, with examples from brachiopods and molluscs. *Paleobiology* **15**: 147-164.
2. Ackerly, S.C. (1989). Shell coiling in gastropods: analysis by stereographic projection. *Palaios* **4**: 374-378.
3. Checa, A., Jimenez-Jimenez, A.P. and Rivas, P. (1998). Regulation of spiral coiling in the terrestrial gastropod Sphincterochila: an experimental test of the road-holding model. *Journal of Morphology* **235**: 249-257.
4. Fowler, D.R., Meinhardt, H. and Prusinkiewicz, P. (1992). Modeling seashells. *Computer Graphics* **26**: 379-387.
5. Gunji, Y-P., Kusunoki, Y. and Ito, K. (1999). Pigmentation of molluscs: how does global synchronisation arise? In Savazzi, E. (ed) *Functional Morphology of the Invertebrate Skeleton*, John Wiley & Sons, Chichester, pp 37-55.
6. Hutchinson, J.M.C. (1989). Control of gastropod shell shape: the role of the preceding whorl. *J. theor. Biol.* **140**: 431-444.
7. Illert, C. (1987). Formulation and solution of the classical seashell problem. *Il Nuovo Ciemento* **9D**: 791-813.
8. Meinhardt, H. (1984). Models for positional signalling, the threefold subdivision of segments and the pigmentation pattern of molluscs. *Journal of Embryology and Experimental Morphology* **83** (supplement): 289-311.
9. Meinhardt, H. (1995). *The Algorithmic Beauty of Sea Shells*. Springer-Verlag, pp i-xi, 1-204.
10. Meinhardt, H. and Klinger, M. (1987). A model for pattern formation on the shells of molluscs. *J. theor. Biol.* **126**: 63-89.
11. Morita, R. (1991). Finite element analysis of a double membrane tube (DMS-tube) and its implication for gastropod shell morphology. *Journal of Morphology* **207**: 81-92.
12. Morita, R. (1991). Mechanical constraints on aperture form in gastropods. *Journal of Morphology* **207**: 93-102.
13. Morita, R. (1993). Developmental mechanics of retractor muscles and the "dead spiral model" in gastropod shell morphogenesis. *Neues Jahrbuch fur Geologie und Palaontologie Abhandlugen* **190**: 191-217.
14. Okamoto, T. (1984). Theoretical morphology of Nipponites (a heteromorph ammonoid). *Kaseki* **36**: 37-51. (in Japanese)
15. Okamoto, T. (1988). Analysis of heteromorph ammonoids by differential geometry. *Palaeontology* **31**: 35-52.
16. Okamoto, T. (1988). Developmental regulation and morphological saltation in the heteromorph ammonite Nipponites. *Paleobiology* **14**: 272-286.
17. Okamoto, T. (1993). Theoretical modelling of ammonite morphogenesis. *Neues Jahrbuch fur Geologie und Palaontologie Abhandlugen* **190**: 183-190.

18. Paul, C.R.C. (1999). Terrestrial gastropods. In Savazzi, E. (ed) *Functional Morphology of the Invertebrate Skeleton*, John Wiley and Sons, Chichester, pp 149-167.
19. Raup, D.M. (1962). Computer as aid in describing form in gastropod shells. *Science* **138**: 150-152.
20. Raup, D.M. (1966). Geometric analysis of shell coiling: general problems. *Journal of Paleontology* **40**: 1178-1190.
21. Raup, D.M. (1967). Geometric analysis of shell coiling: coiling in ammonoids. *Journal of Paleontology* **41**: 43-65.
22. Raup, D.M. and Michelson, A. (1965). Theoretical morphology of coiled shells. *Science* **147**: 1294-1295.
23. Savazzi, E. (1983). Aspects of the functional morphology of fossil and living invertebrates (bivalves and decapods). *Acta Universitatis Upsaliensis, Abstracts of Uppsala Dissertations from the Faculty of Science* **680**: 1-21.
24. Savazzi, E. (1986). Burrowing sculptures and life habits in Paleozoic lingulacean brachiopods. *Paleobiology* **12**: 46-63.
25. Savazzi, E. (1990). Biological aspects of theoretical shell morphology. *Lethaia* **23**: 195-212.
26. Savazzi, E. (1991). Burrowing in the inarticulate brachiopod Lingula anatina. *Palaeogeography Palaeoclimatology Palaeoecology* **85**: 101-106.
27. Savazzi, E. (1994). Functional morphology of burrowing and boring organisms. In Donovan, S.K. (ed) *The Palaeobiology of Trace Fossils*. John Wiley and Sons, London, pp 43-82.
28. Savazzi, E. (1995). Theoretical shell morphology as a tool in constructional morphology. In Aigner, T., Fursich, F., Lutherbacher, H-P., Mosbrugger, V., Reif, W-E. and Westphal, F. (eds) Festschrift A Seilacher, *Neues Jahrbuch fur Geologie und Palaontologie Abhandlungen* **195**: 229-240.
29. Savazzi, E. (1998). The colour patterns of cypraeid gastropods. *Lethaia* **31**: 15-27.
30. Savazzi, E. (1999). Introduction to functional morphology. In Savazzi, E. (ed) *Functional Morphology of the Invertebrate Skeleton*, John Wiley and Sons, Chichester, pp 3-13.
31. Savazzi, E. and Sasaki, T. (in press)S Function and construction of synchronised sculpture in gastropods. *American Malacological Bulletin*.
32. Seilacher, A. (1972). Divaricate patterns in pelecypod shells. *Lethaia* **5**: 325-343.
33. Seilacher, A. (1973). Fabricational noise in adaptive morphology. *Systematic Zoology* **22**:451-465.
34. Seilacher, A. and Gunji, P-Y. (1993). Morphogenetic countdowns in heteromorph shells. *Neues Jahrbuch fur Geologie und Palaontologie Abhandlugen* **190**: 237-265.
35. Stanley, S.M. (1969). Bivalve mollusk burrowing aided by discordant shell ornamentation. *Science* **166**: 634-635.
36. Wrigley, A. (1948). The colour patterns and sculpture of molluscan shells. *Proceedings of the Malacological Society* **27**: 206-217.

Why do Univalve Shells of Gastropods Coil so Tightly? A Head-Foot Guidance Model of Shell Growth and its Implication on Developmental Constraints

Rihito Morita

Natural History Museum and Institute, Chiba. 955-2 Aoba-cho, Chuo-ku, Chiba
2608682, Japan
email: morita@chiba-muse.or.jp

Summary. Tight coiling with whorl overlap is the most frequent mode of shell coiling in gastropods. A new model of shell growth, the "head-foot guidance model", implies that the contact of the head-foot mass with the mantle edge plays the main role in producing tight coiling. Computer simulation of the model suggests that the head-foot guidance mechanism imposes developmental constraints on the evolution of gastropods; 1) bilaterally symmetric coiling is associated with bilaterally symmetric musculature of the foot mass; 2) tight coiling is associated with the contact of the head-foot mass and the mantle edge; 3) insertion of the foot muscle occurs in gastropods with an uncoiled shell.

29.1 Introduction

Organismic forms produced by evolution are not distributed randomly or continuously in a geometrical sense, but in most cases discontinuously and in discrete clusters. Why? Is such a discrete cluster the result of natural selection or historical contingency? Developmental constraints are also possible factors that may explain this discreteness [4]. The concept implies that morphological variation is not generally continuous according to the inflexibility of modification of developmental processes. This means that morphological variation is potentially limited prior to the operation of natural selection. Although there are many hypotheses on developmental constraints to explain some evolutionary phenomena, so far none of them has been tested.

In order to test any hypothesis of developmental constraints, we need the entire catalogue of existing forms of organisms which share a common developmental constraint. The catalogue must also include fossil data. However, fossil records of evolution are generally incomplete. Some phyla of animals have no fossil record. Fossil records provide only a limited number of "stories" out of the diverse evolutionary histories of organisms which have existed. However, in spite of such general incompleteness, patterns of evolution recorded by fossil data can provide important implications to evolutionary biology if we focus on those organisms which have left an almost complete fossil record. Molluscs of the class Gastropoda with univalve shells are one such case. They are one of the most common fossils in marine sediments since their first appearance in the Cambrian.

The geometrical regularity or beauty of shells have also fascinated many scientists. However, the main contribution has been provided by theoretical morphology pioneered by Raup [7, 8]. Although the aim of modern theoretical morphology varies according to the authors [2, 9], the original target of theoretical morphology is to construct a morphospace which subsumes all theoretically possible forms. It is intended to visualize existing forms within a morphospace which covers also all possible but non-existing forms. However, the term 'theoretical' in 'theoretical morphology' is not necessarily equivalent to the term 'developmental' or 'morphogenetic'. This is rather equivalent to 'a rule' or 'kinematics' of shell construction. For example in [7] a logarithmic helicospiral rule is used.

29.2 Head-foot guidance mechanism of shell growth

A marginally growing shell is the sum of many small steps of accretionary growth (Fig. 29.1). Thus the shape of the shell depends on the shape and size of the shell increment at each accretionary step, and also on the ontogenetic change of the latter. For instance, any coiled shell consists of many curved increments whose length is different around the aperture. If the direction of the curve is concordant with the gradient in length of increment, a regularly coiled shell like a helicospiral is generated.

The shape of a shell increment is almost parallel to that of its mantle which is the tissue lining the inner surface of the shell and secreting the shell (Fig. 29.2). The differential growth rate of the mantle tissue along its margin is always concordant with the direction of coiling. That is, the shortest mantle lobe is always located on the inner side of coiling. Furthermore, the shape of any shell increment seems to be almost equivalent to that of the mantle edge poking out from the aperture at the time of shell growth. However, controlling the shape of mantle tissue is not easy because the mantle has only a hydrostatic skeleton. It can be easily deformed when a rigid object contacts it. The head-foot mass is such an object. Actually, in marine creeping snails, the foot always presses the mantle edge on the inner lip while the mantle edge is poking out from the aperture. Is the bending of the mantle edge pressed by the foot included in the morphogenetic system of the growing shell?

It was observed that the developing foot in the pedi-veliger stage had pressed the aperture margin so that shell growth was changed into the direction concordant with the pressure [1]. Morphological rule of apertural outline

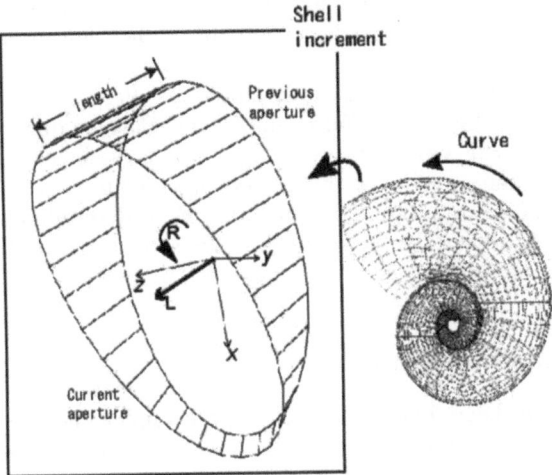

Fig. 29.1. An isometric mode of coiling is generated by a shell increment which curves in a certain direction, the length is reduced in the same direction, and increases size at a constant rate. The incremental growth can be described as an apertural motion in a Cartesian frame (x, y, z) defined on the previous aperture plane. R and L are rotation matrix and translational motion vector respectively in equation (29.1).

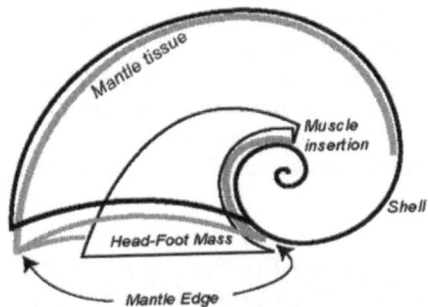

Fig. 29.2. Schematic arrangement of mantle and head-foot mass in an imaginary transparent shell of a creeping snail. The mantle is a skirt-like tissue lining the inner surface of the shell, and its edge pokes out from the aperture at the time of shell growth. In a creeping mode of life, the mantle edge is pressed by the head-foot mass on the inner side of coiling on which the head-foot mass is inserted.

also indicates contact of the foot with the mantle edge on the inner lip of the aperture at the time of shell growth [5]. According to these facts, Morita [6] proposed a model of the mechanism for controlling the direction of shell growth. In this mechanism, shell growth follows the direction of the extension of the head-foot mass which makes contact with the mantle edge. Here, I rename this mechanism the "head-foot guidance" mechanism, instead of the "dead spiral model" in [6].

The head-foot mass does not alone guide the direction of shell growth. There are other possible agents. For example, the spatially differential growth of soft parts confined within the shell may also contribute to determining the direction of shell growth. In this paper, in order to test to what extent the mechanism of head-foot guidance contributes to determining shell shape, in the model, all other factors have been placed in a parameter set called 'intrinsic growth parameters'.

29.3 Computational model

The mechanism of head-foot guidance has been implemented in a computational program (detailed description of the program is in preparation). In the computational model, the effect of head-foot guidance is simplified as rotation of the aperture plane in the direction in which the head-foot mass presses the mantle edge. Furthermore, I assume that apertural rotation due to the contact effect of the head-foot mass is independent of the effect of the intrinsic growth parameters at every accretionary step.

29.3.1 Intrinsic growth model

An intrinsic growth model is a model of shell growth controlled by intrinsic growth parameters alone. The parameters here are defined as kinematic parameters in the following equation of shell growth (Fig. 29.1). The equation expresses the transformation of a point (x,y,z) on the previous apertural plane to the homologous point (X,Y,Z) on the current apertural plane. The equation is related to a local Cartesian coordinate system defined on the previous aperture plane:

$$\begin{pmatrix} X \\ Y \\ Z \end{pmatrix} = E \left[\begin{pmatrix} R_{11} & R_{12} & R_{13} \\ R_{21} & R_{22} & R_{23} \\ R_{31} & R_{32} & R_{33} \end{pmatrix} \begin{pmatrix} x \\ y \\ z \end{pmatrix} + \begin{pmatrix} L_x \\ L_y \\ L_z \end{pmatrix} \right] \tag{29.1}$$

where the matrix $[R_{ij}]$ and vector L denote a rotation matrix and a translational motion vector respectively, and E denotes an expansion rate. In principle, all the regular spirals can be simulated by this equation, and moreover if these parameters are set as a function of an accretionary step, any shell growth can be simulated.

29.3.2 Head-foot guidance model

A head-foot guidance model is a model of shell growth controlled by the effect of head-foot guidance in addition to the intrinsic growth parameters.

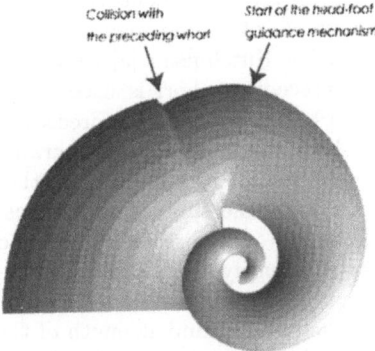

Fig. 29.3. In a planispiral shell with bilaterally symmetric musculature, contact pressure of the head-foot mass always operates on the bilaterally symmetric axis if there is contact. This contact pressure produces much tighter coiling.

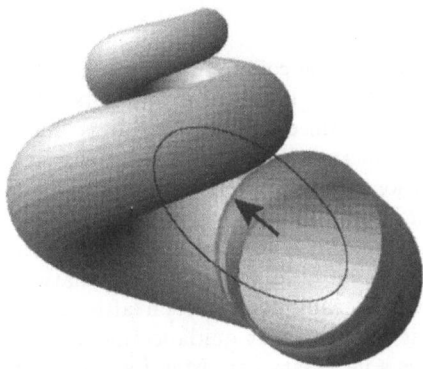

Fig. 29.4. The contact pressure of an idealized head-foot mass (elliptic outline) is assumed to operate perpendicularly on aperture margin near the preceding whorl.

The effect of head-foot guidance is defined as the rotation of the aperture plane in the direction of the contact pressure of the head-foot mass (Fig. 29.3). In actual processes, at each accretionary step, firstly the aperture is expanded, rotated, and translated according to the given intrinsic growth parameters, and thereafter, the aperture is rotated again according to the mechanism of head-foot guidance. The secondary rotation of the aperture moves the aperture margin closer to the preceding whorl, since, in most cases, the contact of the head-foot mass with the mantle is expected to be near the preceding whorl. Ultimately, the aperture margin collides with the preceding whorl. In the computation, I implemented a collision detection so that the aperture margin conforms to the nearest surface of the preceding whorl. This means that the mode of shell growth depends on the shape of the preceding whorl, and the resultant mode of coiling may be the same as was expected by the road-holding model of Hutchinson [3]. According to the Hutchinson model, the shape of the preceding whorl is used as a cue to dictate where upon it the next new whorl attaches onto the preceding one.

The most difficult problem in executing the program of the head-foot guidance model is to determine the place of the foot at the time of shell growth. This is the key information for the simulation, because without the knowledge of the position of the foot, the direction of secondary rotation can not be determined. In order to identify the position of the foot, it is necessary to identify the life position at the time of shell growth. The life position also depends on the mode of shell coiling and strength of the foot. Unfortunately, there is no general rule to figure out the relationship between them. In this paper, I will show some results of computer simulation of the head-foot guidance mechanism only for some modes of coiling where the foot position can be easily inferred: planispiral, helicospiral, and uncoiling modes.

29.3.3 Planispiral shell

A planispiral shell is a mode of coiling in a single horizontal plane. Such shells are common in cephalopods, but rare in gastropods. An important aspect of this bilaterally symmetric coiling is that it is always associated with a bilaterally symmetric musculature of the foot. This fact allows us to assume that the foot is positioned on the plane of bilateral symmetry.

With this assumption, computer simulation of the head-foot guidance model generates a tightly coiled shell with successive whorls overlapping even if the initial coiling is open (Fig. 29.4). The simulation also shows that this state of plane coiling is easily broken soon after the musculature becomes asymmetric. Thus, if the head-foot guidance model is valid, bilaterally symmetric musculature is a necessary condition for keeping the planispiral during growth. Actually, all the known plane coilings have bilaterally symmetric musculature since their first appearance in the Cambrian.

29.3.4 Helicospiral shell

A helicospiral shell is a shell which grows around an axis and at the same time translates along the axis. Among such helicospiral shells, there are creeping snails which carry the shell resting on the back side of the foot. Here, I assume that these shells have the same life position also at the time of shell growth. This assumption means that the region of the aperture margin which is pressed by the foot is always located near the preceding whorl, and the direction of the pressure is almost constant (say always perpendicular to the inner lip, which is close to the preceding whorl, as shown in Fig. 29.3).

The head-foot guidance model with this assumption generates a regularly coiled shell with whorl overlap for a given initial helicospiral shell (Fig. 29.5B).

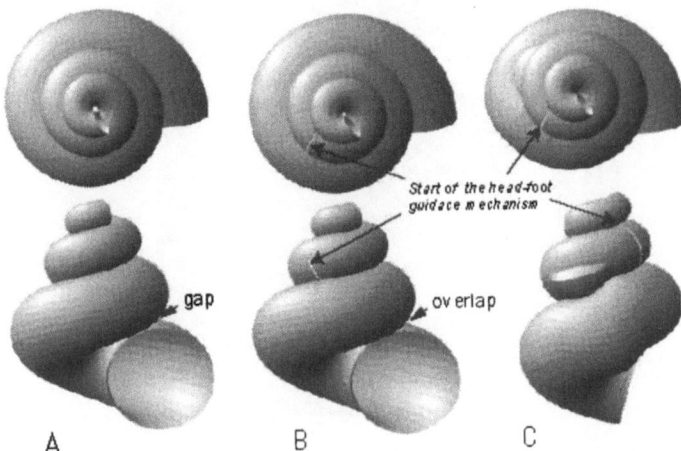

Fig. 29.5. In a helicospiral coiling, contact pressure of the head-foot is assumed to operate perpendicularly to a certain region of the aperture margin near the preceding whorl. (A) A coiling controlled by intrinsic helicospiral growth parameters alone with a gap between successive whorls. (B) A similar coiling to that of (A) but with whorl overlap was generated by activation of the head-foot guidance mechanism starting in mid-growth. (C) The head-foot guidance mechanism can be transferred by projection on the preceding whorl.

This consequence is not only similar to the natural coiling of gastropods, but also shows a similar ability to react to an obstacle attached to the preceding whorl (Fig. 29.6). In a computer simulation with a swollen part on the preceding whorl, the shell growth changed its direction downward when it met the swollen portion (Fig. 29.5C). This reaction is possible because the head-foot guidance model has an ability to rotate the aperture in order to conform the surface of the preceding whorl. Actually, even without the swollen part, the mode of coiling generally depends on the shape of the preceding whorl. For example, the prominent spiral element influences a route of whorl overlapping and its stability. As mentioned above, these properties of the head-foot guidance model are similar to those expected by the "road-holding" model of [3]. Although the latter is a simple kinematic model without any assumption of the mechanism involved, these results of our computer simulations suggest that the head-foot guidance mode is simply a mechanism for road-holding.

29.3.5 Uncoiled shell

The head-foot guidance model is a model under the assumption that the head-foot mass contacts with the aperture margin at the time of shell growth. In other words, if there is no contact, the growth of the shell is controlled only by

Fig. 29.6. A marine snail, Turbo (Batillus) cornutus (Lightfoot) with an attached oyster.

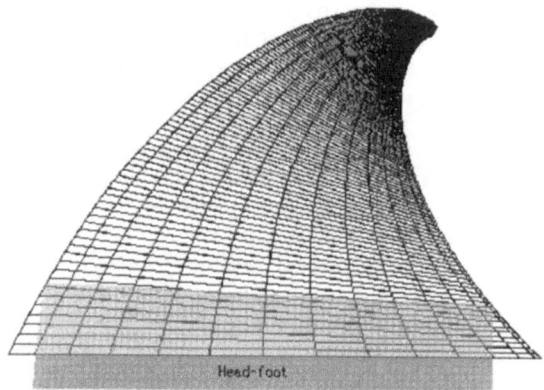

Head-foot

Fig. 29.7. Uncoiling or disjunct coiling is associated with widely extended muscle insertion of the foot. This makes the apertural margin (mantle edge) free from contact with the head-foot mass.

the intrinsic growth parameters. A shell with a wide area of muscle insertions near the aperture is considered in such condition (Fig. 29.7). According to the muscle scars preserved in fossil shells, all known cup- or horn-like uncoiled shells have muscle insertions such that they could avoid the contact of the shell margin with the head-foot mass. Does this mean that intrinsic growth parameters are limited as constraints in those potentially uncoiled modes? This can not be answered yet, but at least it is likely that tight coiling is the inevitable result in head-foot guidance when the head-foot mass lies in contact with the aperture margin.

29.4 Conclusion

The head-foot guidance model implies that shell growth follows the extension of the head-foot mass if the latter presses the mantle margin at the time of shell growth. Thus, according to the model, the resultant shape of a shell depends on the position of the head-foot mass on the aperture plane. If the head-foot mass is located around the center of the aperture and does not contact the aperture margin at the time of shell growth, shell growth has no head-foot guidance and follows the influence of other agents (here called intrinsic growth parameters).

Shell form can not be predicted if shell growth is controlled by the intrinsic growth parameters alone. However, if there definitely is contact of the head-foot mass with the aperture, then all the forms can be predicted to be tightly coiled with whorl overlap, because of the control of the head-foot contact. This prediction is valid at least in the case where the head-foot mass is undoubtedly pressing the aperture margin, such as in the case of creeping modes of life. In addition, the model also predicts the condition when all bilaterally symmetric and uncoiled shells can be generated. Bilaterally symmetric coiling needs the musculature to be bilaterally symmetric. Otherwise, according to the head-foot guidance model, the asymmetric position of the head-foot mass causes the coiling also to be asymmetric. The uncoiling mode needs only the condition that the head-foot mass does not make contact with the aperture margin. This condition occurs only if the musculature has a wide area of insertion near the aperture. All these predictions may be regarded as developmental constraints in the evolution of shell form.

Acknowledgements

I am grateful to Dr. Toshio Sekimura for inviting me to present this work at the International Conference on Morphogenesis and Pattern Formation in Biological Systems in Nagoya, Japan. I also thank Dr. Enrico Savazzi and Dr. Ludmila Titova for reviewing this paper. This work was party supported by Grant-in-Aids for Scientific Research from the Japan Society for the Promotion of Science (No. 12640460).

References

1. Crofts, D. (1937). The development of Haliotis tuberculata, with special reference to the organogenesis during torsion. *Phil. Trans. Roy. Soc. Lond.* B **228**, 219-268.

2. Hickman, C.S. (1993). Theoretical design space: A new program for the analysis of structural diversity. *N.Jb.Geol.Palaeont.Abh.*, **190(2/3)**, 169-182.
3. Hutchinson, J.M.C. (1989). Control of gastropod shell shape; the role of the preceding whorl. *J. Theor. Biol.*, **140**, 431-444.
4. Maynard Smith, J., Burian, R., Kauffman, S., Alberch, P., Campbell, J., Goodwin, B., Lande, R., Raup, D. and Wolpert, L. (1985). Developmental constraints and evolution. *Quart.Rev.Biol.*, **60**, 265-287.
5. Morita, R. (1991). Mechanical constraints on aperture form in gastropods. *J.Morphology*, **207**, 93-102.
6. Morita, R. (1993). Developmental mechanics of retractor muscles and the "dead spiral model" in gastropod shell morphogenesis. *N. Jb. Geol. Palaeont. Abh.*, **190(2/3)**, 191-217.
7. Raup, D. M. (1966). Geometric analysis of shell coiling: general problems. *J. Paleontology*, **40(5)**, 1178-1190.
8. Raup, D. M. and Michelson A. (1965), Theoretical morphology of the coiled shell. *Science*, **147(3663)**, 1294-1295.
9. Stone, J.R. (1996). The evolution of ideas: a phylogeny of shell models. *Am.Nat.*, **148(5)**, 904-929.

Computer Modeling of Microscopic Features of Molluscan Shells

Takao Ubukata

Institute of Geosciences, Shizuoka University, 836 Oya, Shizuoka 422-8529, Japan
e mail: sbtubuk@ipc.shizuoka.ac.jp
telephone: +81-54-238-4797
fax: +81-54-238-0491

Summary. Computer modeling of growth of biominerals and microgrowth increments of the bivalve shell helps us understand the rules or algorithms that lead to the formation of the geometric pattern of the microscopic features of the shell. Computer models of elongation of aggregated minerals on the radial shell section or outer shell surface indicate that the direction of crystal elongation relative to growth rings depends on the rate of crystal growth and its anisotropy. Computer models of crystal expansion on the outer shell surface and comparison of the models with biometric analyses reveal a close relationship between growth rate of crystals or the entire shell and microscopic morphologies such as size, shape and distribution of nucleation sites of aggregated crystals. An analysis of internal microgrowth pattern and its relation to topography of the outer shell surface using computer modeling allows us to reconstruct successive behaviors of mantle when it is secreted from mantle.

Keywords: Bivalvia, computer modeling, microgrowth increments, shell microstructure

30.1 Introduction

A molluscan shell consists of a lot of small carbonate minerals arranged into various distinct fabrics to form several kinds of shell microstructures. In addition, accretionary shell growth is accompanied by formation of a large number of growth rings which are the record of periodic time series of microgrowth increments. These microscopic features of the molluscan shell record some physiological conditions of the animal at the time the shell was formed, and are readily preserved in fossil materials. Therefore, understanding the process of pattern formation of the microscopic features will provide a reliable basis to establish a "paleophysiology" of fossil organisms. Since the middle of the 20th century, taxonomic distribution of bivalve shell microstructure has been studied by many paleontologists and malacologists for inferring phylogenetic evolution [2, 3, 4, 22, 23, 33]. On the other hand, physiological, biochemical and crystallographic studies of molluscan shell formation have also been carried out in the multidisciplinary field of biomineralization [6, 7, 18, 20, 34, 36]. In addition, a variety of theoretical modeling studies has been attempted to investigate gross shell morphology or shell ornamentation [1, 14, 15, 17, 32]. However, these fields have been developed independently, without strong reciprocal interactions. Thus, little attention has been paid to rules or algorithms generating the geometric pattern of the microstructures. Recently, I have studied molluscan shell microstructures from a viewpoint of pattern formation. The goal of this series of studies is to couple geometric properties of microscopic morphology to biomineralogical aspects in molluscan shells. In the present paper, I review some theoretical models on microscopic features of bivalve shells which I have introduced.

30.2 Modeling crystal elongation in radial shell section

A bivalve shell consists of a large number of columnar, fibrous or sheetlike structural units, each of which is not a single crystal but a bundle of aggregations of numerous small minerals. Each structural unit elongates inward and/or toward the growing shell margin. In radial shell section, the elongation axes of structural units appear to be vertical, inclined or horizontal to the outer shell surface (Fig. 30.1A).

Inclined structural units are often curved toward the inner shell surface. In some microstructures, elongation axes of structural units and internal microgrowth increments in radial shell section usually intersect at 90°. In this case, however, structural units occasionally become slightly inclined to the microgrowth increments as crystals grow toward the inner surface [26]. In other cases, structural units are remarkably inclined to the increments throughout crystal growth. I attempted to reproduce the variation of geometric pattern of the microstructures in radial shell section using theoretical modeling [25, 26]. For modeling the process of microscopic growth, information on initial growth stage of structural units helps understand the nature of crystal growth. In many cases, incipient structural units occurring at the growing margin are hemispheric or spherulitic [25, 26]. Then, in radial shell section, each structural unit can be approximated by an enlarging circle or ellipse which is accreted at the shell margin. Although the shape of an incipient microstructural unit is not exactly hemispheric but more or less variable among shell microstructures, the essential nature of the theoretical model does not seriously depend on the shape of the model of each unit. Each structural unit continues to expand on its lateral sides until it comes into contact with a neighboring one, and then elongates towards the inner shell surface or ventral shell margin, forming a boundary between units (Fig. 30.1B).

Total shell growth consists of accretion of new structural units at the shell margin and elongation of each unit. Therefore, a theoretical model of radial shell section can be defined if we specify the lateral expansion rate of each unit

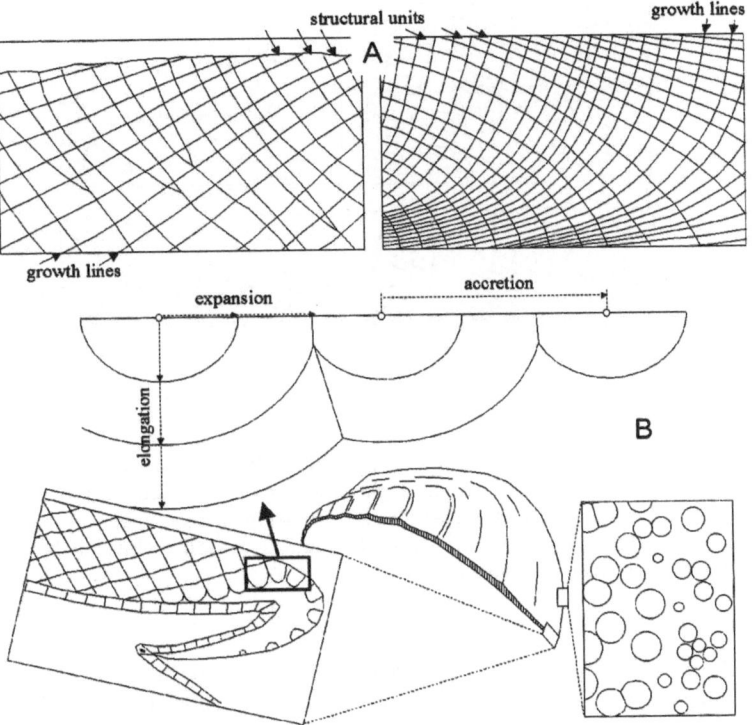

Fig. 30.1. A, Sketches of radial sections of bivalve shells. The outer surface is upward and the ventral side to the right. B, Schematic diagram illustrating growth process of structural units and theoretical modeling of radial shell section exhibited by enlargement of the structural units. In this modeling, shell growth is represented by accretion, lateral expansion, and extension inward of elliptic structural units.

and elongation rate of the unit inward, with respect to accretion rate of new structural units. If each hypothetical structural unit grows as an enlarging circle at a constant rate, the elongation axis of the unit is always perpendicular to any growth rings (Fig. 30.2A). In this case, the degree of inclination of preferred oriented structural units is predominantly determined by the relative enlargment rate of the circular structural unit with respect to the accretion of new structural units. If each crystal is represented by an enlarging ellipse at a steady rate, structural units are considerably inclined both to the outer shell surface and microgrowth increments toward the ventral direction (Fig. 30.2B). If each elliptical structural unit decreases its enlargment rate during growth while maintaining its aspect ratio constant, the elongation axes of structural units are curved and inclined to microgrowth increments in the

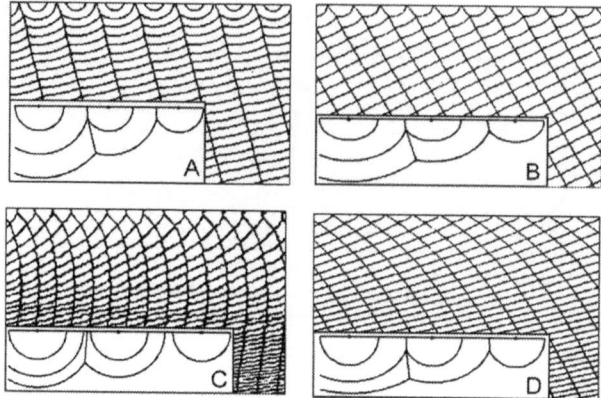

Fig. 30.2. Selected computer models of radial shell sections. The outer surface is upward and the ventral side to the right. Each structural unit is represented by an enlarging circle (A, C) or ellipse (B, D). Enlargment rate is constant (A, B) or decreases isotropically (C), or only the extension rate inward declines with remaining lateral expansion rate constant (D).

dorsal direction (Fig. 30.2C). This pattern is not found in actual bivalves. On the other hand, if each structural unit reduces its rate of elongation inward with lateral expansion rate fixed throughout growth, crystals are also curved but are inclined to the microgrowth increments in the ventral direction (Fig. 30.2D). This pattern represents well the pattern commonly observed in natural bivalves. From comparison of computer models with actual shells, the growth rate of structural units and its anisotropy appear to control the direction of elongation of the structural units. Various patterns observed in actual bivalves can be reproduced using theoretical modeling by changing the relative elongation rate of structural units with respect to lateral expansion of the units and total shell accretion.

30.3 Modeling crystal elongation and microsculptures on the outer shell surface

In some shell microstructures, structural units are strongly reclining or nearly parallel to the outer shell surface. In this case, the elongation axes of their structural units may form a linear structure on the outer shell surface. Hayami and Okamoto [8] studied pattern formation of some oblique microsculptures on the outer surface of bivalve shells using a theoretical model, and pointed out that their architectures are based on orientations of microstructural units.

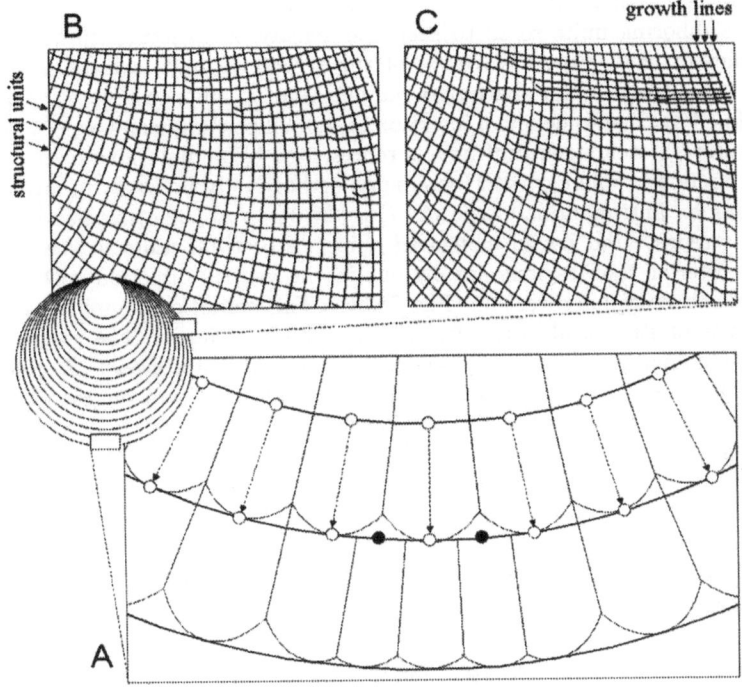

Fig. 30.3. A, Diagram showing a theoretical model of crystal elongation on the outer shell surface. In this model, total shell growth is represented by enlargement of a circle, and each structural unit is inserted and grows at the shell margin. B, C, Selected computer models. Each structural unit is represented by an enlarging circle (B) or ellipse (C).

A similar theoretical model as used on radial shell section is also applicable to the structures on the outer shell surface after some necessary modifications. I have introduced a theoretical model to analyze the microscopic geometry on the outer shell surface in bivalves [28].

In this model, the outline of the shell is represented by an enlarging circle. Initiation of microstructural units occurs along the margin of the circular shell (Fig. 30.3A). Each structural unit is approximated by an enlarging ellipse. A new structural unit is installed with the elongation axis of the ellipse perpendicular to the margin of the circular shell. The direction of the major axis of the elliptic unit is kept throughout the unit growth. Each structural unit grows until it is inscribed in the neighboring unit or the next circle of the shell margin. The center of the elliptic unit shifts towards the growing margin as the circular shell enlarges. A new structural unit is inserted between the neighboring units so as to keep the density of structural units uniform throughout shell growth. Growth of the structural unit is terminated if it loses in competition for space between neighboring units. A theoretical model is defined by the frequency of insertion of new elliptic units and aspect ratio of each unit. If each structural unit is represented by an enlarging circle, structural units and microgrowth increments orthogonally intersect in any region forming an anti-marginal structure (Fig. 30.3B). On the other hand, if each unit is represented by an elongated ellipse, structural units in the anterior and posterior shell regions are oblique in the ventral direction with respect to the anti-marginal direction (Fig. 30.3C). This condition complies well with obliquity of structural units observed in actual specimens. The result shows that actual patterns are reproduced well by the computer models in which the growth rate of each structural unit in the radial direction is larger than that in the transverse direction. Considering the results of the study on radial shell section together, it is concluded that microstructural units of bivalve shells tend to grow more rapidly in the radial direction than any other direction.

30.4 Modeling crystal expansion on the outer shell surface

Among various kinds of shell microstructures, the prismatic structure consists of many parallel arrayed columnar units [3, 4]. Each prism is surrounded and bounded by an organic matrix showing a honeycomb-like appearance on the outer shell surface (Fig. 30.4A, B). Sizes, shapes and distribution patterns of prisms are quite variable among individuals or species. They also often change during growth of a single individual. A seasonal change of size and shape of crystals through growth was reported in other types of microstructure [35]. It has also been suggested that size and shape of crystals may reflect the rate of crystal growth in bivalve molluscs. I have analyzed the relationship among sizes, morphologies and distribution patterns of nucleation sites of aggregated prisms on the outer shell surface using a theoretical model [29, 30]. On the outer surface of the prismatic shell, many circular growth rings are commonly observed within an individual prism (Fig. 30.4B) [25]. Then, in the present model growth of each prism is represented by an enlarging circle at a steady rate. Potential nucleation sites of prisms are regularly arranged within the nucleation zone along the growing shell margin (Fig. 30.4C).

During a short growth step, nucleation of a prism may occur at each potential site with a given probability. Total shell accretion during the period is reflected in a shift of the shell margin and nucleation zone. Meanwhile, each circular prism increases its radius. As the prisms grow, neighboring prisms come closer and finally into contact with one another, resulting in the formation of a boundary between two prisms. A theoretical model can be defined if we specify the following three variables: relative growth rate of prisms with

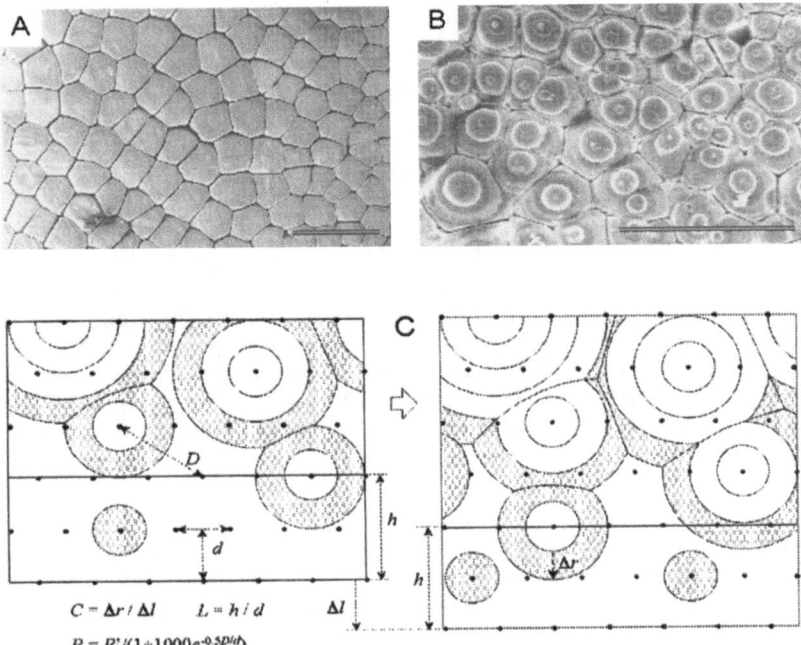

Fig. 30.4. A, B, Scanning electron micrographs of the outer surface of bivalve shells showing prismatic shell structure. Scale bar: 0.1mm. C, Schematic diagram showing a theoretical model of prismatic structure on the outer shell surface. The model is defined by the relative growth rate of prisms with respect to total shell accretion; C, extent of the nucleation zone, L, and probability of nucleation at a potential site during a unit growth step, P.

respect to total shell accretion, extent of nucleation zone, and probability of nucleation at a potential site during a unit growth step. In order to compare theoretical models with actual specimens, sizes, shapes and spatial distribution of nucleation of aggregated prisms were estimated in each theoretical model and each portion of the actual shell. An average size of prisms on a restricted shell portion was represented by the median of the area of prisms in the shell portion [28]. Shapes of prisms were represented by deviation of the geometry from Voronoi division in which any points inside the polygon are closer to its center than to any other centers (Fig. 30.5A) [9, 10]. Distribution of nucleation sites was represented by the index of basic contagion defined by [11] (Fig. 30.5B). It is determined by the relationship between the number of square grids of an arbitrary size and the number of nucleation sites in each square [13]. From the result of biometric analyses, the cellular pattern of

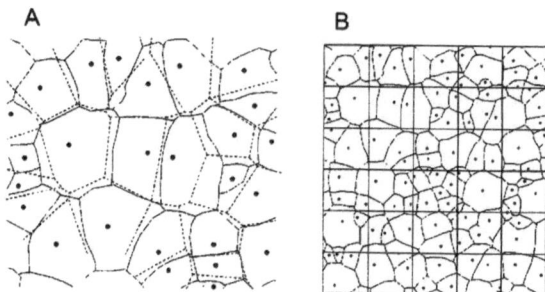

Fig. 30.5. A sketch of a shell portion on the outer surface illustrating boundaries between prisms (solid lines) and nucleation sites (bold points). A, Ideal Voronoi polygons based on the distribution of the nucleation sites are shown by broken lines. B, If the image is subdivided into squares, the number of nucleation sites in each square and the number of squares both determine degree of aggregation of the nucleation sites.

prismatic structures tends closely to Voronoi polygons with increasing sizes of prisms. In the actual shell, nucleation of prisms tends to occur at regular intervals in the case of low density of nucleation. Computer simulations show that, as prisms grow faster, or the nucleation zone becomes wider, prisms tend to have smaller sizes, more non-uniform distribution, and more irregular shapes (Fig. 30.6). In theoretical models, the increase in probability of nucleation results only in reducing sizes of prisms. Since nucleation generally occurs at the growing shell margin, the width of the nucleation zone is considered to be not quite variable. Therefore, from comparison of computer models with actual shells, variation of geometric pattern of prismatic structure on the outer shell surface appears to be controlled mainly by the relative growth rate of prisms with respect to total shell accretion.

30.5 Modeling pattern of microgrowth increments in radial shell section

Internal microgrowth increments of the molluscan shell have been used as a record of periodic time series of environments [12, 16, 21, 24]. They also record mantle kinematics during growth, because the mantle, which secretes shell matter, is in contact with the inner surface of the shell throughout growth. They often show a periodic pattern in terms of their spaces, but it is not always conformable to the commarginal sculpturing pattern on the shell (Fig. 30.7A). I have analyzed the internal microgrowth pattern and its relation to topography of the outer shell surface using a theoretical model, and tried to reconstruct successive behaviors of the mantle when it secreted the microgrowth

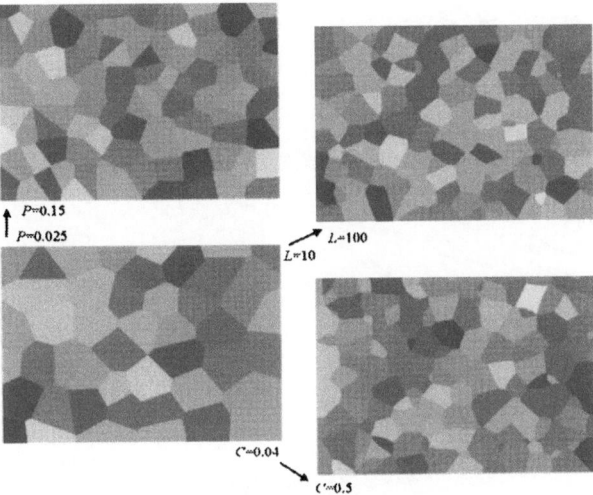

Fig. 30.6. Selected computer models of prismatic structure. As C or L increases, prisms become smaller and more uneven in size, and their geometries deviate from Voronoi division.

increments [27]. Internal microgrowth increments within the outer shell layer are formed by the distal part of the mantle which attaches to the inner shell surface by pallial muscle. In the present modeling, shell growth in radial shell section is exhibited by kinematics of a growth line. The growth line consists of a linear dorsal part and ventral edge of a circular arc. During growth, it translates in the ventral direction and changes its length (Fig. 30.7B). The translation of the growth line represents shell precipitation. Progression of the pallial muscle attachment and extension or shrinkage of the distal part of the mantle, change the total length of the growth line. A theoretical model is defined if periodic functions or constant values are given to the following three vaiables: shell precipitation rate, progression rate of the pallial muscle attachment and rate of extension or shrinkage of the distal part of the mantle. Computer models indicate that a considerable change in any parameters results in the formation of a periodic sculptural pattern on the shell surface, whereas the periodic change of spacing of microgrowth increments requires fluctuation of shell precipitation rate (Fig. 30.7C). The relationship between the pattern of microgrowth increments and phase of sculpture on the shell was analyzed in theoretical models and actual shells. The result of the analysis indicates that periodic change of either shell precipitation rate or progression rate of the pallial muscle attachment does not function to form prominent sculpture on the shell in actual bivalves.

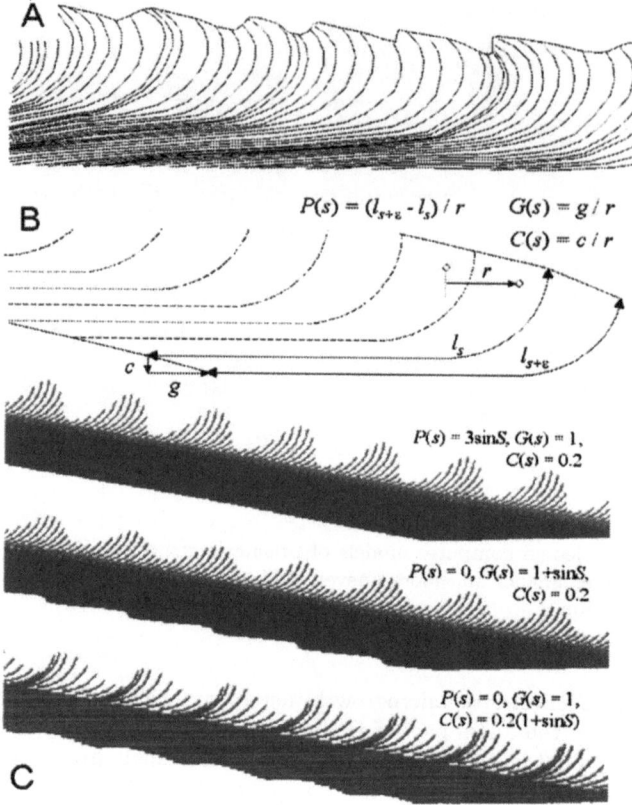

Fig. 30.7. A, Sketch of a radial shell section showing internal microgrowth increments. B, Theoretical model of radial shell section exhibited by kinematics of growth lines. The model is defined by rate of shell precipitation, C, progression rate of the pallial muscle attachment, G, and rate of change in the total length of the growth line, P. C, Selected models in which one of the parameters is periodic.

30.6 Summary and perspectives

In order to understand how to construct microscopic features of molluscan shells, it is necessary to recognize an algorithm for its pattern formation. Analysis of geometric pattern of aggregated biominerals based on theoretical modeling is useful for understanding the architectural aspect [19] of the microstructures. However, the process of microstructure formation is closely related to crystallographic and biochemical aspects of biomineralization which are controlled ultimately by a genetic program. More elaborate models in which these aspects are incorporated will enable us to couple a geometric model with biomineralogy and has the potential to inspire the research field of "paleophysiology". On the other hand, gross shell morphology can be also regarded as a result of stacking of microgrowth increments. Therefore, the direction of total shell growth, which defines coiling geometry, is at least partly related to the process of microscopic growth. Recently, I have introduced a theoretical modeling of bivalve shell growth based on the balance in rates of growth components such as shell accretion, mantle growth and pallial muscle translation [31]. In this modeling, the balance of the growth components determines the shapes of stacking of microgrowth increments, which define the shell form in radial section. Although it only defines the direction of shell growth in two dimensions, this approach may be extended to link microscopic pattern formation with macroscopic shell morphology in bivalve molluscs.

Acknowledgements

This study is a result of a fellowship of the Bilateral Program for Scientist Exchange between the Royal Swedish Academy of Sciences and the Japan Society for Promotion of Science. I thank E. Savazzi for his kind help during my stay at Uppsala. A fund for this work was in part provided by Grant-in-Aids for Scientific Research from the Japan Society for the Promotion of Science (No. 14340267 in 2002).

References

1. Bayer, U. (1978). Morphogenetic programs, instabilities, and evolution - a theoretical study. *N. Jb. Geol. Paleont. Abh.*, **156**:226-261.
2. Bøggild, O.B. (1930). The shell structure of mollusks. *D. Kgi. Danske Vidensk. Selsk. Skrifter, Naturvidensk. og Mathem. Afd. 9*, **2**:231-326.
3. Carter, J.G. and Clark, G.R. II. (1985). Classification and phylogenetic significance of molluscan shell microstructure. In: Broadleaf, T.W. (ed) *Mollusks: Notes for a Short Course*. Department of Geological Science, Studies in Geology, B., University of Tennessee, Knoxville, pp 15-71.
4. Carter, J.G. (1991). Evolutionary significance of shell microstructure in the Palaeotaxodonta, Pteriomorphia and Isofilibranchia (Bivalvia: Mollusca). In: Carter J.G. (ed) *Skeletal Biomineralization: Pattern, Process and Evolutionary Trends*, vol 1. Van Nostrand, New York, pp 135-296.
5. Carter, J.G., Bandel, K., de Buffrénil, V., Carlson, S., Castanet, J., Dalingwater, J., Francillon-Vieillot, H., Géraudie, J., Meunier, F.J., Mutvei, H., de Ricqlès, A., Sire, J.Y., Smith, A., Wendt, J., Williams, A. and Zylberberg, L. (1991). Glossary of skeletal biomineralization. In: Carter J.G. (ed) *Skeletal Biomineralization: Pattern, Process and Evolutionary Trends*, vol 1. Van Nostrand, New York, pp 337-399.

6. Crenshaw, M.A. (1972). The inorganic composition of molluscan extrapallial fluid. *Biological Bulletin*, Marine Biological Laboratory, Woods Hole, Massachusetts, **143**:506-512.

7. Greenfield, E.M., Wilson, D.C. and Crenshaw, M.A. (1984). Inorganic nucleation of calcium carbonate by molluscan matrix. *Am. Zool.*, **24**:925-932.

8. Hayami, I. and Okamoto, T. (1986). Geometric regularity of some oblique sculptures in pectinid and other bivalves: recognition by computer simulations. *Paleobiology*, **12**:433-449.

9. Honda, H. (1978). Description of cellular patterns by dirichlet domains: the two dimensional case. *J. theor. Biol.*, **72**:523-543.

10. Honda, H. (1983). Geometrical analysis of cells becoming organized into a tensile sheet, the blastular wall, in the starfish. *Differentiation*, **25**:16-22.

11. Iwao, S. (1968). A new regression method for analyzing the aggregation pattern of animal populations. *Res. Popul. Ecol.*, **10**:1-20.

12. Jones, D.S., Thompson, I. and Ambrose, W. (1978), Age and growth rate determinations for the Atlantic surf clam *Spisula solidissima* (Bivalvia: Mactracea), based on internal growth lines in shell cross-sections. *Marine Biology*, **47**:63-70.

13. Lloyd, M. (1967). Mean crowding. *Journal of Animal Ecology*, **36**:1-30.

14. Meinhardt, H. and Klingler, M. (1987). A model for pattern formation on the shell of molluscs. *J. theor. Biol.*, **126**:63-89.

15. Okamoto, T. (1988). Analysis of heteromorph ammonoids by differential geometry. *Palaeontology*, **31**:35-52.

16. Pannella, G. and MacClintock, C. (1968). Biological and environmental rhythms reflected in molluscan shell growth. *J. Paleontol.*, Paleontological Society Memoir, **2**:64-80.

17. Raup, D.M. (1966). Geometric analysis of shell coiling: general problems. *Journal of Paleontology*, **40**:1178-1190.

18. Sarashina, I. and Endo, K. (1998). Primary structure of the soluble matrix protein of scallop shell: implications for calcium carbonate biomineralization. *American Mineralogists*, **83**:1510-1515.

19. Seilacher, A. (1970). Arbeitskonzept zur Konstruktions-Morphologie. *Lethaia*, **3**:393-396.

20. Simkiss, K. (1976). Cellular aspects of calcification. In: Watabe, N., Wilbur, K.M. (eds) *The Mechanism of Biomineralization in the Inverbrates and Plants*. University of South Carolina Press, Columbia, pp 1-31.

21. Tanabe, K. (1988). Age and growth rate determination of an intertidal bivalve, Phacosoma japonicum, using internal shell increments. *Lethaia*, **21**:231-241.

22. Taylor, J.D., Kennedy, W.J. and Hall, A. (1969). The shell structure and mineralogy of the Bivalvia. Introduction, Nuculacea-Trigonacea. *Bull, Br. Mus. Nat. Hist., Zool. Suppl.*, **2**:1-125.

23. Taylor, J.D., Kennedy, W.J. and Hall, A. (1973). The shell structure and mineralogy of the Bivalvia. II. Lucinacea-Clavagellacea. Conclusions. *Bull, Br. Mus. Nat. Hist., Zool. Suppl.*, **22**:253-294.

24. Tojo, B., Ohno, T. (1999). Continuous growth-line sequences in gastropod shell. *Palaeogeogr. Palaeoclimetol. Palaeoecol.*, **145**:183-191.

25. Ubukata, T. (1994). Architectural constraints on the morphogenesis of prismatic structure in Bivalvia. *Palaeontology*, **37**:241-261.

26. Ubukata, T. (1997). Microscopic growth of bivalve shells and its computer simulation. *The Veliger*, **40**:165-177.

27. Ubukata, T. (1997). Mantle kinematics and formation of commarginal shell sculpture in Bivalvia. *Paleontological Research*, **1**:132-143.
28. Ubukata, T. (2000). Theoretical morphology of composite prismatic, fibrous prismatic and foliated shell microstructures in bivalves. *Venus*, **59**:297-305.
29. Ubukata, T. (2001). Geometric pattern and growth rate of prismatic shell structures in Bivalvia. *Paleontogical Research*, **5**:33-44.
30. Ubukata, T. (2001). Nucleation and growth of crystals and formation of cellular pattern of prismatic shell microstructure in bivalve molluscs. *Forma*, **16**:141-154.
31. Ubukata, T. (2002). Stacking increments: a new model and morphospace for the analysis of bivalve shell growth. *Historical Biology*, **15**:303-321.
32. Ubukata, T. and Nakagawa, Y. (2000). Modelling various sculptures in the Cretaceous bivalve *Inoceramus hobetsensis*. *Lethaia*, **33**:313-329.
33. Uozumi, S. and Suzuki, S. (1981). The evolution of shell structure in the Bivalvia. In: Habe, T., Omori, M. (eds) *Studies of Molluscan Paleobiology*, Professor Omori Memorial Volume. Publication Committee, Niigata University, Niigata, Japan, pp 63-77.
34. Wada, K. (1961). Crystal growth of the molluscan shell. *Bull. Natl.Pearl Res. Lab.*, **7**:703-838.
35. Wada, K. (1972). Nucleation and growth of aragonite crystals in the nacre of bivalve molluscs. *Biomineralization*, **6**:41-159.
36. Weiner, S. and Traub, W. (1984). Macromolecules in the mollusc shell and their functions in biomineralization. *Phil. Trans. R. Soc. Lond.*, **B 304**:425-434.

How *Anomalocaris* Swam in the Cambrian Sea; A Theoretical Study Based on Hydrodynamics

Yoshiyuki Usami, Keiko Kamono[1] and Kayako Kawamura[1]

Institute of Physics, Kanagawa University
Rokkakubashi 3-27-1, Kanagawa-ku, Yokohama 221-8686, Japan
e-mail; usami-yoshiyuki@nifty.com, http://www. museum.fm
[1] Department of Mathematical and Physical Sciences, Japan Women's University
Mejirodai 2-8-1, Bunkyo-ku, Tokyo 12-8681, Japan

Summary. Swimming motion of *Anomalocaris* is studied based on hydrodynamic calculations. *Anomalocaris* was the largest predator in the age of the Cambrian explosion, 530 million years ago. It is considered to swim using 14 paired lateral lobes. The relating dynamics of the lobes and surrounding water is calculated by the moving particle method. Various types of swimming styles are calculated, and swimming speed and energy are discussed. As a result, a certain waving pattern for the series of lobes is calculated as the most energy saving motion.

Keywords: *Anomalocaris*, locomotion, simulation, moving particle method

31.1 Introduction

It is widely stated that the Cambrian Explosion was the sudden appearance of the major animal groups that occurred during the Cambrian period (544-505 millions years ago) [12]. A rapid increase of fossils of hard shell creatures has been observed from the early Cambrian in contrast to the remains of soft-bodied fauna from the Precambrian. This phenomenon is of interest not only to paleontologist but also to researchers in biological evolution. Historically, limited fossil resources have prevented detailed discussion on what happened to the ecological system, however, recent paleontological discoveries may shed light on this problem [3, 10, 13]. The new data may help in connect the relationships between the Edicaran fauna and that of the early Cambrian [8, 9]. It is noted that many new discoveries on Cambrian creatures reported from the fossils of Chengian, China [3], including the predator *Anomalocaris* [1, 2, 4, 5, 6]. *Anomalocaris* is a large Cambrian predator, sometimes stated to be the largest. *Anomalocaris* is a name of genera, the other two genera are *Leggania* and *Amplectobelua*, and these are a group of *Anoma-*

locarids. Anomalocarid has been found from Cambrian fossil localities around
the world, including the Burgess Shale formation in Canada, Chengjiang formation in China and the Emu Bay Shale formation in Australia. Each genera
has morphological differences from each other, however, Anomalocaris has
common morphological characteristics of lateral lobes, two head appendages,
mouth beneath the head and prominent eyes, etc. It has been reported recently
that Anomalocaris has legs beneath lobes, however, well-developed lobes of
Anomalocaris make it possible to swim in water. In this paper we study the
locomotion of Anomalocaris based on numerical simulation of hydrodynamics.
Anomalocaris is considered as a group of arthropods, hence the body shape
is completely different from well-studied vertebrates such as fish. It has 14
lateral lobes, and problem here is how Anomalocaris swam in the Cambrian
ocean using their lobes.

31.2 Hydrodynamical calculations based on the moving particle method

From the viewpoint of hydrodynamics, calculating the water environment surrounding a moving biological object is a difficult problem. The boundary is
usually complicated and varies with time. The Navier-Stokes equation is a
faundamental and useful one for studying fluid mechanics, however, considering a vector field on a fixed coordinate space is not suitable for the problem
of changing boundaries in time.

Instead of obtaining the velocity vector field by solving the Navier-Stokes
equation, the moving particle method has been developed to study the interaction between a moving object and water. In the moving particle method,
fluid is represented by moving particles. Calculating the motion of particles
allows us to determine characteristics of water such as velocity and pressure.
It is an advantage that computation at errors such as numerical diffusion,
never appear in the moving particle method, in contrast to the finite difference method.

At first, let us describe the Navier-Stokes equation, which is given by the
Lagrange differential for velocity field $\frac{D\vec{u}}{Dt}$ as,

$$\frac{D\vec{u}}{Dt} = -\frac{1}{\rho}\nabla P + \nu\nabla^2\vec{u} + \vec{F} \tag{31.1}$$

,where P and $\vec{F}$ are pressure and external force, and ρ and ν are density
and viscosity constant, respectively. Recently, Koshizuka and Oka proposed
equivalent dynamics of the Navier-Stokes equation by solving the particle
method. In this method, one calculates the motion of moving particles instead
of solving vector field equations. We describe briefly the MPS (moving particle
semi-implicit) method, which is a model of the particle method proposed by
Koshizuka and Oka [11, 14, 15, 16].

In the MPS a particle has a weight function $\omega(r)$ to restrict the particle interaction within a finite radius r_e,

$$\omega(r) = \begin{cases} r_e - 1 & (r < r_e) \\ 0 & (r_e < r) \end{cases},$$ (31.2)

where r is the distance between two particles. Fluid density is represented by the particle number density n at the particle i with the use of a weight function of particle i and its neighbor j as,

$$n_i = \sum_{j \neq i} w(|\vec{r}_j - \vec{r}_i|).$$ (31.3)

When we consider a certain scalar quantity φ, a gradient vector at the particle i is given as the weighted average of neighbor particle j as,

$$\nabla \varphi_i = -\frac{d}{n_i} \sum_{j \neq i} \frac{\varphi_j - \varphi_i}{|\vec{r}_j - \vec{r}_i|^2}(\vec{r}_j - \vec{r}_i)w(|\vec{r}_j - \vec{r}_i|),$$ (31.4)

where d is the number of the space dimension.

Diffusion is modeled by the distribution of part of a quantity from a particle i to its neighbor particle j by the use of the weight function as follows,

$$\nabla^2 \vec{u}_i = \frac{2d}{\lambda n^0} \sum_{j \neq i} (\vec{u}_j - \vec{u}_i)w(|\vec{r}_j - \vec{r}_i|),$$ (31.5)

where a parameter λ is introduced to fit with the analytical solution as $\lambda = -\sum_{j \neq i} |\vec{r}_j - \vec{r}_i|w(|\vec{r}_j - \vec{r}_i|)/\sum_{j \neq i} w(|\vec{r}_j - \vec{r}_i|)$. Let $\vec{u}_i^n$ and $\vec{r}_i^n$ are velocity and position, respectively, at n-th time step, then particles move as,

$$\vec{u}_i^* = \vec{u}_i^n + \vec{F}\Delta t$$ (31.6)
$$\vec{r}_i^* = \vec{r}_i^n + \vec{u}_i^* \Delta t,$$ (31.7)

where $\vec{u}_i^*$ and $\vec{r}_i^*$ are temporal velocity and position, respectively. By this calculation, number density becomes temporal one as n^* Then we need to correct the number density to n^0 as,

$$n^0 = n_i^* + n_i',$$ (31.8)

where n' is a correction such that the system keeps the number density n^0. Note that if we let $\vec{u}^{n+1}$ be the next time step velocity, then a correction $\vec{u}_i'$ is also needed as,

$$\vec{u}_i^{n+1} = \vec{u}_i^* + \vec{u}_i'.$$ (31.9)

This correction $\vec{u}_i'$ is calculated by the pressure gradient as,

$$\vec{u}_i' = -\frac{\Delta}{\rho}\nabla P^{n+1}.$$ (31.10)

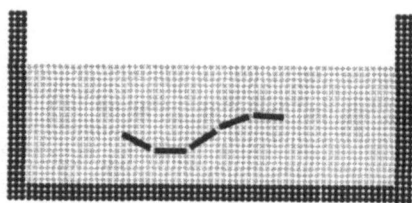

Fig. 31.1. This graph shows the set up for the simulation. A fixed boundary is assumed, shown as a black wall. The *Anomalocaris* lobe is modeled as a board. Simulation is accomplished in two-dimensional space, then *Anomalocaris'* lobe is represented as a black line in the middle of this container. Gray circle filled in the container shows a moving particle.

The momentum conservation law requires the following,

$$\frac{n_i'}{\Delta t} + \nabla(n^0 \vec{u}_i') = 0. \tag{31.11}$$

Then we have an expression of Poisson equation in the particle method as

$$\nabla^2 P_i^{n+1} = -\frac{\rho}{\Delta t^2} \frac{n_i^* - n^0}{n^0}. \tag{31.12}$$

In this study, the two-dimensional space of vertical and horizontal direction is considered. The boundary is shown as a black wall in Fig. 31.1 Moving particles are filled inside the boundary shown as gray circles. Each particle experiences gravitational force, which keeps the particle inside the boundary. Lobes of *Anomalocaris* are modeled as a series of boards which are a set of lines in two-dimensional space shown as thick black lines in the middle of Fig. 31.1.

Let us now note how a solid in water is described in the MPS method. A solid is represented by a group of moving particles which has the same weight function of water. However, particles which consist of the solid keep the same relative position. This model of liquid and solid has been studied by Koshizuka and Oka, and they reported that this model simulates the actual system. They discussed the comparison between experiment and simulation, and found good agreement between them. On the motion of lobes of *Anomalocaris*, we give the same distance for each board in the horizontal direction.

A board is assumed to change its vertical position according to a periodic function. Let Y_i be the vertical position of the i -th board; Y_i changes its value as,

$$Y_i = A\sin(\omega t + i\theta), \tag{31.13}$$

where A, ω, θ, and t are the amplitude, frequency, phase and time, respectively. One of the authors (Usami) has studied the motion including the first two terms of the Fourier expansion in the previous study. It is found that the first term contributes mostly to the motion of the *Anomalocaris* lobes. Note that hydrodynamic effects were not taken into account in the study. In this work we include hydrodynamical effects of the water environment, and examine how a model *Anomalocaris* behaves with the change of phase θ .

31.3 Computational Results

We vary the phase and calculate the motion of the model *Anomalocaris*. Fig. 31.2 displays a snapshot of swimming pattern for the case of θ =0.5 Radians. The other parameters used in the simulations are summarized at the end of this paper. Basically, each parameter is fitted into the one of real system of Cambrian creature. The moving particle velocity is shown as a line inside the boundary in the figure. The thick bold line represents high speed motion. On the contrary, dots represent inactive particles. From this figure we can observe that the model *Anomalocaris* swims to the right side. In Fig. 31.3. and Fig. 31.4, the swimming speed and total energy of moving particles are shown, respectively, with the change of phase parameter . Fig. 31.3 indicates head position of *Anomalocaris* versus time. We can observe from this figure that the swimming speed has a peak around $\theta = 0.25$ and decreases with the increase of θ . In contrast, from Fig. 31.4 we observe monotone decrease of energy level with the increase of phase parameter θ. This means that when θ increases from zero, the induced water energy of moving particles gradually decreases. However, the swimming speed increases from $\theta = 0$ and has a maximum speed at a certain value of θ, and then decreases. The behavior of swimming speed and induced energy with different θ are summarized in Fig. 31.5.

We can observe from Fig. 31.5 that the swimming speed attains its maximum value for a wide θ range from 0.25 to 0.4. However, the energy of the moving particles decreases rapidly with increasing θ. In Fig. 31.4 we can still observe the decrease of energy from $\theta = 0.375$ to $\theta = 0.5$. In fact, oscillatory behavior for the energy is observed for $\theta > 0.375$. We make two observations for the locomotion of *Anomalocaris*. If *Anomalocaris* prefers to swim faster, a phase difference around $\theta = 0.375$ is suitable. But if *Anomalocaris* chooses energy saving locomotion, motion with $\theta = 0.5$ is more suitable. The motion with $\theta = 0.5$ is shown in Fig. 31.2. There is a possibility that *Anomalocaris* controlled the phase difference θ. In the case of high speed motion, small θ is chosen, however, *Anomalocaris* might usually swim in energy saving mode of large θ. It is noted that for large θ value, the motion becomes snail-like, however, small θ motion results in tuna-like motion.

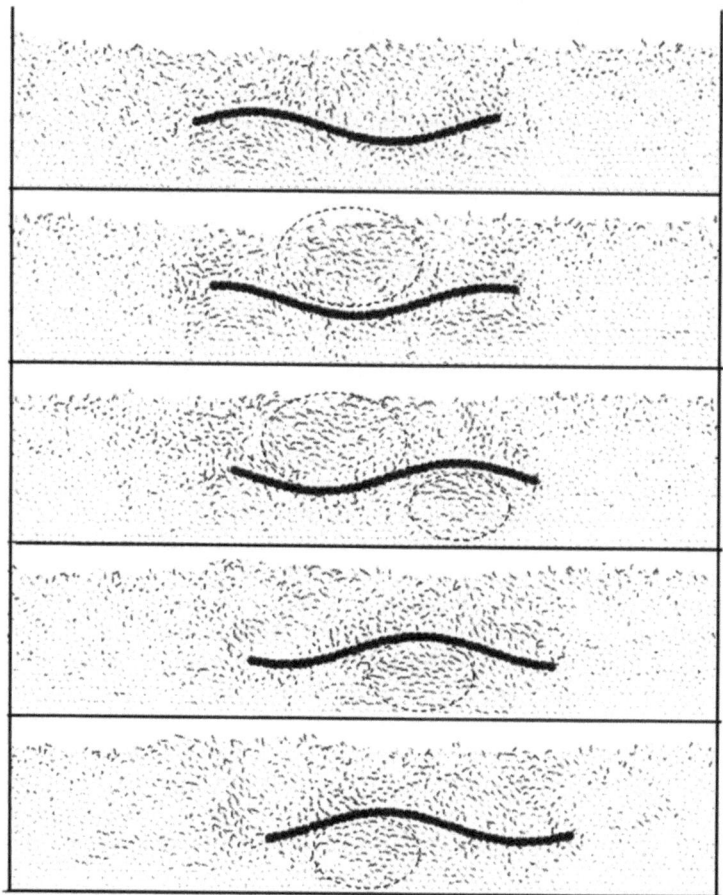

Fig. 31.2. Motion of model *Anomalocaris* and surrounding moving particles. The Anomalocaris lobe is modeled as a solid particle in the MPS method, which is shown as black circles in the middle. The velocity of the moving particles is shown as line. Dotted circles show collective motion of water generated by wave motion of a series of lobes.

Finally, we summarize the parameters used in the simulation, and discuss each quantity. A diameter of moving particle is 1cm on an average, and assumed *Anomalocaris* size is 42cm. The simulation was performed in two dimensional space, then energy used in the figure has a dimension of Joule per

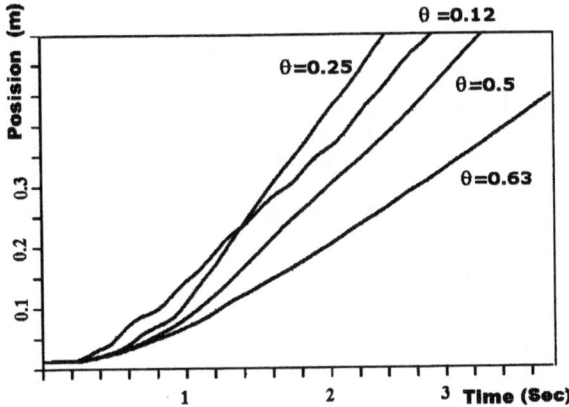

Fig. 31.3. Head position of model *Anomalocaris*. The gradient of the slope gives the swimming speed. Around speed has maximum.

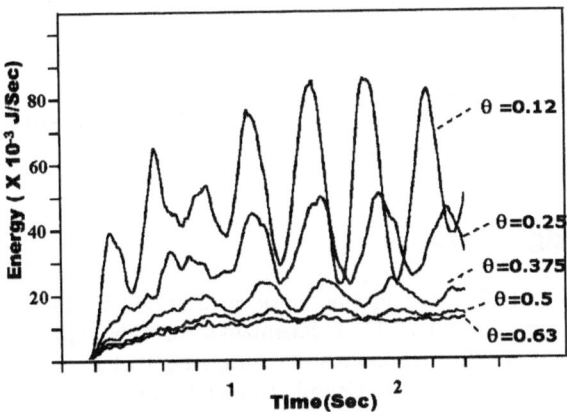

Fig. 31.4. Total energy of moving particles. Small θ brings high energy motion of surrounding water. Large θ motion does not disturb the water so much.

centimeter and second [J/cm· sec]. If we give both lobes width of *Anomalocaris* as 10cm, the induced energy of moving particles becomes 0.1-0.3 J / sec. The frequency of each lobe is given as 2.3/sec. The observed swimming speed is around 0.2m-0.3m/sec. Nagai pointed out that carp swims at 1m/sec by expending energy of 0.5 J/sec [25]. Considering the fact that carp is expected to be heavier than *Anomalocaris*, and is a vertebrate, the value of 0.2-0.3 m/sec

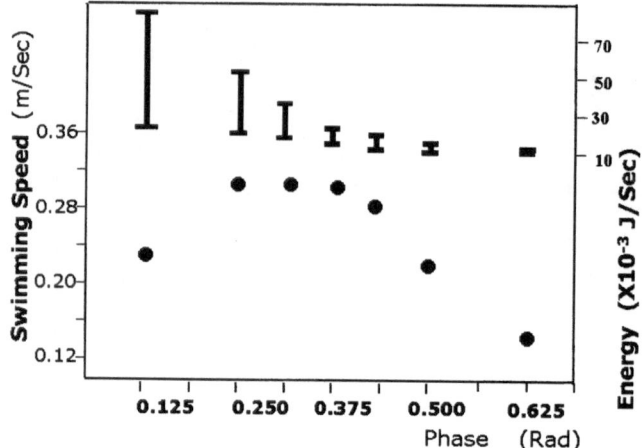

Fig. 31.5. Black circle shows swimming speed. Thick line shows total energy of moving particles. Swimming speed shows a plateau maximum around θ =0.25-0.4. However, water energy decreases monotonically with the increase in θ.

swimming speed with the induced energy 0.1-0.3 J/sec is appropriate for an invertebrate arthropod such as *Anomalocaris*.

Acknowledgements

The program of MPS used in this work is provided by Prof. S. Koshizuka. The authors are grateful to Prof. Koshizuka for this support. The authors are also grateful to the continuous encouragement of Prof. Ono throughout this work. One of the authors (Usami) thanks to a curator Mike Everhart for useful discussions on swimming motion of ancient vertebrates.

References

1. Nedin, C. (1999). *Anomalocaris* predation on mineralized and non-mineralized trilobites. *Geology*, **27**, 987-990.
2. Collins, D. (1996). The 'Evolution' of *Anomalocaris* and its classification in the arthropod class Dinocardia (NOV.) and order Radiodonta (NOV.). *Journal of Paleontology*, **70**(2), 280-293.
3. Shu, D-G., Luo, H-L., Conway, Morris, S., Zhang X-L., Hu, S-X., Chen, L., Han, J. and Zhu, M. (1990). Lower Cambrian vertebrates from south China, Nature, **402**, 42-46.

4. Graham, E.B. (1998). Arthropod body-plan evolution in the Cambrian with an example from *Anomalocaridid* muscle. *Lethaia*, **31**, 197-210.
5. Bergstr, J. (1987). The Cambrian Opabinia and *Anomalocaris*. *Lethaia*, **20**, 187-188.
6. Chen, J.Y., Ramskold, L., Gui-qing, Z. (1994). Evidence for monophyly and arthropod affinity of Cambrian giant predators. *Science*, **264**, 1304-1308.
7. Nagai, M. *Hydrodynamics Learned from Dolphin* (in Japanese), Ohm Publisher.
8. McMenamin, M. A. (1998). *The Garden of Ediacara*, Columbia University Press.
9. Nedin, C. and Jenkins, R.J.F. (1998). First occurrence of the Ediacaran fossil Charnia from the Southern Hemisphere. *Alcheringa*, **22**: 315-316.
10. Gould, S.J. (1990). *Wonderful Life*, Century Hutchinson.
11. Koshizuka, S. (1997). *Numerical Methods for Hydrodynamics*, (Baifu-kan). "Su-chi Ryu-tai Rikigaku" (in Japanese).
12. Bowring, S.A., Grotzinger, J.P., Isachsen, C.E., Knoll, A.H., Pelechaty, S.M. and Kolosov, P. (1993). Calibrating rates of Early Cambrian evolution. *Science*, **261 (3)**, 1293-1298.
13. Morris, S.C. and Whittington, H.B. (1979). The animals of the Burgess Shale. *Scientific American*, **241**, 122-133.
14. Koshizuka, S., Nobe, A. and Oka, Y. (1998). Numerical analysis of breaking. waves using the moving particle semi-implicit method. *Int. J. Numer. Meth Fluids*, **26**, 751-769.
15. Koshizuka, S., Tamako, H. and Oka, Y. (1995). A particle method for incompressible viscous flow with fluid fragmentation. *Comput. Fluid Dynamics J.*, **4**, 29-46.
16. Koshizuka, S. and Oka, Y. (2001). Application of moving particle semi-implicit method to unclear reactor safety. *Comput. Fluid Dynamics J.*, **9**, 366-375.
17. Usami, Y., Saburo, H., Inaba, S. and Kitaoka, M. (1998). Reconstruction of extinct animals in the computer. In Adami, C., Belew, R., Kitano, H. and Taylor, C. (eds.). *Artificial Life* IV, MIT Press, 173-177.

Morphology and Function of Cuticular Terraces in Stomatopoda (Crustacea) and Mantodea (Insecta)

Enrico Savazzi

Hagelgränd 8, 75646 Uppsala, Sweden
e-mail: enrico@savazzi.net

32.1 Introduction

Relief patterns consisting of sets of subparallel ridges with a distinctly asymmetrical cross-section have been described in a variety of marine and marine-derived invertebrates (see below). In the literature, these ridges usually are called terraces, terrace-lines or terrace-sculptures [25, 26, 27, 28, 29, 31] because of their superficial similarity with agricultural terraces on sloping terrains. Typically, the cross-section of a terrace is asymmetrically triangular, with a very steep face and a gently sloping opposite face, and a sharp edge (either straight or crenulated) delimiting the distal end of the steep face.

Observations and experiments on living organisms [17, 21, 22, 23, 26, 27, 29, 30, 32, 33, 34, 35] , inferential reasoning [12, 13, 17, 27, 28, 31] and experiments with artificial terrace patterns [16, 24, 36] showed that these features are functional in the context of increasing friction between the exoskeleton and a substrate. Because of the asymmetrical cross-section of terraces, friction is higher in the direction faced by the steeper sides of terraces (see above references).

The adaptive significance of these features to the organism varies. Two general adaptive patterns have been recognised:

(1) In organisms that burrow through loose sediments by repeating a push-pull sequence of movements, terraces reduce back-slippage, while offering a lower friction against the sediment when the sculptured surfaces move in the burrowing direction. This enhances the effectiveness of the burrowing process [12, 13, 16, 17, 18, 19, 20, 21, 22, 23, 26, 27, 28, 29, 31, 32, 33, 34, 36, 37]. In these organisms, terraces are approximately perpendicular to the direction of burrowing, and their steep faces are directed away from the burrowing direction (see also Savazzi, this volume).

(2) In a few crevice-dwelling and burrow-dwelling organisms, the increased friction provided by terraces is functional in preventing the organism from

being dislodged from a shelter by predators or competitors [19, 26]. Terraces in these taxa are oriented to provide maximal friction in a direction toward the mouth of the crevice or tunnel.

Seilacher [29] discussed several morphological properties that characterise terrace sculptures. Two of these properties are especially relevant to the present paper: (1) perpendicular orientation to the burrowing or wedging direction, and (2) asymmetry of cross-section, related to the directionality of terrace-friction. As a consequence of these properties, the terraces forming a pattern are subparallel to each other, and their steep faces are oriented in the same direction (i.e., toward the direction of highest friction).

The literature on terrace patterns (see references above) records the presence of these sculptures in a broad range of marine burrowers, and in marine and a few subaerial crevice- and burrow-dwellers (these subaerial forms occur only among decapod crustaceans). There are no studies of terraces with other functions, nor of terraces occurring in typically freshwater or terrestrial groups like insects. Therefore, it is legitimate to ask if this lack of records stems from constraints that prevented the evolution of terraces in these groups, and/or in association with different functions and life habits.

Among arthropods, raptorial appendages, used to catch and hold a prey, occur mainly in stomatopod crustaceans and in several groups of insects (see below). A functional requirement of this type of appendage is that it must prevent the prey from slipping away once it is caught. Therefore, structures that increase friction between the appendage and the prey can be expected to be advantageous. In addition, the shape of the appendage and its mode of functioning are likely to require that this friction be applied selectively in a specific direction. This directional friction, in principle, could be provided by terraces placed in suitable positions and orientations on the raptorial appendages. Thus, their function would be comparable with that of terraces in burrowing and crevice-dwelling organisms. In order to verify these assumptions, arthropods that possess raptorial appendages were examined, looking for terrace sculptures. Instances were found in the Stomatopoda (Crustacea) and Mantodea (Insecta). This provides the opportunity for testing the validity of inferential techniques used to assess the adaptive value of more "conventional" terrace patterns in living and fossil invertebrates.

32.2 Materials and methods

Insect specimens illustrated in this paper are stored at the Dept. of Entomology, Swedish Museum of Natural History, Stockholm, Sweden. All stomatopod specimens were collected by, and are in the collection of, the author. The behaviour of living stomatopods was observed in the field and in temporary tanks set up during field work. The behaviour of Mantodea is discussed only on the basis of the literature. Localities are indicated in the figure captions. Specimens were gold-sputtered prior to SEM observations.

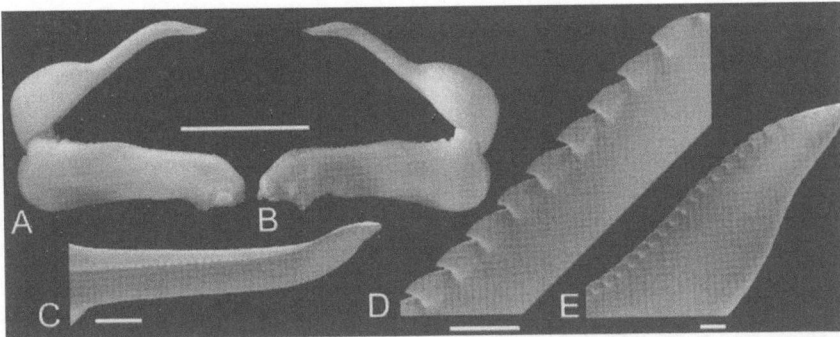

Fig. 32.1. Gonodactylus cf. falcatus Forskal, Tayud, Cebu Island, the Philippines, left second maxilliped, left lateral (A) and right lateral (B) views, and details of dactylial tip (C-E). Scale bars represent 10mm (A-B), 1mm (C) and 100mm (D-E).

32.3 Observations

32.3.1 Stomatopoda

The Stomatopoda possess a pair of raptorial appendages, which are modified and enlarged second maxillipeds. These appendages catch and hold a prey between the propodus and the dactylus ([3, 4], pers. obs.), which is kept folded against the dorsal surface of the propodus when at rest. Thus, the appendage opens dorsally (Fig. 32.1A-B, 32.2B-C, E) in a manner reminiscent of a switchblade.

The Stomatopoda can be subdivided into two groups, called smashers and spearers, based on the morphology of the raptorial appendages [4]. Smashers possess a dactylus with a single spine and a thickened, swollen proximal portion (Fig. 32.1A-B). This swelling is used to impact and break the exoskeletons of invertebrates ([4], pers. obs.). The spine can be used to stab soft-bodied prey (see above references) and can also hold a stunned or killed prey against the propodus (pers. obs.). Spearers possess several spines with subparallel tips (Fig. 32.2A-E), which are used to spear soft-bodied prey (see above references; pers. obs.). Spearers lack a swelling comparable to that of smashers, although the ventral proximal portion of the dactylus may be broadened and thickened (Fig. 32.2D).

The dorsal (when extended) edge of the dactylus of smashers bears a serrated row of asymmetrical, terrace-like barbs the tips of which are directed toward the dactylus-propodus articulation (Fig. 32.1C-E). The propodus displays no terraces or terrace-like sculptures, but it may carry pseudosetae (Fig. 32.1A-B) on a ridge flanking a groove that houses the folded dactylus.

The spines on the dactylus of spearers bear a well-developed terrace-pattern (Fig. 32.2G-H). The terraces are especially well developed along the

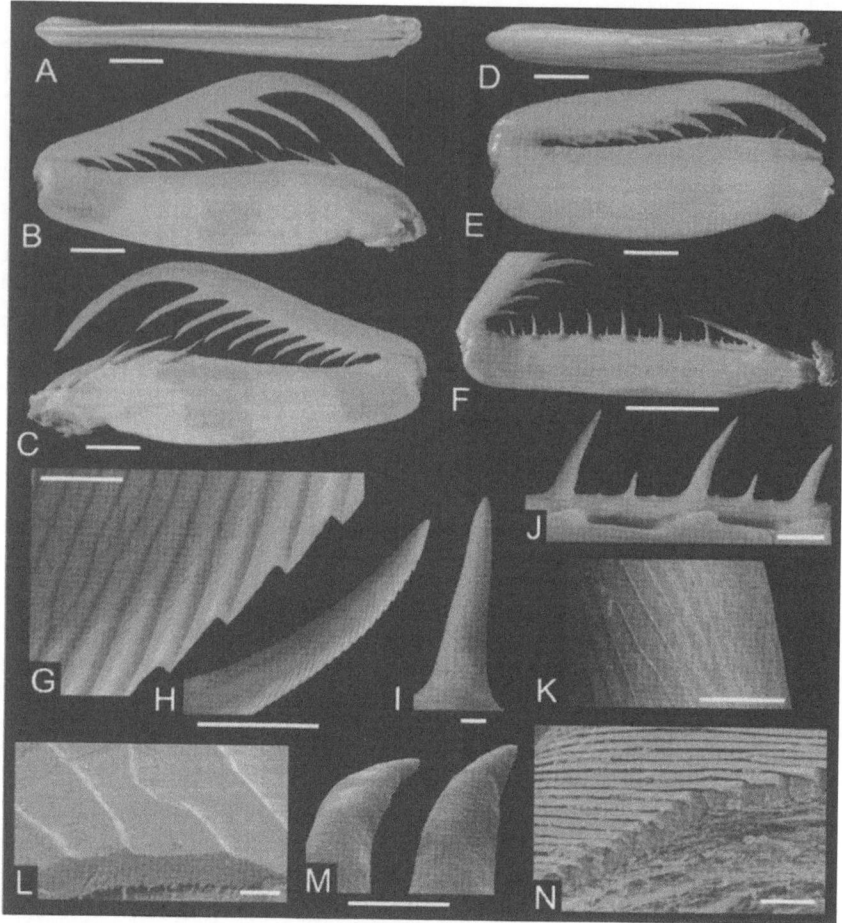

Fig. 32.2. . A-C. Lysiosquillina maculata (Fabricius), Hilotongan Island, the Philippines, left second maxilliped, left second maxilliped, dorsal (A), left lateral (B) and right lateral (C) views. D-E. Oratosquilla cf. oratoria (de Haan), Hilotongan Island, the Philippines, left second maxilliped, dorsal (D) and left lateral (E) views. F, I-L. Harpiosquilla harpax (de Haan), Ibo, Mactan island, the Philippines, right second maxilliped, left lateral views (F, J) and detail of propodal fixed spine (I, K-L). G-H, M-N. Squilla mantis (Linnaeus), Chioggia, Italy, details of second dactylial spine (G-H) and propodal pseudosetae (M-N). Scale bars represent 10mm (A-C, D-F), 1mm (H, J), 100mm (G, I, M), 50mm (K) and 10mm (L, N).

dorsal and ventral edges of the spines, while the pattern may display irregularities on the lateral regions (Fig. 32.2G). Terraces are also shallower in the lateral regions. The cross-section of the spines typically is flattened sideways (i.e., where the irregular regions of the terrace-pattern are located). The terraces are oriented roughly perpendicularly to the direction of the piercing strike, i.e., approximately perpendicularly to the length of the dactylus but slightly oblique with respect to the length of the spines, which are inserted obliquely on the dorsal side of the dactylus. The steep faces of the terraces are directed toward the dactylus-propodus articulation.

At rest, the dactylus folds against the propodus, with the spine tips sunk within loose fitting sockets on the dorsal face of the dactylus (Fig. 32.2F, J). A ridge on one side of these sockets typically carries a row of pseudosetae (Fig. 32.2B-C, E, M-N) that come into contact with a speared prey when the dactylus is folded. In a few taxa, the pseudosetae are supplemented or substituted by a row of fixed spines, often arranged in hierarchies of different sizes (Fig. 32.2F, J). The largest among these spines bear a terrace-pattern finer than that on the dactylus (Fig. 32.2I, K-L). Unlike on the dactylus, these terraces are roughly parallel to the length of the spines (i.e., perpendicular to the length of the propodus), and their steep faces are directed toward the dactylus-propodus articulation (Fig. 32.2I, K-L).

The pseudosetae on the propodus bear a very fine surface relief superficially similar to terraces perpendicular to the length of the pseudosetae (Fig. 32.2M), but observations of their internal structure shows that this relief is associated with a collar-in-collar stacking of epithelial structures (Fig. 32.2N), and therefore is not related to terraces. The pseudosetae are relatively soft structures, and probably the nested collars make them slightly compressible along their length. Pseudosetae and related features in stomatopods have been studied in detail [11].

The holding action of the terraced dactylial spines and of the propodal pseudosetae and fixed spines is likely enhanced by a few mobile spines present near the proximal end of the propodus. These spines can be raised into a direction approximately opposite to that of the dactylial spines (Fig. 32.2B-C, F).

32.3.2 Mantodea

The first pair of legs in the Mantodea (Fig. 32.3) is modified into raptorial appendages that catch and hold a prey between tibia and femur [1, 2, 5, 6, 7, 9, 14]. Unlike in stomatopods, the appendage folds in the ventral direction. The appendage usually is kept folded when at rest. The prey of Mantodea consists mostly of insects and spiders [8, 15, 38], but also includes vertebrates (for areview, see [8]).

The opposing surfaces of both tibia and femur carry two longitudinal rows of fixed spines. Typically, the number of spines on the tibia is higher than on the femur, but the spines on the femur can be longer and stouter. The tibial

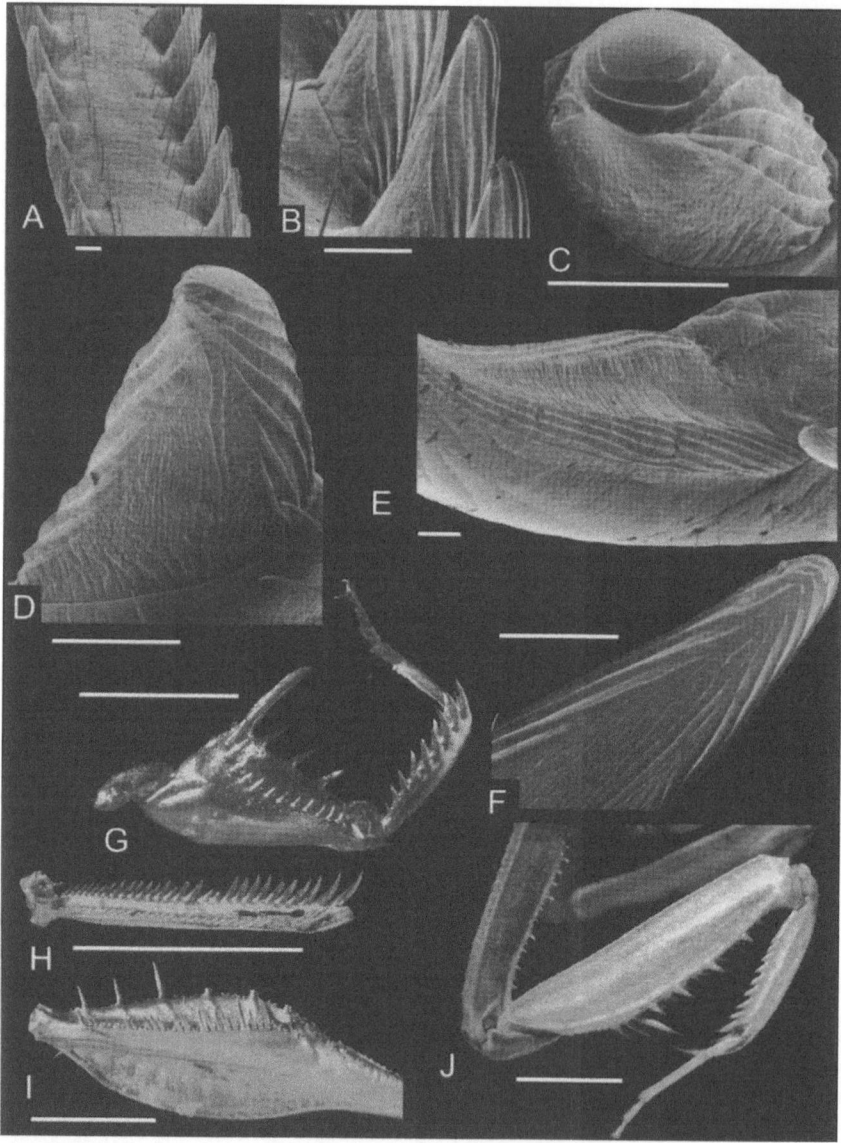

Fig. 32.3. A-F. Mantis religiosa Linnaeus, left front tibia, fixed spines (A-D) and
tibial hook (E-F). G. Metallyticus splendidus Westwood, Sumatra, Indonesia, right
front leg, left lateral view. H. Chaetessa caudata Saussure, Brazil, right front tibia,
left lateral view. I. Hestiasula phyllopus de Haan, Kina Balu, Borneo, front left
femur, right lateral view. J. Polyspylota variegata Oliv, unknown locality, right front
leg (with left front leg partly visible in background), right lateral view. Scale bars
represent 5mm (J), 400mm (G-I) 200mm (A-E) and 50mm (F) (all specimens except
J turned upside down).

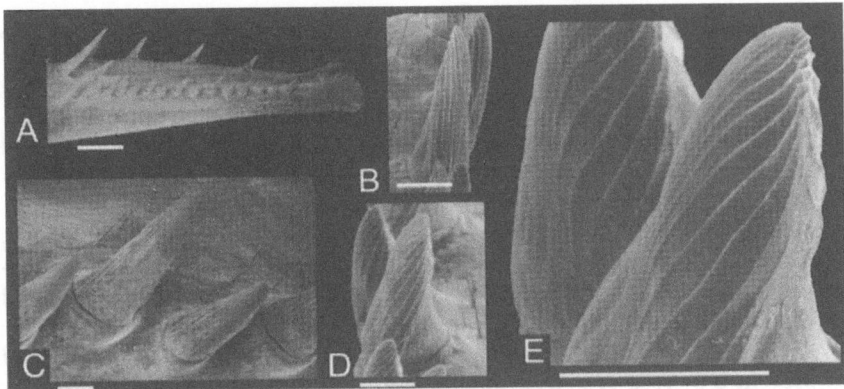

Fig. 32.4. Mantis religiosa Linnaeus, Italy, right front femur, left lateral view (A) (specimen turned upside down) and femoral fixed spines (B-E). Scale bars represent 200mm (A) and 50mm (B-E).

and femoral spines often mesh with each other, and this in turn causes the tips of the femoral spines to be staggered laterally (Fig. 32.4B-E). A few mobile spines are present on the proximal portion of the femur. The fixed spines of Mantodea are not an integral part of the appendage as in the Stomatopoda, but instead are connected to the appendage by a slightly flexible basis, visible in Figs. 32.3A-D, 32.4B-D as a discontinuity in the exoskeletal surface.

The femoral and tibial fixed spines carry a conspicuous terrace-pattern, with the terraces subparallel to the length of the spines (Fig.32.3A-D).

The steep faces of the terraces are oriented toward the medial direction (i.e., they are oriented in opposite directions in the two parallel rows of spines). The tips of the spines usually (albeit not always) are blunt, and in this case the terraces often continue across their tips (Fig. 32.3C-D).

The distal extremity of the tibia carries a long curved hook. The sides of the tibial hook carry a terrace-pattern similar to that present on the spines (Fig. 32.3E-F). The genus *Metallyticus* also possesses a large, fixed femoral spine projecting perpendicularly from the ventral surface of the femur in its proximal portion. The sides of this spine are terraced in a way similar to the tibial hook.

32.4 Discussion

Raptorial legs function by catching a prey, either by piercing it (e.g. in spearing stomatopods) or by pressing it between two opposing sclerites. It is crucial to the function of raptorial appendages that the prey does not break loose. The adaptive significance of the terrace-patterns observed in Stomatopoda and Mantodea can be analysed in this context. One may state, as a working hypothesis, that terraces in these groups are functional by increasing friction between appendage and prey. This hypothesis can be tested by checking whether the distribution and orientation of terraces is optimal for this function. The requirements of distribution are obvious: terraces should be placed where they touch the prey. The requirements of orientation need a more careful analysis. As discussed in the Introduction, terraces provide the highest friction in the direction opposite to the orientation of their steep faces. Thus, terraces should be oriented perpendicularly to the likely direction of escape-movement of the prey. In addition, their steep faces and sharp edges should be oriented to exert their maximal friction against this movement.

In smashing Stomatopoda, the short terraces on the dactylus are oriented to prevent a pierced prey from slipping out of the tip of the dactylus. In this respect, the terraces can be compared in function to the barbs of a harpoon. That they are present only in a narrow longitudinal band on the dorsal side of the dactylus likely depends on the fact that this region of the dactylus can be pressed against the propodus, thus making the terraces more effective. The tip of the dactylus is bent obliquely toward the propodus, which further reduces the likelihood of a pierced prey slipping loose.

In spearing Stomatopoda, terraces are present on most of the surfaces of the dactylial spines, but their pattern is organised more precisely on the dorsal and ventral sides of the spines. Their orientation indicates a function in preventing a speared prey from slipping toward the tips of the spines. The better orientation, higher steep faces and more clearly defined sharp edges of the terraces in the dorsal and ventral regions (i.e., the regions facing adjacent spines) suggests that the friction of the terraces here is particularly important. One of the causes of this is the stretching of the pierced prey as the spine penetrates it. This is due to the diameter of the spine's cross-section increasing from tip to basis. Stretching is effected mostly by the dorsal and ventral edges of a spine, because the cross-section of the spine is laterally compressed. Another factor is compression of the pierced prey between adjacent spines, as the prey gradually becomes wedged deeper between adjacent spines. Compression increases because the spacing between adjacent spines decreases from tip to basis. The slightly oblique (rather than perpendicular as expected from the hypothesised function) orientation of the terraces on the spines can be explained by the fact that the spines are correspondingly oblique to the dactylus and to the direction of spearing (see above). It has been argued [10] that small prey can be trapped between adjacent spines of spearers, rather than pierced. In this case, the terraces are not suitably oriented to hold a prey, which could slip sideways. Therefore, the terraces are not optimised for holding such small prey. The spines of Mantodea are not used for spearing, as indicated by their generally blunt tips, and by the occurrence of terraces on both sides and tips of the spines. The orientation of the terraces indicate that the prey is being contained between the two parallel rows of spines on the tibia (and the two corresponding rows on the femur), and prevented from slipping sideways. The terraces do not hinder slippage in the distal or prox-imal directions. In these directions, slippage is hindered by the tibial hook and/or additional projections, including mobile spines. While the presence of terraces in stomatopods is mentioned in passing in the literature (e.g. [4]), the writer found no mention of terraces in the literature on Mantodea. This could be due to the fact that terraces in these insects are finer than those in stomatopods (which are visible to the naked eye), and therefore more difficult to notice, although clearly visible with a dissection microscope.

32.5 Concluding remarks

The inferential criteria used to assess the function of terraces in burrow-ing, crevice-dwelling and burrow-dwelling invertebrates prove successful in analysing the function of terraces on the raptorial appendages of Stomatopoda and Mantodea. The results obtained from the application of these criteria support the idea of a frictional function of terraces in these organisms. The differences in orientation and distribution of terraces in these two groups are explained by the correspondingly different construction and function of the raptorial appendages, i.e., sharp spines used for piercing a soft-bodied prey in spearing stomatopods, versus parallel rows of typically blunt-tipped spines holding a hard-bodied prey in the Mantodea. This study also shows that no constraints prevent the evolution of terrace-patterns in invertebrate groups and for functions other than those examined in past papers. It is entirely possible that further instances of terrace-patterns have remained undetected in other invertebrate groups. Earlier studies have identified a very frequent, albeit not constant, association between a burrowing function of terraces in soft sediments and an allometric growth-pattern of the terraces (in particu-lar, the secondary introduction of new terraces among existing ones during growth; see [16] for a discussion and an exception to this trend). In contrast, terrace-patterns functional for crevice-wedging tend to display an isometric growth and a constant number of terraces during ontogeny [19, 26, 27]. The mode of growth of terraces in Stomatopoda and Mantispidae remains to be studied.

Acknowledgments

The writer thanks the organisers of the MPB 2002 symposium and the Swedish Research Council for providing financial support for attendance at the congress.

388   Enrico Savazzi

References

1. Ass, M.J. (1973). Die Fangbeine der Arthropoden, ihre Entstehung, Evolution und Funktion. *Deutsche Entomologische Zeitschrift* **1-3**:127-152.
2. Bremond, J. (1974). Remarques sur le phenomene du convergence des members prehensiles chez la Mante religieuse et le Crustace Squilla mantis. *Entomologiste* **30**:183:188.
3. Burrows, M. (1969). The mechanics and neural control of the prey capture strike in the mantid shrimps, Squilla and Hemisquilla. *Zeitschrift fur vergleichende Physiologie* **62**:361-381.
4. Caldwell, R.L. and Dingle, H. (1975). Stomatopods. *Scientific American* **234**:81-89.
5. Cleal, K.S. and Prete, F.R. (1996). The predatory strike of free ranging praying mantises, Sphodromantis lineola (Burr.) II: Strikes in the horizontal plane. *Brain Behaviour and Evolution* **48**:191-204.
6. Copeland, J. and Carlson, A.D. (1979). Prey capture in mantids: a non-stereotyped component of lunge. *Journal of Insect Physiology* **25**:263-269.
7. Corrette, B.J. (1990). Prey capture in the praying mantis Tenodera aridifolia sinensis: coordination of the capture sequence and strike movements. *Journal of Experimental Biology* **148**:147-180.
8. Ehrmann, R. (1992). Vertebrates as food for praying mantids (Mantodea). *Entomologische Zeitschrift* **102**:153-161.
9. Gray, P.T. and Mill, P.J. (1983). The mechanics of the predatory strike of the praying mantis Hierodula membranacea. *Journal of Experimental Biology* **107**:245-275.
10. Hamano, T. and Matsuura, S. (1986). Optimal prey size for the Japanese mantis shrimp from structure of the raptorial claw. *Bulletin of the Japanese Society of Scientific Fisheries* **52**: 1-10.
11. Jacques, F. (1989). Pseudosetal formations of maxillipeds in Stomatopoda. In Ferrero E.A. (ed.): *Biology of stomatpopods. Selected Symposia and Monographs U.Z.I.* **3**: 133-139.
12. Jefferies, R.P.S. (1984). Locomotion, shape, ornament and external ontogeny in some mitrate calcichordates. *Journal of Vertebrate Paleontology* **4**:292-319.
13. Kohn, A.J. (1986). Slip-resistant silver-feet: shell form and mode of life in Lower Pleistocene Argyropeza from Fiji. *Journal of Paleontology* **60**:1066-1074.
14. Loxton, R.G. and Nicholls, I. (1979). The functional morphology of the praying mantis forelimb (Dictyoptera:Mantodea). *Zoology of the Linnean Society* **66**:185-203.
15. Reitze, M. and Nentwig, W. (1991). Comparative investigations into the feeding ecology of six Mantodea species. *Oecologia* **86**:568-574.
16. Savazzi, E. (1981). Functional morphology of the cuticular terraces in Ranina (Lophoranina) (brachyuran decapods; Eocene of NE Italy). *Neues Jahrbuch fur Geologie und Palaontologie Abhandlungen* **162**:231-243.
17. Savazzi, E. (1982). Burrowing habits and cuticular sculptures in Recent sand-dwelling brachyuran decapods from the Northern Adriatic Sea (Mediterranean). *Neues Jahrbuch fur Geologie und Palaontologie Abhandlungen* **163**:369-388.
18. Savazzi, E. (1985). Adaptive themes in cardiid bivalves. *Neues Jahrbuch fur Geologie und Palaontologie Abhandlungen* **170**:291-321.
19. Savazzi, E. (1985). Functional morphology of the cuticular terraces in burrowing terrestrial brachyuran decapods. *Lethaia* **18**:147-154.

20. Savazzi, E. (1986). Burrowing sculptures and life habits in Paleozoic lingulacean brachiopods. *Paleobiology* **12**:46-63.
21. Savazzi, E. (1989). Burrowing mechanisms and sculptures in Recent gastropods. *Lethaia* **22**:31-48.
22. Savazzi, E. (1991). Burrowing sculptures as an example in functional morphology. *Terra Nova* **3**:242-250.
23. Savazzi, E. (1994). Adaptations to burrowing in a few Recent gastropods. *Historical Biology* **7**:291-311.
24. Savazzi, E. and Pan, H. (1994). Experiments on the frictional properties of terrace sculptures. *Lethaia* **27**:325-336.
25. Savazzi, E., Jefferies, R.P.S. and Signor, P.W.III. (1982). Modification of the paradigm for burrowing ribs in various gastropods, crustaceans and calcichordates. *Neues Jahrbuch fur Geologie und Palaontologie Abhandlungen* **164**:206-217.
26. Schmalfuss, H. (1978). Structure, Patterns, and function of cuticular terraces in Recent and fossil arthropods. I. Decapod crustaceans. *Zoomorphologie* **90**:19-40.
27. Schmalfuss, H. (1978). Constructional morphology of cuticular structures in crustaceans. *Neues Jahrbuch fur Geologie und Palaontologie* **156**:155-159.
28. Schmalfuss, H. (1981). Structure, patterns and function of cuticular terraces in trilobites. *Lethaia* **14**:331-341.
29. Seilacher, A. (1973). Fabricational noise in adaptive morphology. *Systematic Zoology* **22**:451-465.
30. Seilacher, A. (1984). Constructional morphology of bivalves: evolutionary pathways in primary versus secondary soft-bottom dwellers. *Palaeontology* **27**:207-237.
31. Seilacher, A. (1985). Trilobite palaeobiology and substrate relationships. *Transactions of the Royal Society of Edinburgh* **76**:231-237.
32. Signor, P.W.III. (1982). Constructional morphology of gastropod ratchet sculptures. *Neues Jahrbuch fur Geologie und Palaontologie Abhandlungen* **163**:349-368.
33. Signor, P.W.III. (1983). Burrowing and the functional significance of ratchet sculpture in turritelliform gastropods. *Malacologia* **23**: 313-320.
34. Stanley, S.M. (1969). Bivalve mollusk burrowing aided by discordant shell ornamentation. *Science* **166**:634-635.
35. Stanley, S.M. (1970). Relation of shell form to life habits of the Bivalvia. *Geological Society of America Memoir* **125**:1-296.
36. Stanley, S.M. (1977). Coadaptation in the Trigoniidae, a remarkable family of burrowing bivalves. *Palaeontology* **20**:869-899.
37. Stanley, S.M. (1981). Infaunal survival: alternative functions of shell ornamentation in the Bivalvia (Mollusca). *Paleobiology* **7**:384-393.
38. Suckling, D.M. (1984). Laboratory studies on the praying mantis Orthodera ministralis (Mantodea: Mantidae). *New Zealand Entomologist* **8**:96-101.

Subject Index